Lecture Notes in Control and Information Sciences

Edited by A.V. Balakrishnan and M. Thoma

43

Stochastic Differential Systems

Proceedings of the 2nd Bad Honnef Conference
of the SFB 72 of the DFG at the University of Bonn
June 28 – July 2, 1982

Edited by M. Kohlmann and N. Christopeit

Springer-Verlag
Berlin Heidelberg New York 1982

Series Editors
A. V. Balakrishnan · M. Thoma

Advisory Board
L. D. Davisson · A. G. J. MacFarlane · H. Kwakernaak
J. L. Massey · Ya. Z. Tsypkin · A. J. Viterbi

Editors
M. Kohlmann
Universität Hamburg
Institut für Mathematische Stochastik
Bundesstr. 55
D-2000 Hamburg 13

N. Christopeit
Institut für Ökonometrie und Operations
Research der Universität Bonn
Ökonometrische Abteilung
Adenauerallee 24–42
D-5300 Bonn 1

ISBN 3-540-12061-0 Springer-Verlag Berlin Heidelberg NewYork
ISBN 0-387-12061-0 Springer-Verlag NewYork Heidelberg Berlin

Printed in Germany

Printing and binding: Beltz Offsetdruck, Hemsbach/Bergstr.
…0-543210

Herrn Professor Dr.Walter Vogel

anläßlich seines 60.Geburtstags

von seinen Schülern gewidmet.

PREFACE

Since 1979 when the first Bad Honnef Conference on Stochastic Control and Stochastic Differential Systems took place the art has developed a lot.Problems in partially observed nonlinear control e.g. forced the control people to become familiar with filter theory and those working in filter theory tried to give results which were applicable in control theory.

So the main idea for the second Bad Honnef Conference was to bring together filter and control specialists for an exchange of ideas and a discussion of new results.We gratefully acknowledge the generous support of the Sonderforschungsbereich 72 of the Deutsche Forschungsgemeinschaft at the University of Bonn which made this idea become a conference in June/July 1982.

This book presents nearly all contributions to this conference. Following the concept of the conference the book is divided into three sections : The first presents contributions to the general theory and the next two contain recent results in filter and control theory.We are very grateful to all lecturers for the presentation of their works and for their willingness to explain and discuss their results. Furthermore,we would like to thank them for the careful preparation of their manuscripts which made a rapid publication possible.

We also want to express our thanks to our co-organizers R.J.Elliott and M.Hazewinkel and many of our students for doing a lot of organizational work before,during,and after the conference.

The conference took place in the Elly Hölterhoff-Böcking Stiftung of the Deutsche Physikalische Gesellschaft in Bad Honnef.We thank the staff there for their manifold efforts towards a successs of the meeting.

Bonn,July 1982 N.Christopeit and M.Kohlmann

LIST OF PARTICIPANTS

A. AL-HUSSAINI
University of Alberta
Edmonton, Canada

M. ARATO
SZAMKI
Budapest, Ungarn

A.V. BALAKRISHNAN
University of California
Los Angeles, USA

V.E. BENEŠ
Bell Laboratories
Murray Hill, USA

A.BENSOUSSAN
INRIA
Le Chesnay, France

J.M. BISMUT
Université de Paris-Sud
Paris,France

B. BOBROWSKY
ETH Zürich
Zürich ,Switzerland

R.K. BOEL
University of Ghent
Ghent,Belgium

N. CHRISTOPEIT
Universität Bonn
Bonn, BRD

N.J. CUTLAND
University of Hull
Hull, England

M.H.A. DAVIS
Imperial College
London, England

M. DEISTLER
TU Wien
Wien,Austria

G. DEL GROSSO
Universita di Roma
Roma, Italy

M. DEMPSTER
Balliol College
Oxford, England

G.B. DI MASI
LADSEB-CNR
Padova, Italy

R.J. ELLIOTT
University of Hull
Hull, England

A. FRIEDMAN
Northwestern University
Evanston, Illinois, USA

H. GLÖCKNER
Universität Bonn
Bonn, BRD

K. GOOSSEN
Universität Essen
Essen,BRD

J. GROH
Friedrich-Schiller-Universität
Jena, DDR

U. HAUSSMANN
University of British Columbia
Vancouver, Canada

M. HAZEWINKEL
Erasmus University
Rotterdam, The Netherlands

K. HELMES
Universität Bonn
Bonn, BRD

Y. KABANOV
CEMI
Moskow, USSR

G. KALLIANPUR
University of North Carolina
Chapel Hill, USA

I. KARATZAS
Columbia University
New York, USA

V. KOHA
Universität Bonn
Bonn, BRD

M. KOHLMANN
Universität Bonn
Bonn, BRD

F. KONECNY
Universität für Bodenkultur
Wien, Austria

H.J. KUSHNER
Brown University
Providence, USA

J.P. LEPELTIER
Université du Maine
Le Mans Cedex, France

D. LEPINGLE
Université d'Orléans
Orléans, France

R. LIPTSER
Institute of Control Science
Moscow, USSR

A. LINDQUIST
University of Kentucky
Lexington,USA

B. MARCHAL
Université Paris
Paris, France

G. MAZZIOTTO
C.N.E.T.
Les Moulineaux, France

A. MELNIKOV
V.A. Steklov Mathematikal Institute
Moscow, USSR

V.J. MIZEL
Carnegie-Mellon University
Pittsburgh, USA

D. PAPPAS
Universität Kaiserslautern
Kaiserslautern, BRD

E. PARDOUX
Université de Provence
Marseille, France

M.A. PINSKY
Northwestern University
Evanston, USA

S.R. PLISKA
Northwestern University
Evanston, USA

M. PONTIER
Université d'Orléans
Orléans, France

H. PRAGARAUSKAS
Academy of Sciences of the Lithuanian SSR
Vilnius, USSR

R. RISHEL
University of Kentucky
Lexington, USA

M. SCHEUTZOW
Universität Kaiserslautern
Kaiserslautern, BRD

Z. SCHUSS
Tel-Aviv University
Ramat Aviv, Israel

J. SZPIRGLAS
C.N.E.T.
Les Moulineaux, France

L. STETTNER
Inst. of Mathem. Polish Academy
of Sciences, Warsaw, Poland

B. STORB
Universität Bonn
Bonn, BRD

D. TALAY
Marseille
France

L. TUBARO
Libera Universita'Degli Studi di Trento
Trento, Italy

J. VAN GELDEREN
Delft University of Mathematics
Delft, Holland

J.H. VAN SCHUPPEN
Mathematisch Centrum
Amsterdam, The Netherlands

D. VERMES
Computer and Automation Inst. of the
Hungarian Acad. Sci. , Budapest,
Hungary

B. WALTER
Universität Bonn
Bonn, BRD

H. WITTING
Institut für Mathematische
Stochastik, Freiburg, BRD

T. WITTING
Universität Bonn
Bonn, BRD

J. ZABCZYK
Institute of Mathem., Polish Acad.
of Sciences, Warsaw, Poland

M. ZAKAI
Israel Institute of Technology
Haifa, Israel

J. ZIENTEK
Universität Bonn
Bonn, BRD

CONTENTS:

I. GENERAL THEORY

STOCHASTIC DIFFERENTIAL SYTEMS

PART I : GENERAL THEORY

*Holzmacher und Waldhüter kennen die Wege.
Sie wissen, was es heißt, auf einem Holzweg
zu sein.*

Martin Heidegger: Holzwege

RADON-NIKODYM DERIVATIVES IN CASE OF RATIONAL SPECTRAL DENSITIES

Mátyás Arató[1]

1. Introduction.

A stationary Gaussian process $\xi(t)$ has rational spectral density function if and only if it is an autoregressive moving average type process, or the component of a multidimensional Markov process. In this paper we give explicite formulae for the Radon-Nikodym derivatives of measures related to this process and defined in functional spaces. These processes are the most important from the statistical point of view because they can be represented as the components of stochastic differential equations with constant coefficients. It turns out that the Riccati type differential equations in the Kalman filtering problem can be solved and the solution of the conditional expectation enables us to calculate the explicite Radon-Nikodym derivatives.

The examination of this result gives that the autoregressive moving average processes, in the general case, does not allow non-trivial system of sufficient statistics.

Furthermore, using the same method for solving the Riccati equation, we can answer some problem of estimation theory with noise. In the beginning, for the continuous time parameter the Radon-Nikodym derivative of a stationary process, $\xi(t)$, with rational spectral density was calculated only for autoregressive processes (see [24], [16], [2], [11]). In this case a set of sufficient statistics exists (see [1]). The case when the nominator is not constant was investigated first heuristically by Pisarenko [15] and later by Hajek [8], Rozanov [18], and Pisarenko [17]. In all attempts special methods, e.g. as the linear Hilbert space theory (with reproducing kernels) and the Wiener-Hopf theory were used. Here we shall give the representation with stochastic differential equation and we shall use Girsanov's theorem and optimal linear filtering developped by Liptser and Shiryaev [13] (Chapters 7 and 13), see also [9], [10].

In the following we shall use the filtering equations examined by Kalman and Bucy in the late fifties. We consider a $(k + \ell)$ dimensional Gaussian random process $(\underline{\theta}(t), \underline{\xi}(t))^* = [(\theta^1(t), \ldots, \theta^k(t)), (\xi^1(t), \ldots, \xi^\ell(t))]$, $0 \leq t \leq T$, satisfying the stochastic differential equations

---

[1] Visiting Professor at UCLA, Department of System Science, Los Angeles, CA 90024, and Budapest, Eotvos University, Department of Probability.

(1.1) $d\underline{\theta}(t) = [\underline{a}_0(t) + a_1(t)\ \underline{\theta}(t) + a_2(t)\ \underline{\xi}(t)]dt + b_1\ d\underline{w}_1(t) + b_2\ d\underline{w}_2(t)$

(1.2) $d\underline{\xi}(t) = [\underline{A}_0(t) + A_1(t)\ \underline{\theta}(t) + A_2(t)\ \underline{\xi}(t)\ dt + B_1\ d\underline{w}_1(t) + B_2\ d\underline{w}_2(t)$,

where $\underline{w}_1^*(t) = (w_1^1(t), \ldots, w_1^k(t)), \underline{w}_2^*(t) = (w_2^1(t), \ldots, w_2^\ell(t))$ are independent Wiener processes. $\underline{\theta}(0)$, $\underline{\xi}(0)$ are Gaussian and independent of the processes $\underline{w}_1(t)$, $\underline{w}_2(t)$. The measurable deterministic functions $a_i(t)$, $A_i(t)(i = 0,1,2)$ are square integrable. Let

$$\underline{m}(t) = E(\underline{\theta}(t)|F_t),\ \gamma(t) = E(\underline{\theta}(t) - \underline{m}(t))(\underline{\theta}(t) - \underline{m}(t))^*$$

be the conditional expectation vector and covariance matrix, respectively. Using the notation

(1.3) $$b \circ b = b_1 B_1^* + b_2 B_2^*$$

we have the following statement (see e.g. [13]).

Theorem 1. *The functions m(t), γ(t) satisfy the equations*

(1.4) $$d\underline{m}(t) = [\underline{a}_0(t) + a_1(t)\ \underline{m}(t) + a_2\ \underline{m}(t)]\ dt + [(b \circ B) + \gamma(t)\ A_1^*(t)] \times$$
$$\times (B \circ B)^{-1}\ [d\underline{\xi}(t) - (\underline{A}_0(t) + A_1(t)\ \underline{m}(t) + A_2\ \underline{\xi}(t))\ dt],$$

(1.5) $$\dot{\gamma}(t) = a_1(t)\ \gamma(t) + \gamma(t)\ a_1^*(t) + b \circ b - [(b \circ B) + \gamma(t)\ A_1^*(t)] \times$$
$$\times (B \circ B)^{-1}\ [b \circ B + \gamma(t)\ A_1^*(t)]^*,$$

with the initial conditions.

$$\underline{m}(0) = E(\underline{\theta}(0)|\underline{\xi}(0),\quad \gamma(0) = (\gamma_{ij}(0)).$$

The following two lemmas we shall use many times and they can be checked by direct calculations.

Lemma 1. *The one dimensional homogeneous equation.*

(1.6) $$\dot{\gamma}(t) = -A\gamma(t) - B\gamma^2(t)\ ,\ (A \geq 0\ ,\ B \geq 0)\ ,$$

has the solution

(1.7) $$\gamma(t) = e^{-At}\ [c_o + \frac{B}{A}(1 - e^{-At})]^{-1}\ ,\quad \gamma(0) = \frac{1}{c_o}\ ,\ t \geq 0,$$

The inhomogeneous equation

(1.8) $$\dot{\gamma}(t) = -A\gamma(t) - B\gamma^2(t) + b\ (A \geq 0\ ,\ B \geq 0\ ,\ b \geq 0),$$

has the solution

(1.9) $$\gamma(t) = e^{-At}\ [c_o + \frac{B}{A}(1 - e^{-At})]^{-1} + c\ ,\quad t \geq 0\ ,$$

where c *is the root of the equation (a particular and stationary solution).*

(1.10) $B c^2 + A c - b = 0$,

and

(1.11) $\gamma(0) = \frac{1}{c_0} + c$.

In the case $B = 0$ *we have*

(1.12) $\gamma(t) = \frac{1}{c_0} e^{-At} + c , \quad c = \frac{b}{A}$.

Lemma 2. *Let a, b, A, B be matrices, then the matrix solution* $\gamma(t)$ *of the following Euler-Riccati type equation*

(1.13) $\dot{\gamma}(t) = a\,\gamma(t) + \gamma(t)\,a^* + b\,b^* - \gamma(t)\,A^*\,(B\,B^*)^{-1}\,A\,\gamma(t)$

is given by

(1.14) $\gamma(t) = e^{at}\,[c_0 + \int_0^t e^{a^*u}\,A^*\,(B\,B^*)^{-1}\,A\,e^{au}\,du]^{-1}\,e^{a^*t} + c,$

where c *is the solution (a particular solution of the differential equation) of the following matrix equation*

(1.15) $a\,c + c\,a^* + b\,b^* - c\,A^*\,(B\,B^*)^{-1}\,A\,c = 0.$

The above explicite solutions of the Euler-Riccati equations first we shall use for the determination of Radon-Nikodym derivatives (see Arato [3]) and then for estimators in the presence of different types of noises. These results seem very useful because even in textbooks it is tacitly considered that no explicite solutions can be given for Riccati equations (see e.g. Liptser-Shiryaev [13], Sections 15.3, 16.2, 16.3, 17.1).

The estimation of parameters of stationary Gaussian Markov processes in the presence of noise was proposed to the author in the late fifties by A.N. Kolmogorov (see e.g. [4]).

2. Representation by Stochastic Differential Equations

Let $\xi(t)$ be a regular Gaussian process with spectral density

(2.1) $$f_\xi(\lambda) = \frac{|b_0(i\lambda)^q + b_1(i\lambda)^{q-1} + \dots + b_q|^2}{|(i\lambda)^p - [a_1(i\lambda)^{p-1} + \dots + a_p]|^2} = \frac{|Q(i\lambda)|^2}{|P(i\lambda)|^2} ,$$

where $P(z) = z^p - [a_1 z^{p-1} + \dots + a_p]$ has roots with negative real parts, $Q(z) = b_0 z^q + b_1 z^{q-1} + \dots + b_q$, $q < p$, and both polynomials have real coefficients.

The spectral representation gives

(2.2) $$\xi(t) = \int_{-\infty}^{\infty} e^{i\lambda t}\,\Phi_\xi\,(d\lambda) = \int_{-\infty}^{\infty} e^{i\lambda t}\,\frac{Q(i\lambda)}{P(i\lambda)}\,\Phi\,(d\lambda) ,$$

where $\phi(d\lambda)$ is an orthogonal spectral measure, $E\,\Phi(d\lambda) = 0$, $E|\Phi(d\lambda)|^2 = \frac{d\lambda}{2\pi}$

Such a process we shall call an ARMA (autoregressive-moving average) Gaussian process. We prove that an ARMA process is one component of a p-dimensional elementary Gaussian process $\underline{\xi}^*(t) = (\xi^1(t), \ldots, \xi^p(t))$, i.e.,

(2.3) $$d\,\underline{\xi}(t) = A\,\underline{\xi}(t)\,dt + B_w^{\frac{1}{2}} \cdot d\underline{w}(t),$$

where $\underline{w}(t)$ is the standard Wiener process.

Theorem 2. (*Representation A.*) (*see Liptser-Shiryaev* [13] § 15.3). *The stationary Gaussian ARMA process* $\xi(t) = \xi^1(t)$ *permitting the spectral representation given by (2.2), is the first component of the p-dimensional stationary Gaussian process* $\underline{\xi}^*(t) = (\xi^1(t), \ldots, \xi^p(t))$ *satisfying the linear stochastic equations*

(2.4) $$d\xi^j(t) = \xi^{j+1}(t)\,dt + \beta_j dw(t), \quad = 1,2, \ldots, p-1,$$
$$d\xi^p(t) = \sum_{k=0}^{p-1} a_{p-k}\,\xi^{k+1}(t)\,dt + \beta_p\,dw(t),$$

with the Wiener process

(2.5) $$w(t) = \int_{-\infty}^{\infty} \frac{e^{i\lambda t}-1}{i\lambda}\,\Phi(d\lambda),$$

and the coefficients $\beta_1, \beta_2, \ldots, \beta_p$ *are given by*

(2.6) $$\beta_p = [a_1\,\beta_{p-1} + \ldots + a_{p-1}\,\beta_1] + b_q,$$
$$\beta_{p-j} = \sum_{i=1}^{p-j-1} \beta_{p-j-i}\,a_i + b_{q-j}, \quad j = 1,2, \ldots, p-1,$$

(*where* $b_{-1} = \ldots = b_{-(p-q-1)} = 0$, $q \leq p-1$, $\beta_1 = b_{q-(p-1)}$) .

The components $\xi j(t)$ *are given by*

(2.7) $$\xi^j(t) = \int_{-\infty}^{\infty} e^{i\lambda t}\,W_j(i\lambda)\,\Phi(d\lambda), \quad \gamma = 1,2, \ldots, p,$$

where

(2.8) $$W_j(z) = \frac{1}{Z}[W_{j+1}(Z) + \beta_j], \quad j = 1,2, \ldots, p-1,$$
$$W_p(Z) = \frac{1}{Z} \sum_{k=0}^{p-1} a_{p-k}\,W_{k+1}(Z) + \beta_p,$$

and $E\,\xi^j(0)\,w(t) = 0$ $(t \geq 0,\ j = 1,2, \ldots, p)$.

For computational and other purposes there exists a more simple representation, which can be easily proved (see [3]).

Theorem 2'. (*Representation B*). *The stationary Gaussian ARMA process* $\xi(t)$ *permitting the spectral representation given by (2.2) is the first component of the p-dimensional stationary Gaussian process* $\underline{\xi}^*(t)$ *satisfying the linear stochastic equations (for simplicity let* $P(z)$ *have distinct roots* λ_i*,* $Re\lambda_i < 0$*).*

$$\frac{d\xi^1(t)}{dt} = \xi^2(t), \qquad \xi^1(t) = \xi(t),$$

$$\vdots$$

$$\frac{d^{p-q-1}\xi^1(t)}{dt^{p-q-1}} = \xi^{p-q}(t),$$

$$d\,\xi^{p-q}(t) = \sum_{i=1}^{p} d_i \cdot \xi^i(t)dt + b_o\, dw(t), \tag{2.9}$$

$$d\,\xi^{p-q+1}(t) = \lambda_{p-q+1}\,\xi^{p-q+1}(t)dt + dw(t),$$

$$\vdots$$

$$d\,\xi^p(t) = \lambda_p \xi^p(t)dt + dw(t),$$

where

$$\xi^1(t) = \xi(t) = \int_{-\infty}^{\infty} e^{i\lambda t}\left[\frac{Q_1}{i\lambda - \lambda_1} + \dots + \frac{Q_p}{i\lambda - \lambda_p}\right]\Phi(d\lambda) =$$

$$= \int_{-\infty}^{t}\left[Q_1 e^{\lambda_1(t-s)} + \dots + Q_p e^{\lambda_p(t-s)}\right] dw(s) =$$

$$= Q_1\eta_1(t) + \dots + Q_p\,\eta_p(t),$$

$$\xi^2(t) = \frac{d\xi^1(t)}{dt} = \int_{-\infty}^{\infty} \frac{i\lambda Q(i\lambda)}{P(i\lambda)}\Phi(d\lambda) = \int_{-\infty}^{\infty} e^{i\lambda t}\left[\frac{Q_1'}{i\lambda - \lambda_1} + \dots + \frac{Q_p'}{i\lambda - \lambda_p}\right]\Phi(d\lambda)$$

$$= Q_1'\,\eta_1(t) + \dots + Q_p'\,\eta_p(t), \tag{2.10}$$

$$\vdots$$

$$\xi^{p-q}(t) = \frac{d^{p-q-1}\,\xi^1(t)}{dt^{p-q-1}} = \int_{-\infty}^{\infty} e^{i\lambda t}\,\frac{(i\lambda)^{p-q-1}\,Q(i\lambda)}{P(i\lambda)}\,\Phi(d\lambda) =$$

$$= Q_1^{(p-q-1)}\eta_1(t) + \dots + Q_p^{(p-q-1)}\eta_p(t),$$

$$\xi^{p-q+1}(t) = \eta_{p-q+1}(t),$$

$$\vdots$$

$$\xi^p(t) = \eta_p(t),$$

and

(2.11) $$d\,\eta_j(t) = \lambda_j\,\eta_j(t)\,dt + dw(t)\,,$$

$$E\,\eta_j(0)\,w(t) = 0 \quad (t \geq 0,\ j = 1,2,\ \ldots,\ p).$$

Proof. Immediately follows from (2.10) and Lemma 15.3 in [13].

Remark 2.1. From (2.9) we see that in case $q = 0$ ($Q(z) = \text{const} = b_o$) we have an autoregressive, AR, process with "observable" components $\xi(t)$, $\xi'(t)$, ..., $\xi^{(p-1)}(t)$ and one stochastic equation

(2.12) $$d\,\xi^{(p-1)}(t) = \sum_{k=0}^{p-1} a_{p-k}\,\xi^{(k)}(t)\,dt + dw(t)\,,$$

with the solution

(2.13) $$\xi(t) = \int_{-\infty}^{\infty} e^{i\lambda t}\,\frac{1}{P(i\lambda)}\,\Phi(d\lambda) = \int_{-\infty}^{\infty} e^{i\lambda t}\left[\frac{C_1}{i\lambda - \lambda_1} + \ldots + \frac{C_p}{i\lambda - \lambda_p}\right]\Phi(d\lambda) =$$

$$= \int_{-\infty}^{t} \left[C_1\,e^{\lambda_1(t-s)} + \ldots + C_p\,e^{\lambda_p(t-s)}\right] dw(s).$$

Remark 2.2. If $q \geq 1$ the components $\xi^{p-q+1}(t)$, ..., $\xi^{p}(t)$ in (2.9) are unobservable and they are part of the equation of $\xi^{p-q}(t)$. The coefficients d_i are calculated by inverting the system of equations (2.10)

(2.10') $$\underline{\xi}(t) = \tilde{Q}'\,\underline{\eta}(t)$$

In case of complex roots of $P(z)$ the d_i may be complex.

To find the Radon-Nikodym derivative of AR processes, or elementary Gaussian processes of the form

$$d\,\xi^1(t) = \sum_{i=1}^{p} a_{1i}\,\xi^i(t)\,dt + dw_1(t)\,,$$

$$\vdots$$

$$d\,\xi^p(t) = \sum_{i=1}^{p} a_{pi}\,\xi^i(t)\,dt + dw_p(t)\,,$$

with independent Wiener processes $w_1(t)$, ..., $w_p(t)$ is known (see [2], [13], § 7.2). In case (2.10) the process $\xi^{p-q}(t)$ is an Ito Gaussian process and the Radon-Nikodym derivative has to be calculated by the help of filtering.

Example. (See [8], [6], [7], [15].) Let the spectral density have the form

(2.14) $$f_\xi(\lambda) = K\,\frac{\lambda^2 + 2\theta^2}{\lambda^4 + 4\theta^4}$$

Then representation (A) gives

$$(2.15)\qquad \begin{aligned} d\xi^1(t) &= \xi^2(t)\,dt + \sqrt{K}\,dw(t)\,, \\ d\xi^2(t) &= -(2\,\theta^2\xi^1(t) + 2\,\theta\xi^2(t))\,dt + \theta(\sqrt{2}-2)\,\sqrt{K}\,dw(t)\,, \end{aligned}$$

while representation (B) gives

$$(2.16)\qquad \begin{aligned} d\xi^1(t) &= [(i\theta-\theta)\xi^1(t) + \theta(1-i(1-\sqrt{2}))\,\xi^2(t)]\,dt + \sqrt{K}\,dw(t)\,, \\ d\xi^2(t) &= -(i\theta+\theta)\,\xi^2(t)\,dt + \sqrt{K}\,dw(t). \end{aligned}$$

3. The Radon-Nikodym Derivative

We shall use the following theorem, which can be deduced from Girsanov's Theorem (see [13], Th. 7.15). Let us consider the Ito process $\xi(t)$, $0 \le t \le T$, with the differential

$$(3.1)\qquad d\,\xi(t) = \beta(t,\omega)\,dt + dw(t)\,,$$

and let $P_\xi(\cdot)$, $P_w(\cdot)$ be the measures corresponding to the processes $\xi(t)$ and $w(t)$ respectively.

Theorem 3. Let $\beta(t,\omega)$ be a continuous (in the mean square) Gaussian process. Then $P_\xi \sim P_w$ and

$$(3.2)\qquad P\left\{\int_0^t \alpha^2\,(t,\xi)\,dt < \infty\right\} = P\left\{\int_0^t \alpha^2(t,w)\,dt < \infty\right\} = 1\,,$$

$$(3.3)\qquad \frac{dP_\xi}{dP_w}\,(t,\xi) = \exp\left\{\int_0^t \alpha\,(s,\xi)\,d\xi(s) - \frac{1}{2}\int_0^t \alpha^2\,(s,\xi)\,ds\right\}\,,$$

$$(3.4)\qquad \frac{dP_w}{dP_\xi}\,(t,w) = \exp\left\{-\int_0^t \alpha\,(s,w)\,dw + \frac{1}{2}\int_0^t \alpha^2(s,w)\,ds\right\}\,,\quad \alpha = E(\beta(t,w)\,|\,\mathcal{F}_t^\xi).$$

On the basis of Theorem 2, and representation (2.9) we distinguish between cases, a) the process $(\xi^1(t), \xi^2(t), \ldots, \xi^p(t))$ has only observable components, b) there are some unobservable components. In case b) for simplicity we assume that the unobservable components are $\xi^{p-q+1}(t), \ldots, \xi^p(t)$ with the representation

$$(3.5)\qquad \begin{aligned} d\,\xi^{p-q}(t) &= \sum_{i=1}^{p} d_i\,.\,\xi^i(t)\,dt + dw(t)\,, \\ d\,\xi^i(t) &= \lambda_i\,\xi^i(t)\,dt + dw(t),\quad i = p-q+1, \ldots, p, \end{aligned}$$

where $d_i \ne 0$ $(i = p-q, \ldots, p)$.

To calculate

$$(3.6)\qquad \alpha_i(t) = E\,(\xi^i(t)\,|\,\mathcal{F}_t^{\xi^{p-q}})\,,\quad i = p-q+1, \ldots, p,$$

(3.7) $\quad \gamma_{ij}(t) = E(\alpha_i(t) - \xi^i(t))(\alpha_j(t) - \xi^j(t));\ i,j = p-q+1, \ldots, p,$

we use Theorem 1. in this special case. We have the following equations for $\underline{\alpha}(t)^* = (\alpha_{p-q+1}(t), \ldots, \alpha_p(t))$ and $\gamma(t) = (\gamma_{ij}(t))$,

$$d\,\underline{\alpha}(t) = a_1\,\underline{\alpha}(t)\,dt + [\underline{b}_1 + \gamma(t)\,\underline{A}_1]\,[d\,\xi^{p-q}(t) - (\underline{A}_1^*\,\underline{\alpha}(t) + d_{p-q}\cdot\xi^{p-q}(t))\,dt] =$$
$$= [a_1 - \underline{b}_1\underline{A}_1^* - \gamma(t)\,\underline{A}_1\underline{A}_1^*]\,\underline{\alpha}(t) - [\underline{b}_1 + \gamma(t)\,\underline{A}_1]\,d_{p-q}\,\xi^{p-q}(t)\,dt + [\underline{b}_1 + \gamma(t)\,\underline{A}_1]\,d\,\xi^{p-q}(t), \tag{3.8}$$

$$\gamma(t) = \frac{d\gamma(t)}{dt} = 2\,[a_1 - \underline{b}_1\underline{A}_1^*]\,\gamma(t) - \gamma(t)\,\underline{A}_1\underline{A}_1^*\,\gamma(t)\ , \tag{3.9}$$

where

$$a_1 = \begin{pmatrix} \lambda_{p-q+1} & 0 & \cdots & 0 \\ 0 & \lambda_{p-q+2} & \cdots & 0 \\ \vdots & & & \\ 0 & & & \lambda_p \end{pmatrix}_{q\times q},\ \underline{b}_1 = \begin{pmatrix} 1 \\ \vdots \\ 1 \end{pmatrix},\ \underline{A}_1 = \begin{pmatrix} d_{p-q+1} \\ \\ d_p \end{pmatrix}_{q\times 1}$$

Equation (3.9) is an Euler type with solution (see (1.14))

$$\gamma(t) = e^{-2[\underline{b}_1\underline{A}_1^* - a_1]t}\,[\gamma^{-1}(0) + \underline{A}_1\underline{A}_1^*\int_0^t e^{-2[\underline{b}_1\underline{A}_1^* - a_1]u}\,du]^{-1}\ , \tag{3.10}$$

Further for $\underline{\alpha}(t)$ we get, as it can easily be done directly

$$\underline{\alpha}(t) = e^{\int_0^t [a_1 - \underline{b}_1\underline{A}_1^* - \gamma(s)\underline{A}_1\underline{A}_1^*]\,ds}\,\{\,\underline{\alpha}\,(0) + \int_0^t e^{\int_0^s [\underline{b}_1\underline{A}_1^* + \gamma(u)\underline{A}_1\underline{A}_1^* - a_1]\,du}\,[-(\underline{b}_1 + \gamma(s)\underline{A}_1)d_{p-q}\cdot\xi^{p-q}(s)\,ds + (\underline{b}_1 + \gamma(s)\,\underline{A}_1)\,d\,\xi^{p-q}(s)]\,\}\ , \tag{3.11}$$

with the initial values.

$$\underline{\alpha}(0) = E\left(\begin{matrix}\xi^{p-q+1}(0)\\ \vdots \\ \xi^p(0)\end{matrix}\ \middle|\ \begin{matrix}\xi^{p-q}(0)\\ \cdot\end{matrix}\right)\ , \tag{3.12}$$

(3.13) $\quad \gamma(0) = (E\,(\xi^{p-q+i}(0) - \alpha_i(0))\,(\xi^{p-q+j}(0) - \alpha_j(0)))$.

In order to find these values in (3.12) and (3.13) we shall take again advantage of elementary Gaussian process $(\xi^{p-q}(t), \ldots, \xi^{p}(t)$ that $B(0) = (E\,\xi^{p-q+i}(0)\,\xi^{p-q+j}(0))$ is the unique solution of the system of equations

$$A\,B(0) + B(0)\,A^{*} = -\,B_w$$

where

$$A = \begin{pmatrix} d_{p-q} & d_{p-q+1} & \cdots & d_p \\ 0 & \lambda_{p-q+1} & \cdots & 0 \\ \vdots & & & \\ 0 & 0 & & \lambda_p \end{pmatrix}, \quad B_w = \begin{pmatrix} 1 & 1 & \cdots & 1 \\ 1 & 1 & \cdots & 1 \\ \vdots & & & \\ 1 & 1 & & 1 \end{pmatrix}$$

Using formula (3.3) we get that in the general case no sufficient statistic exists which we can state in the following.

Theorem 4. *Let the stationary Gaussian process* $\xi(t)$ *permit the spectral representation given by (2.2), then if* $Q(z) \neq$ *const no sufficient statistic exists for the set of unknown parameters* $(b_o, \ldots, b_q), (a_1, \ldots, a_p)$.

4. Estimation problems

In this part we assume that $\theta(t)$ is a one dimensional first order autoregressive type Gaussian process, i.e.,

(4.1) $\quad d\theta(t) = -\,\alpha\theta(t)\,dt + \sqrt{c_1}\,dw_1(t)$,

where $w_1(t)$ is a Wiener process. Let $\xi(t)$ be the observable process

(4.2) $\quad \xi(t) = \theta(t) + \varepsilon(t)$,

where

(4.3) $\quad d\varepsilon(t) = -\,\beta\varepsilon(t)\,dt + \sqrt{c_2}\,dw_2(t)$,

and $w_1(t), w_2(t)$ are independent Wiener processes, and independent of $\xi(0)$, $\theta(0)$.

Theorem 1 gives the following filtering equations for $m(t) = E(\theta(t)\,|\,F_t^{\xi})$ and $\gamma(t) = E(\theta(t) - m(t))^2$,

(4.4) $$d\,m(t) = -\,\alpha m(t)\,dt + \frac{c_1 + (\beta-\alpha)\,\gamma(t)}{c_1 + c_2}\,[d\xi(t) - ((\beta-\alpha)\,m(t) - \beta\xi(t))\,dt] = -\,[\tfrac{A}{2} + B\gamma(t)]\,m(t)\,dt + \frac{\beta}{\beta-\alpha}\,[A/2 - \alpha + B\gamma(t)]\,\xi(t)\,dt + \frac{1}{\beta-\alpha}\,[A/2 - \alpha + B\gamma(t)]\,d\xi(t),$$

$$(4.5)\quad \dot{\gamma}(t) = -2\alpha\gamma(t) + c_1 - \frac{1}{c_1+c_2}(c_1 + (\beta-\alpha)\gamma(t))^2 =$$

$$= -A\gamma(t) - B\gamma^2(t) + b$$

where

$$(4.6)\quad A = 2\frac{\alpha c_2 + \beta c_1}{c_1 + c_2}, \quad B = \frac{(\beta-\alpha)^2}{c_1+c_2}, \quad b = \frac{c_1 c_2}{c_1+c_2}.$$

Using Lemma 1 we get that $\gamma(t)$ has the form

$$(4.7)\quad \gamma(t) = e^{-At}[c_o + \frac{B}{A}(1 - e^{-At})]^{-1} + c$$

where

$$(4.8)\quad c = \frac{-A+\sqrt{A^2+4Bb}}{2B}, \quad \gamma(0) = \frac{c_1 c_2}{2(\alpha c_2 + \beta c_1)},$$

$$\frac{1}{c_o} = \gamma_o - c .$$

The optimal estimator $m(t)$ can be given in the following way

$$(4.9)\quad m(t) = e^{-\int_o^t [A/2+B\gamma(s)]\,ds}\{m(0) + \int_o^t e^{\int_o^s [A/2+B\gamma(u)]du}\{\frac{\beta}{\beta-\alpha}[A/2 +$$

$$+ B\gamma(s) - \alpha]\xi(s)\,ds + \frac{1}{\beta-\alpha}[a/2+B\gamma(s) - \alpha]\,d\xi(s)\} =$$

$$= e^{-\int_o^t [A/2+B\gamma(s)]\,ds}\{m(0) + e^{\int_o^t [A/2+B\gamma(s)]\,ds}\frac{1}{\beta-\alpha}[A/2+B\gamma(t)-\alpha]\xi(t)$$

$$- [A/2+B\gamma(0)-\alpha]\xi(0) + \int_o^t e^{\int_o^s [A/2+B\gamma(u)]\,du}\frac{1}{\beta-\alpha}[A\frac{\beta-\alpha}{2} - \beta\alpha - \frac{A^2}{4} +$$

$$- bB + B\gamma(s)(\beta - 2A + \alpha) - \gamma^2(s)B^2]\xi(s)\,ds\},$$

where

$$(4.10)\quad \int_o^t [A/2 + B\gamma(s)]\,ds = (A/2 + c)\,t - \frac{1}{B}\ln\frac{c_o}{c_o + B/A - B/A\,e^{-At}}.$$

Putting $\beta \to \infty$, $c_2 \to \infty$ in such a way that

$$(4.11)\quad \frac{\beta^2}{c_2} \to \beta_o$$

holds, one can get that

$$A \to 2\alpha,\ B \to \beta_o\ ,\ c \to \frac{-\alpha + \sqrt{\alpha^2 + b\beta_0}}{\beta_0} = \tilde{c}\ ,$$

(4.12)

$$\gamma_0 \to \frac{c_1}{2\alpha}\ ,\ \frac{1}{c_0} \to \gamma_0 - \tilde{c} = \frac{1}{\tilde{c}_0}\ .$$

Then for $\tilde{\gamma}(t)$ and $\tilde{m}(t)$ we have

(4.13) $$\tilde{\gamma}(t) = e^{-2\alpha t}\ [\tilde{c}_0 + \frac{\beta 0}{2\alpha}(1 - e^{-2\alpha t})]^{-1} + \tilde{c}$$

and, using (4.3) and (4.10),

$$\tilde{m}(t) = e^{-(\alpha + \tilde{c})\,t} \left(\frac{\tilde{c}_0}{\tilde{c}_0 + \frac{\beta_0}{2\alpha} - \frac{\beta_0}{2\alpha} e^{-2\alpha t}} \right)^{\beta_0} \{\ m(0) +$$

$$+ \beta_0 \int_0^t e^{(\alpha+\tilde{c})s} \left(\frac{\tilde{c}_0}{\tilde{c}_0 + \frac{\beta_0}{2\alpha} - \frac{\beta_0}{2\alpha} e^{-2\alpha t}} \right)^{-\beta_0} \left[e^{-2\alpha s} (\tilde{c}_0 + \frac{\beta 0}{2\alpha} - \frac{\beta 0}{2\alpha} e^{-2\alpha s})^{-1} + \tilde{c} \right] \xi(s)ds$$

If we are interested in the estimation of the unknown drift parameter in the presence of noise the above method can be used. Let us assume that the noise is a Wiener process, i.e., the two dimensional process $(\theta(t), \xi(t))$ is given by the equations

(4.15) $$d\ \theta(t) = -\ \alpha\ \theta(t)\ dt + \sqrt{c_1}\ dw_1(t)\ ,$$

(4.16) $$d\ \xi(t) = d\ \theta(t) + d\ \varepsilon(t) = -\ \alpha\ \theta(t)\ dt + \sqrt{c_1}\ dw_1(t) + \sqrt{c_2}\ dw_2(t),$$

where $w_1(t)$, $w_2(t)$ are independent Wiener processes. The filtering equations are given by Theorem 1 in the form

(4.17) $$d\ m(t) = -\ \alpha\ m(t)\ dt + \frac{c_1 - \alpha\gamma(t)}{c_1 + c_2}(d\ \xi(t) + \alpha\ m(t)\ dt) =$$

$$= -\frac{\alpha}{c_1 + c_2}[c_2 + \alpha\ \gamma(t)]\ m(t)\ dt + \frac{c_1 - \alpha\ \gamma(t)}{c_1 + c_2}\ d\ \xi(t)$$

(4.18) $$\dot{\gamma}(t) = -\ 2\alpha\ \gamma(t) - \frac{[c_1 - \alpha\gamma(t)]^2}{c_1 + c_2} + c_1 =$$

$$= -\frac{2\alpha\ c_2}{c_1+c_2}\gamma(t) - \frac{\alpha^2}{c_1+c_2}\gamma^2(t) + \frac{c_1 c_2}{c_1+c_2}\ ,$$

with the initial conditions

$$m(0) = E(\theta(0)|\xi(0)), \quad \gamma(0) = E(\theta(0) - m(0))^2.$$

Using Lemma 1 one can get

$$(4.19)\qquad \gamma(t) = e^{-\frac{2\alpha c_2}{c_1+c_2}} + [c_0 + \frac{\alpha}{2c_2}(1 - e^{-\frac{2\alpha c_2}{c_1+c_2}}]^{-1} + c$$

where

$$c = \frac{1}{\alpha}[c_2 + \sqrt{c_2^2 + c_1 c_2}], \quad \frac{1}{c_0} = \gamma(0) - c.$$

Taking advantage of Girsanov's theorem (Theorem 3.) we have

$$(4.20)\qquad \frac{dP_\xi}{dP_w}(\xi(t)) = \exp\{\int_0^t \alpha(s,\xi)\, d\xi - \frac{1}{2}\int_0^t \alpha^2(s,\xi)\, ds\},$$

where

$$(4.21)\qquad \alpha(t,\xi) = E(-\alpha\,\theta(t)|F_t^\xi) = -\alpha\, E(\theta(t)\,|\,F_t^\xi) = -\alpha\, m(t)$$

From (4.17), after tedious calculations, we get

$$(4.22)\qquad m(t) = e^{-\frac{\alpha}{c_1+c_2}\int_0^t [c_2+\alpha\gamma(s)]\, ds}\{m(0) + e^{\frac{\alpha}{c_1+c_2}\int_0^t [c_2+\alpha\gamma(s)]\, ds}\cdot$$

$$\cdot \frac{c_1 - \alpha\gamma(t)}{c_1 + c_2}\xi(t) - \frac{c_1 - \alpha\gamma(0)}{c_1 + c_2}\xi(0) - \int_0^t e^{\frac{\alpha}{c_1+c_2}\int_0^s [c_2+\alpha\gamma(u)]\, du}\frac{\alpha^2}{c_1+c_2}\gamma(s)\xi(s)ds\}.$$

where

$$(4.23)\qquad \int_0^t [c_2 + \alpha\gamma(s)]\, ds = (c_2 + c_1)t - \frac{c_2+c_2}{\alpha^2}\ \ln \frac{c_0}{c_0 + \frac{\alpha}{2c_2} - \frac{\alpha}{2c_2}e^{-\frac{2\alpha c_2}{c_1+c_2}t}}$$

From (4.19) - (4.23) one can get the maximum likelihood estimator of α, but note that no finite system of sufficient statistics exists in this case.

If $c_2 \to 0$ one can get the maximum likelihood estimator of α in the first order autoregressive process, which was discussed by the author (see e.g. in [4]).

REFERENCES

[1] M. Arató. On the sufficient statistics of stationary Gaussian random processes, Theory of Probability of Applic., 6 (1961) 199-201.

[2] M. Arató. Exact formulas for density measure of elementary Gaussian processes, Studia. Sci. Math. Hung, 5 (1970) 17-27.

[3] M. Arató. On sufficient statistics of Gaussian processes with rational spectral density function (submitted to Analysis Mathematia), (1982).

[4] M. Arato. Linear Stochastic Systems with Constant Coefficients (manuscript, to appear in Lecture Notes in Control and Information Sciences, Springer) (1982).

[5] M. Arató, A. Benczúr, A. Krámli, F. Pergel. Statistical problems of the elementary Gaussian processes, MTA Sztaki Tanulmanyok, 41 (1975) 1-65.

[6] K. O. Dzhaparidze. On the estimation of the spectral parameters of a Gaussian stationary process with spectral density, Theory Probability and Applic., 15 (1970) 531-538.

[7] K. O. Dzhaparidze. On methods for obtaining aymptotically efficient spectral parameter estimates for a stationary Gaussian process with rational spectral density, Theory Probability and Applic., 16 (1971) 550-554.

[8] J. Hajek. On linear statistical problems in stochastic processes, Czechoslovak Math. Fourn., 12 (1962) 404-444.

[9] T. C. Kailath. The structure of Radon-Nikodym derivatives with respect to Weiner and related measures, Annals Math. Stat., 42 (1971) 1054-1067.

[10] T. C. Kailath, M. Zakai. Absolute continuity and Radon-Nikodym derivatives for certain measures relative to Wiener measure, Annals Math. Stat., 42 (1971) 130-140.

[11] A. LeBreton. Parameter estimation in a vector linear stochastic differential equation, Trans. Seventh Prague Conf., (1974) Vol. A, 353-366.

[12] A. LeBreton. On continuous and discrete sampling for parameter estimation in diffusion type processes, Math. Programming Studies 5 (1976) 124-144.

[13] R. Liptser, A. Shiryaev. Statistics of random processes (in Russian, 1974), Springer, (1978).

[14] E. Parzen. Efficient estimation of stationary time series mixed schemes, Bull. Internat. Stati. Inst., 44, Book 2, (1971) 315-319.

[15] V. Pisarenko. On the problem of discovering a random signal in noisy background (in Russian), Radiotechnica & Elektronika, 6 (1961) 515-528.

[16] V. Pisarenko. On the estimation of paramaters of stationary Gaussian process with spectral density $|p(i\lambda|^{-2}$, Litovsk. Mat. Sbornik, 2 (1963) 159-167.

[17] V. Pisarenko. On the computation of the likelihood ratio for Gaussian processes with a rational spectrom, Theory Probability and Applic., 10 (1965) 299-303.

[18] V. Pisarenko, F. Rozanov. On certain problems for stationary processes leading to integral equations related to the Wiener-Hopf equation, Problemi Peredachi Informacii, 14 (1963) 113-135.

[19] J. Rozanov. Stationary Random Processes (in Russian, 1962) Holden-Day San Francisco (1967)

[20] J. Rozanov. Infinite dimensional Gaussian distributions, Moscow, Nauka (1971).

[21] P. Shaman. On the inverse of the covariance matrix of a first-order moving average, Biometrika, 56 (1969) 595-600.

[22] P. Shaman. On the inverse of the covariance matrix for an autoregressive-moving average process, Biometrika, 60 (1973) 193-196.

[23] A. Shiryaev. Probability (in Russian), Nauka, Moscow, (1980).

[24] Ch. Striebel. Densities for stochastic processes, Annals Math. Stat., 30 (1959) 559-567.

[25] P. Whittle. Estimation and information in stationary time series, Arkiv Math., 2 (1953) 423-434.

DIFFERENTIATION OF MEASURES RELATED TO STOCHASTIC PROCESSES

A.N. AL-HUSSAINI
Department of Statistics and Applied Probability
University of Alberta
Edmonton, Alberta
Canada T6G 2G1

0. SUMMARY

Vector measures induced by stochastic processes, especially martingales, have been discussed by several authors ([2], [3], [5]), primarily in the context of stochastic integration. Our purpose here is to invetigate some of their properties, including differentiation and the Radon-Nikodym property. Our approach combines methods drawn from the existing literature with several new techniques.

I. INTRODUCTION

Here we collect definitions and some theorems from vector measure theory. The basic reference is [4].

Def. 1.1. A function F from an algebra $\mathcal{F}$ of subsets of a set Ω to a Banach space X is called a vector measure if whenever E_1, E_2 are disjoint members of $\mathcal{F}$, then

$$F(E_1 \cup E_2) = F(E_1) + F(E_2) .$$

Def. 2.1. A vector measure F from $\mathcal{F}$ above to X above is called countably additive vector measure if whenever $E_1, E_2, \ldots$ are disjoint members of $\mathcal{F}$, such that $\cup E_i \in \mathcal{F}$, then:

$$F(\bigcup_i E_i) = \sum_i F(E_i) ,$$

in the norm topology of X. Let $\| \ \|$ denote this norm.

Def. 3.1. Let $F:\mathcal{F} \to X$ be a vector measure. The variation of $|F|$ is the extended nonnegative real valued $\mathcal{F}$ whose value on a set $A \in \mathcal{F}$ is given by:

$$|F|(A) = \sup_{\pi} \sum_{C\in\pi} \|F(C)\|$$

where the supremum is taken over all partition π of A into a finite number of pairwise disjoint members of F . F is of bounded variation if $|F|(\Omega) < +\infty$.

To introduce the next definition, let X^* denote the dual of X . Its norm is denoted by $\| \ \|$ also.

Def. 4.1. The semivariation of F is the extended nonnegative real valued function $\|F\|$ whose value on a set $A \in F$ is given by:

$$\|F\| (A) = \sup\{|x^*F| : x^* \in X^*, \|x^*\| \leq 1\}$$

where $|x^*F|$ is the variation of the real valued measure x^*F .

An equivalent definition (PROPOSITION 11, page 4, [4]) is the following:

Def. 5.1. Let $F:F \to X$ be a vector measure. Then the semivariation $\|F\|$ of F on $A \in F$ is defined by:

$$\|F\| (A) = \sup\{ \|\Sigma \, \varepsilon_n \, F(A_n)\| \}$$

where the supremum is taken over all partitions π of A into finitely many disjoint members of F and all finite $\{\varepsilon_n\}$ satisfying $|\varepsilon_n| \leq 1$.

If $\|F\| (\Omega) < +\infty$, then F is of bounded semivariation or simply F is bounded.

Def. 6.1. Let $F:F \to X$ be a vector measure. F is strongly additive if it is countably additive i.e. if it satisfies def. 2.1.

F is called weakly countably additive if whenever $E_1, E_2, \ldots$ are disjoint members of F , such that $\cup E_i \in F$, then:

$$x^*F(\bigcup_i E_i) = \sum_i x^*F(E_i) ,$$

for every $x^* \in X^*$.

The following two theorems are drawn from [4].

Theorem 1.1. (Caratheodory-Hahn-Kluvanek Extension Theorem). Let F be an algebra of subsets of a set Ω , and let Σ be the σ-algebra generated by F . Any one of the following four statements about a bounded weakly countably additive vector measure $F:F \to X$ implies all others:

(i) F has a unique countably additive extension $F:\mathcal{F}\to X$.

(ii) There exists a nonnegative real valued countably additive measure μ on $\mathcal{F}$ such that $F \ll \mu$. Here $\ll$ means F is absolutely continuous relative to μ, i.e. $F(C) = 0$ whenever $|\mu|(C) = 0$, where again $|\mu|(C)$ is the variation of μ at C .

(iii) F is strongly addtive.

(iv) $F(\mathcal{F})$ is a relatively weakly compact subset of X.

Theorem 2.1 (Rybakov). Let Σ be a σ-algebra of subsets of a set Ω . If $F:\Sigma\to X$ is a countably additive vector measure. Then there is $x^*\in X^*$ such that $F \ll |x^*F|$.

We need two kinds of integration; they are Bochner integral and integral relative to a vector measure. Details of these are found in [4]. Also we need the concept of Radon-Nikodym property which we introduce:

Def. 7.1. Let (Ω,Σ,μ) a finite measure space. A Banach space X has the Radon-Nikodym property relative to (Ω,Σ,μ) if for each μ-continuous vector measure $G:\Sigma\to X$ of a bounded variation there exists $g\in L_1(\mu,X)$ such that $G(A) = \int_A g d\mu$ for all $A\in\Sigma$.

Here $L_1(\mu,X)$ is the space of Bochner integrable functions relative to μ with values in X .

II. MEASURES RELATED TO STOCHASTIC PROCESSES.

Although the measures we are about to define can be attached to more general processes, we specialize to Martingale set up. Let $(\Omega,\Sigma,P,\Sigma_t)$ be a filtered probability sapce, that is (Ω,Σ,P) is a probability space and Σ_t is an increasing sub-σ-algebras of Σ satisfying the "usual conditions". Thus Σ_0 contains all P-negligible sets and Σ_t is right continuous. Consider:

$$M = \{M: M \text{ is a right continuous martingale relative to } (\Sigma_t), M_0 = 0, \text{ and } EM_\infty^2 < +\infty\}.$$

Here E denotes expectation with respect to P , and $M_\infty = \lim_{t\to\infty} M_t$. Let $\|M\|_2^2 = EM_\infty^2$, then M is a Hilbert space.

For the sake of simplicity we will deal with the time parameter in a finite interval. Specifically we take the time interval to be the unit interval $[0,1]$. Now let $\mathcal{B}$ be the Borel sets of $[0,1]$, let $\mathcal{P}$ be the sub-σ-algebra of $\Sigma\otimes\mathcal{B}$ generated by all Σ_t-adapted left continuous processes or equivalently by sets of the form $C_0\otimes\{0\}$, where $C_0\in\Sigma_0$, and $C\otimes(r,s]$ the Cartesian product of C with the interval $(s,t]$ where $C\in\Sigma_r$. This $\mathcal{P}$ is known as the predictable sets. To continue

Let:

$$X_t^2 = M_t + A_t$$

be the Doob-Meyer decomposition of $X \in M$. Define:

$$X[C \otimes (r,s]) = 1_C[X(s \wedge t) - X(r \wedge t)] , \quad t \geq 0$$

and extend X linearly to the algebra P_0 of all finite disjoint unions of sets of the form $C \otimes (s,t]$. Thus X is a vector measure from P_0 to M .

By Doob-Meyers decomposition and martingale property

$$\|X[C \otimes (r,s]]\|^2 = E[1_C(X_s - X_r)^2]$$

$$\int_C (X_s - X_r)^2 dP = \int_C (X_s^2 - X_r^2) dP$$

$$= E[1_C(A_s - A_r)]$$

$$= E\int_{C \otimes (r,s]} dA_t$$

It is easy to see that:

$$\|X(A)\|^2 = E\int_A dA_s \qquad *$$

holds for all $A \in P_0$. It should have been mentioned a bit earlier that $E\int dA_s$, defines a finite measure on $\Sigma \otimes B$. The same is true on P and P_0 of course. From the continuity from above at the empty set ϕ of $\nu(A) = E\int_A dA_s$, $A \in P_0$ it follows the same for μ , where:

$$\mu(A) = X(A) \qquad A \in P_0 .$$

Finally we use the right hand side of * to extend μ to P by approximating elements of P by elements of P_0 .

By either([4], PROPOSITION 11, pp. 4) or by a simple argument applied to the semivariation of X (def.5.1) it is easy to see that X is bounded, and therefore X enjoys all properties (i) through (iv) stated in Theorem 1.1.

III. INTEGRATION

Let:

$$L = \{f : f \text{ is } P\text{-measurable and } E\int f^2 dA_s < +\infty\} .$$

If $f = \sum_{i=1}^{n} a_i 1_{A_i}(w,t)$ where a_i, $1 \le i \le n$ is real, A_i's are disjoint, of the form $A_i = F_i \otimes (r_i,s_i]$, F_i is Σ_{r_i}-measurable. Define:

$$\int f dX = \sum_{i=1}^{n} a_i 1_{F_i}[X(s_i \wedge t) - X(r_i \wedge t)] , \quad t \ge 0$$

In general for $f \in L$, approximate f by simple functions f_n of the form above and define:

$$\int f dX = \lim_{r \to \infty} \int f_n dX ,$$

using Cauchy's criteria.

IV. DIFFERENTIATION

Recall $\nu(A) = E\int dA_s$ and $\mu(A) = \int_A dX$.

Theorem 1.4. $\mu(A) = \int_A f d\nu$ for some Bochner integrable function f if and only if μ is of bounded variation.

Proof. As μ is a Hilbert space valued vector measure, the representation follows from ([4], pp. 218). On the other hand if $\mu(A) = \int_A f d\nu$, then $|\mu|(\bar{\Omega}) \le \int_{\bar{\Omega}} \|f\| d\nu < +\infty$ which proves the necessary part. Here $\bar{\Omega} = \Omega \otimes [0,1]$.

Remark 1.4. Important and interesting martingales induce unbounded variation measures. See examples at the end. In view of this remark we have:

Theorem 2.4. Let $X \in M$. If EA_t is continuous, then μ is of unbounded variation, provided that $EA_1 \ne 0$.

Proof. Now $\|(A)\|_2 = (E \int_A dA_s)^{\frac{1}{2}}$. Write $\bar{\Omega} = \Omega \otimes [0,1] = \Omega \otimes \{0\} \cup \Omega \otimes (t_i,t_{i+1}]$.

Thus:

$$|\mu|(\bar{\Omega}) \geq E[\int_{\Omega\otimes\{0\}} dA_s]^{\frac{1}{2}} + \sum_i [E(A_{t_{i+1}} - A_{t_i})]^{\frac{1}{2}}$$

which tends to plus infinity by choosing (using the continuity of EA_t) $E(A_{t_{i+1}} - A_{t_i})$ comparable to $\frac{1}{i^2}$.

The scope of this theorem is apparent from the following corollary. First:

Def. 1.4. Suppose $\{X_t\}$ is a right continuous uniformly integrable supermartingale. Then $\{X_t\}$ is called regular if, for every predictable stopping time T , $E(X_{T-}) = E(X_T)$.

Corollary 1.4. If $X \in M$ is regular, then the induced vector measure by X is of unbounded variation.

Proof. A_t in the Doob-Meyer decomposition is continuous by [1], which easily implies that EA_t is continuous.

Theorem 3.4. Let $X \in M$. Then:

$$\nu(A) = \int_A f d\mu ,$$

where f is a random linear function on M .

Proof. A glance at definition 5.1 shows that $\nu(A) = 0$, whenever $\|\mu\|(A) = 0$ implying by (Def. 4.1) and Theorem 2.1 that

$$\nu \ll |x^*\mu| ,$$

for some $x^* \in M^* = M$. So by Radon-Nikodym theorem:

$$\begin{aligned}\nu(A) &= \int_A h d|x^*\mu| \\ &= \int_A hx^*(1_{A_1} - 1_{A'_1})d\mu \\ &= \int_A f d\mu .\end{aligned}$$

A_1 is the positivity set for $x^*\mu$ and A' = complement of A .

Remark 3.4. The proof above is suggested by ([4] pp. 96) but watch $x^*\mu(A \cap A'_1)$ is missing!

V EXAMPLES

1. Brownian motion: Here $A_t = t$, so even without Corollary 1.4 the corresponding measure is of unbounded variation.
2. Poisson process. Let X_t be a Poisson process, say with density equal to one then $X_t - t$ is a martingale, whose $A_t = t$ also. So again an explicit example of an unbounded variation measure μ .

References

(1) Dellacherie, C.: Capacities et Processus Stochastiques. Ergebnisse der Math. 67, Springer-Verlag (1972).

(2) Metivier, M.: The stochastic integral with respect to processes with values in a relexive Banach space. Theory of probability and its application (1974).

(3) Pellaumail, J.: Une nouvelle construction de l'integrale stochastique. Asterisque #9.

(4) Diestel, J. and Uhl, J.J.: Vector Measures. Amer. Math. Soc. (1977).

(5) Walsh, J.B.: Vector Measures and the Ito Integral. Lecture Notes in Math. 645. Springer-Verlag (1970).

DYNKIN GAMES

M. ALARIO-NAZARET
Faculté des Sciences et Techniques
de l'Université de Besançon
Département de Mathématiques
BESANCON

J.P. LEPELTIER
Université du Maine
Département de Mathématiques
Route de Laval
72017 LE MANS Cédex

B. MARCHAL
Université de Paris XII
LA VARENNE - St HILAIRE

Introduction. We study the existence of a value and a saddle-point for Dynkin games whose cost is :

$$E(X_T \mathbb{1}_{(T\leq T')} - X'_{T'} \mathbb{1}_{(T'<T)})$$

where X ans X' are optionals of class (D).

This problem introduced by Dynkin-Yushkevich [5] has been studied for the first time in General Theory by Bismut [2], [3] in the right continuous case. Alwais under the same assumptions Stettner [7] proved the existence of a value. In this paper we generalize their results and in particular we prove the existence of a a value in the upper right semi-continuous case, and the existence of a saddle-point in the upper semi-continuous case ; for this we use sophisticated results about optimal stopping of N. El Karoui [5].

§1 - Zero-Sum games

Définition 1-1. A zero-sum game is described by :

- a probability space (Ω, F, P)
- two sets U and V, U(resp. V) beeing the set of policies for the first (resp. second) player.
- a cost J(u,v), i.e. a family of random variables indexed by UxV.

The rule of the game is as follows : the first (resp. the second) player tries to maximize (resp. minimize) J(u,v), using u(resp.v), in the sense of P-ess sup (resp. P-ess inf).

Any zero-sum game is written :

$$(\Omega, \underline{\underline{F}}, \underline{P}, U, V, (J(u,v))_{(u,v) \in U \times V}) \quad (*)$$

<u>Définition 1-2.</u> (u^*, v^*) is a saddle-point for the game (*) if :

$$\forall (u,v) \in U \times V \qquad J(u,v^*) \leq J(u^*,v^*) \leq J(u^*,v) \quad \text{P.a.s.}$$

<u>Définition 1-3.</u> The upper (resp. lower) value of the game (*) is :

$$V^+ = \text{P-ess}\inf_{v \in V} \quad \text{P-ess}\sup_{u \in U} \quad J(u,v)$$

$$(\text{resp. } V^- = \text{P-ess}\sup_{u \in U} \quad \text{P-ess}\inf_{v \in V} \quad J(u,v))$$

When $V^+ = V^-$ P.a.s., we shall say that the game has a value $V = V^+ = V^-$ P.a.s.

It is well known that the game has a value as soon as there exists a saddle point. More precisely we have :

<u>Proposition 1-4.</u> If (u^*, v^*) is a saddle-point for the game, the game has a value V realized by (u^*, v^*), i.e., :

$$V = J(u^*, v^*) \quad \text{P.a.s.}$$

Nevertheless the next result is useful to prove the existence of a value, without the existence of a saddle-point.

<u>Proposition 1-5.</u> If for all $\varepsilon > 0$, there are u_ε, v_ε such that :

$$J(u, v_\varepsilon) - \varepsilon \leq J(u_\varepsilon, v) + \varepsilon \quad \text{P.a.s.} \quad \forall (u,v) \in U \times V$$

then the game has a value.

§ 2 - <u>Dynkin games</u>

<u>Définition 2-1.</u> By Dynkin game we mean the zero-sum game :

$$(\Omega, \underline{\underline{F}}_\infty, \underline{\underline{F}}_t, P, \mathcal{T}, \mathcal{T}, (J(S,S'))_{(S,S') \in \mathcal{T} \times \mathcal{T}}) \text{ where :}$$

- $(\Omega, \underline{\underline{F}}_\infty, P)$ is a probability space, $(\underline{\underline{F}}_t)_{t \geq 0}$ a filtration of $\underline{\underline{F}}_\infty$, with "usual conditions" [4].
- $\mathcal{T}$ is the set of $\underline{\underline{F}}_t$ - stopping times.
- the cost $J(S,S')$ is :

$$J(S,S') = E(X_S\, 1_{(S\le S')} - X'_{S'}\, 1_{(S'<S)})$$

where X, X' are optional processes of class (D) defined at infinity by $X_\infty = X'_\infty = 0$. Given a stopping time T, we define a T-conditional Dynkin game :

Definition 2-2. By T-conditional Dynkin game we mean the zero-sum game :

$$(\Omega, \underline{\underline{F}}_\infty, \underline{\underline{F}}_t, P, \mathcal{T}_T, \mathcal{T}_T, (J^T(S,S'))_{(S,S')\in \mathcal{T}_T\times\mathcal{T}_T}) \text{ where}$$

- $\mathcal{T}_T$ is the set of $\underline{\underline{F}}_t$ - stopping times $\geq T$
- $J^T(S,S') = E(X_S\, 1_{(S\le S')} - X'_{S'}\, 1_{(S'<S)} \mid \underline{\underline{F}}_T)$ P.a.s.

We now construct (under proper assumptions) the value of the T-conditional Dynkin game $(Z - Z')_T$ where (Z,Z') is solution of :

$$(I) \quad \begin{cases} Z = R(X + Z') \\ Z' = R(X' + Z) \end{cases}$$

(R is the snell envelope-operator)

The equation (I) has always a solution under Mokobodski's assumption. More precisely we have :

Theorem 2-3. There is a solution of (I) $Z, Z' \geq 0$ if and only if the Mokobodski's assumption is satisfied :

(H) $\exists\, \tilde{Z}, \tilde{Z}'$ positive supermartingales of class (D) with :
$X \leq \tilde{Z} - \tilde{Z}' \leq -X'$

Proof : If (H) is true, we construct the double sequence :

$$Z^1 = R(X)\ ,\ Z'^1 = R(X')$$

$$Z^{n+1} = R(X+Z'^n)\ ,\ Z'^{n+1} = R(X'+Z^n) \quad \forall n \geq 1$$

Then it is easy to prove that the sequences Z^n and Z'^n are positive, increasing. Let $Z = \lim\nearrow Z^n$, $Z' = \lim\nearrow Z'^n$. (Z,Z') is solution of (I).

Conversely if (Z,Z') is solution of (I) we have by the definition of Snell's envelope :

$$Z \geq Z' + X\ ,\ Z' \geq Z + X'$$

which implies (H) whith $\tilde{Z} = Z$, $\tilde{Z}' = Z'$.

From the existence of solutions for (I), we deduce a sufficient condition for the existence of a saddle-point.

<u>Theorem 2-4.</u> If (Z,Z') is solution of (I) and :

- S^* optimal for the optimal stopping problem about $Z + X'$
- S'^* optimal for the optimal stopping problem about $Z' + X$,

then (S^*,S'^*) is a saddle-point for the game.

<u>Proof</u> : By [6] we have :

$Z_{S^*} = X_{S^*} + Z'_{S^*}$, $Z'_{S'^*} = X'_{S'^*} + Z_{S'^*}$, and the processes $Z_{t\wedge S^*}$ and $Z'_{t\wedge S'^*}$ an $\underline{\underline{F}}_t$ martingales. Therefore :

$$Z_0 = E(Z_{S^*}) = E(Z_{S^*\wedge S'^*}) \ , \ Z'_0 = E(Z'_{S'^*}) = E(Z'_{S^*\wedge S'^*}) \quad \text{and :}$$

$$Z_0 - Z'_0 = E((Z-Z')_{S^*\wedge S'^*}) = E((Z-Z')_{S^*} 1_{(S^*\le S'^*)} + (Z-Z')_{S'^*} 1_{(S'^*<S^*)}) =$$

$$E(X_{S^*} 1_{(S^*\le S'^*)} - X'_{S'^*} 1_{(S'^*<S^*)}) = J(S^*,S'^*).$$

Let S be any $\underline{\underline{F}}_t$ - stopping time. We have :

$Z_0 \ge E(Z_{S\wedge S'^*})$ (Z is a supermartingale) , $Z'_0 = E(Z'_{S\wedge S'^*})$ and then :

$$\begin{aligned} Z_0 - Z'_0 &\ge E((Z - Z')_{S\wedge S'^*}) \\ &\ge E((Z - Z')_S 1_{(S\le S'^*)} + (Z - Z')_{S'^*} 1_{(S'^*<S)}) \\ &\ge E(X_S 1_{(S\le S'^*)} - X'_{S'^*} 1_{(S'^*<S)}) = J(S,S'^*) \quad \text{since } Z \ge Z' + X. \end{aligned}$$

A similar argument proves that given any $\underline{\underline{F}}_t$ stopping time S', $Z_0 - Z'_0 \le J(S^*,S')$, so that : $J(S,S'^*) \le J(S^*,S'^*) \le J(S^*,S')$

§3 - <u>The defect of value and saddle-point</u>

By analogy with stopping time theory [6] , [2] , we introduce the set of "divided" stopping times, and we prove that there exists always a saddle-point on this set.

<u>Définition 3-1.</u> A system $\sigma = (S,H^-,H,H^+)$ is a divided stopping time if S is a $\underline{\underline{F}}_t$ - stopping time, H^-,H,H^+ are $\underline{\underline{F}}_S$, and such that :

- $H^- \cap (S = 0) = \emptyset$ and S_{H^-} predictable
- $H^+ \cap (S = +\infty) = \emptyset$
- (H,H^-,H^+) is a partition of Ω.

If Y is an optional process, and σ is divided stopping time we write Y_σ for :

$$\underline{Y}_S 1_{H^-} + Y_S 1_H + \overline{Y}_S 1_{H^+} \ ,$$ where $\underline{Y}$ (resp. $\overline{Y}$) is the left

(resp. right) upper limit of Y.

The set of $\underline{\underline{F}}_t$ - stopping times $\mathcal{T}$ is included in the set $\mathcal{F}$ of divided stopping times by the identification :

$$T \equiv (T, \emptyset, \Omega, \emptyset)$$

We are able to define on $\mathcal{F}$ an order which preserves the supermartingale property ; more precisely if σ and τ are in $\mathcal{F}$, $\sigma \leq \tau$, for all positive $\underline{\underline{F}}_t$ - supermartingale Z we have :

$E(Z_\tau / \underline{\underline{F}}_\sigma) \leq Z_\sigma$ P.a.s., where $\underline{\underline{F}}_\sigma$ is the σ-algebra produced by the set of $A \in \underline{\underline{F}}_S$ such that $A \cap H^- \in \underline{\underline{F}}_{S-}$. Finally if σ and τ are any divided stopping times, we are able do define $\sigma \wedge \tau$.

For any $\underline{\underline{F}}_t$ - stopping time T, and any set $\mathcal{J} \subseteq \mathcal{F}$, we introduce the T-conditional Dynkin game on $\mathcal{J}$ by $(\Omega, \underline{\underline{F}}_\infty, \underline{\underline{F}}_t, \mathcal{J}_T, \mathcal{J}_T, (J_T(\sigma,\sigma'))_{(\sigma,\sigma') \in \mathcal{J}_T \times \mathcal{J}_T})$, where

$\mathcal{J}_T = \{\sigma \in \mathcal{J} \mid \sigma \geq T\}$

$J_T(\sigma,\sigma') = E(X_\sigma \mathbb{1}_{(\sigma \leq \sigma')} - X'_{\sigma'} \mathbb{1}_{(\sigma' < \sigma)} / \underline{\underline{F}}_T)$

We shall use the main results of optimal stopping:

Proposition 3-2. [6]

Let Y an optional bounded process ; given any $\underline{\underline{F}}_t$ - stopping time T, let D_T^ε be the beginning after T of the set

$$A^\varepsilon = \{(\omega,t) \,.\, [R(Y)]_t(\omega) \leq Y_t(\omega) + \varepsilon\}, \quad \varepsilon > 0,$$

and set : $\delta_T^\varepsilon = (D_T^\varepsilon, \emptyset, (D_T^\varepsilon \circledS A^\varepsilon), (D_T^\varepsilon \notin A^\varepsilon))$.

Then δ_T^ε is a divided stopping time $\geq T$ such that :

$$[R(Y)]_{\delta_T^\varepsilon} \leq Y_{\delta_T^\varepsilon} + \varepsilon \quad \text{P.a.s., and}$$

$$[R(Y)]_T = E([R(Y)]_{\delta_T^\varepsilon} / \underline{\underline{F}}_T) \quad \text{P.a.s.}$$

When $\varepsilon \to 0$, $D_T^\varepsilon \nearrow \overline{D}_T$: let :

$H_T^- = \{D_T^\varepsilon < \overline{D}_T \ \forall \varepsilon > 0\}$, $H_T = (H_T^-)^c \cap (Y_{\overline{D}_T} \geq [R(Y)]_{\overline{D}_T})$,

$H_T^+ = (H_T^-)^c \cap \{Y_{\overline{D}_T} < [R(Y)]_{\overline{D}_T}\}$

Then $\bar{\delta}_T = (\bar{D}_T , H^-_T , H_T , H^+_T)$ is a divided stopping time such that :

$$(R(Y))_{\bar{\delta}_T} = Y_{\bar{\delta}_T} \quad P.a.s. \quad \text{and}$$

$$(R(Y))_T = E(R(Y)_{\bar{\delta}_T} / \underline{\underline{F}}_T) \quad P.a.s.$$

From these results, with the assumption that $\tilde{Z}$, $\tilde{Z}'$ are bounded we deduce :

Theorem 3-3. For all $\underline{\underline{F}}_t$ - stopping time T, the T-conditional Dynkin game on $\mathcal{F}^r = \{\sigma \in \mathcal{F} , H^- = \emptyset\}$ has a value which is $Z_T - Z'_T$, where (Z,Z') is solution of (I).

Proof : Let D^ε_T (resp. D'^ε_T) be the beginning after T of the set :

$A^\varepsilon = \{(\omega,t) : Z_t(\omega) \leq (X+Z')_t(\omega) + \varepsilon\}$ (resp. $A'^\varepsilon = \{(\omega,t) : Z'_t(\omega) \leq (X'+Z)_t(\omega) +\varepsilon\}$),

and $\delta^\varepsilon_T = (D^\varepsilon_T, \emptyset, (D^\varepsilon_T \in A'^\varepsilon), (D^\varepsilon_T \notin A^\varepsilon))$ (resp. $\delta'^\varepsilon_T = (D'^\varepsilon_T, (D'^\varepsilon_T \in A'^\varepsilon), (D'^\varepsilon_T \notin A'_\varepsilon))$

If δ_T is any divided stopping time of $\mathcal{F}^r_T$, we have $Z_T \geq E(Z_{\delta_T \wedge \delta'^\varepsilon_T} / \underline{\underline{F}}_T)$, since Z has the supermartingale property on divided stopping times. Using proposition 3-2 ; we deduce $Z'_T = E(Z'_{\delta_T \wedge \delta'^\varepsilon_T} / \underline{\underline{F}}_T)$. Hence :

$$Z_T - Z'_T \geq E((Z-Z')_{\delta_T \wedge \delta'^\varepsilon_T} / \underline{\underline{F}}_T) \quad P.a.s.$$

$$\geq E((Z-Z')_{\delta_T} \mathbb{1}_{(\delta_T \leq \delta'^\varepsilon_T)} + (Z-Z')_{\delta'^\varepsilon_T} \mathbb{1}_{(\delta'^\varepsilon_T < \delta_T)} / F_T) \quad P.a.s.$$

But : $(Z-Z')_{\delta_T} \geq X_{\delta_T}$ $\quad (Z = R(X+Z'))$, and

$$(Z-Z')_{\delta'^\varepsilon_T} \geq -X'_{\delta'^\varepsilon_T} - \varepsilon \text{ (prop. 3-2) , so that :}$$

$$Z_T - Z'_T \geq E(X_{\delta_T} \mathbb{1}_{(\delta_T \leq \delta'^\varepsilon_T)} - X'_{\delta'^\varepsilon_T} \mathbb{1}_{(\delta'^\varepsilon_T < \delta_T)} / \underline{\underline{F}}_T) - \varepsilon \quad P.a.s.$$

$$\geq J(\delta_T , \delta'^\varepsilon_T) - \varepsilon \quad P.a.s.$$

A similar argument shows that for any divided stopping time δ'_T of $\mathcal{F}^r_T$:

$$Z_T - Z'_T \leq J(\delta^\varepsilon_T , \delta'_T) + \varepsilon \quad P.a.s.$$

Then the proposition 1-5 concludes the proof.

Theorem 3-4. For all $\underline{\underline{F}}_t$ stopping time T, the T-conditional Dynkin game on F has a saddle-point.

Proof : With the notations of theorem 3-3, let :

$\bar{\delta}_T = (\bar{D}_T, H_T^-, H_T, H_T^+)$, $\bar{\delta}'^T = (\bar{D}'_T, H'^-_T, H'_T, H'^+_T)$ with

$\bar{D}_T = \lim_{\varepsilon\to 0} \nearrow D_T^\varepsilon$, $\bar{D}'_T = \lim_{\varepsilon\to 0} \nearrow D'^\varepsilon_T$ and

$H_T^- = \{D_T^\varepsilon < \bar{D}_T \ \forall\varepsilon>0\}$, $H_T = (H_T^-)^c \cap (Z_{\bar{D}_T} \le (X+Z')_{\bar{D}_T})$, $H_T^+ = (H_T^-)^c \cap (Z_{\bar{D}_T} > (X+Z')_{\bar{D}_T})$

$H'^-_T = \{D'^\varepsilon_T < \bar{D}'_T \ \forall\varepsilon>0\}$, $H'_T = (H'^-_T) \cap (Z'_{\bar{D}'_T} \le (X'+Z)_{\bar{D}'_T})$, $H'^+_T = (H'^-_T)^c \cap (Z'_{\bar{D}'_T} > (X'+Z)_{\bar{D}'_T})$

Using the results of Proposition 3-2, and the same technic as in the proof of Theorem 2-4, we prove that $(\bar{\delta}_T , \bar{\delta}'_T)$ is a saddle point for the T-conditional Dynkin game on $\mathcal{F}$.

§4 - Existence results

We prove in this last section that when X,X' are upper semi-continuous (resp. right upper semi-continuous) there is a saddle-point (resp. a value) for the game.

1 - The upper right semi-continuous case

Suppose that $\bar{X} \le X$ and $\bar{X}' \le X'$. From the theorem 3-3 we deduce :

Theorem 4-1. For any $\underline{\underline{F}}_t$ - stopping time T the T-conditional Dynkin game has a value which is $(Z - Z')_T$.

Proof : Since Z (resp. Z') is Snell's envelope of X + Z' (resp. X' + Z) we have by [7] :

$$Z_{D_T^\varepsilon} \le (Z' + X)_{D_T^\varepsilon} \vee \overline{(Z' + X)}_{D_T^\varepsilon} + \varepsilon \quad \text{P.a.s.} ,$$

$$Z'_{D'^\varepsilon_T} \le (Z + X')_{D'^\varepsilon_T} \vee \overline{(Z + X')}_{D'^\varepsilon_T} + \varepsilon \quad \text{P.a.s.} .$$

If X,X' are upper right semi-continuous, X + Z' and X + Z' are also upper right semi-continuous (since Z and Z' are supermartingales), and then :

$$Z_{D_T^\varepsilon} \le (Z' + X)_{D_T^\varepsilon} + \varepsilon , \ Z'_{D'^\varepsilon_T} \le (Z + X')_{D'^\varepsilon_T} + \varepsilon \ \text{P.a.s.}$$

Hence the beginning after T of A^ε (resp. A'^ε) belongs to A^ε (resp. A'^ε). The divided stopping times δ_T^ε and δ'^ε_T in theorem 3-2, are in this case stopping times.

We have : $J(\sigma, \delta'^{\varepsilon}_T) - \varepsilon \leq (Z - Z')_T \leq J(\delta^{\varepsilon}_T, \tau) + \varepsilon$ P.a.s.

for any σ, τ $\underline{F}_t$ - stopping times $\geq T$. The final result follows from proposition 1-5.

2 - The upper semi-continuous case

We shall use that by [6], if Y is upper semi-continuous, i.e., $\overline{Y} \leq Y$, $\underline{Y} \leq {}^pY$ (${}^pY$ is the predictable projection of Y), the optimal stopping problem relative to Y has solutions ; more precisely the beginning D_o of the set $(Y = R(Y))$ and the beginning S_o of the set $(R(Y) \neq M)$, where M is the martingale part of R(Y), are optimal stopping times. Moreover any optimal stopping time S^* is such that $D_o \leq S^* \leq S_o$.

Suppose that the processes X,X' of the game are upper semi-continuous, and $(X = -X')$ evanescent. Then we have :

Lemma 4-2. Let (Z,Z') be solution of (I). Then Z + X' and Z' + X are upper semi-continuous.

Proof : The right upper semi-continuity of Z + X', Z' + X comes from the supermartingale property of Z and Z' which implies their right upper semi-continuity.

Now, briefly, let $Z = M - A^- - B = M - C$, $Z' = M' - A'^- - B' = M' - C$ be Mertens' decompositions of the supermartingales Z and Z'.

We know by [6] that the jumps of A (resp. A') are included in the set $\{Z = Z' + X\}$ (resp. $(Z' = Z + X')$) and those of B (resp. B') in $\{Z^- = Z'^- + \underline{X}\}$ (resp. $(Z'^- = Z^- + \underline{X}')$. Therefore : $E(Z_{T-} \, 1_{(\Delta B_T > o)}) = E((Z'_{T-} + \underline{X}_T) \, 1_{(\Delta B_T > o)})$.

The set $(X = -X')$ is evanescent. Then :

$$E(Z_{T-} \, 1_{(\Delta B_T > o)}) = E((Z'_T + \underline{X}_T) \, 1_{(\Delta B_T > o)})$$

By the assumption $\underline{X} \leq {}^pX$, and $Z \geq Z' + X$, we obtain gradually :

$$E(Z_{T-} \, 1_{(\Delta B_T > o)}) \leq E((Z'_T + X_T) \, 1_{(\Delta B_T > o)})$$
$$\leq E(Z_T \, 1_{(\Delta B_T > o)})$$

and so $E(Z_{T-}) \leq E(Z_T)$. The supermartingale Z is regular, therefore left upper semi-continuous.

Remark. Under the same assumptions we can prove the unicity of solution of (I), by a generalization of the proof of [3], in the right continuous case, with $Z_\infty = Z'_\infty = 0$.

Using the result of theorem 2-4 we prove the next result avout the existence of a saddle-point in the Dynkin game.

Theorem 4-3. Let X, X' be upper semi-continuous, and $(X = -X')$ evanescent. Then if D_o (resp. D'_o) is the beginning of $(Z - Z' = X)$ (resp. $(Z - Z' = -X')$) and so (resp. S'_o) the beginning of $(Z \neq M)$ (resp. $(Z' \neq M')$) where M (resp. M') is the martingale part of Z (resp. Z') in the Mertens decomposition, the pairs (D_o, D'_o), (S_o, S'_o) are saddle-points for the game. Moreover for any (S^*, S'^*) saddle-point for the game we have :

$$D_o \wedge D'_o \leq S^* \wedge S'^*$$

and $\quad (S^* > S_o, \; S'^* > S'_o)$ or $(S^* \leq S_o, \; S'^* \leq S'_o)$

Proof : The first part of this theorem is straight for ward from Theorem 2-4 and the results of optimal stopping.

For the second part, since $Z_T - Z'_T$ is the value of the T-conditional Dynkin game we deduce easily that $(Z-Z')_{S^* \wedge S'^*} = X_{S^*} 1_{(S^* \leq S'^*)} - X'_{S'^*} 1_{(S'^* < S^*)}$

and $(Z - Z')_{t \wedge S^* \wedge S'^*}$ is a martingale, if (S^*, S'^*) is a saddle-point for the game. From this and the definitions of D_o, D'_o, S_o, S'_o we obtain the final result.

Conclusion. This study can be easily generalized to X, X' $\wedge$-processes, where $\wedge$ is a Meyer σ-algebra. Sec [1] for details.

Bibliography

[1] M. ALARIO-NAZARET
Jeux de Dynkin - Thèse de 3e cycle. Université de Besançon - 1982.

[2] J.M. BISMUT
Temps d'arrêt optimal et quasi-temps d'arrêt C.R. Acad. Sci. Paris, Ser. A. 284, 1519 - 1521 - 1977.

[3] J.M. BISMUT
Contrôle de processus alternants et applications Z. Wahr. Veruv. Geb.47, 247 - 288 - 1979.

[4] C. DELLACHERIE

Capacités et processus stochastiques. Springer n° 67, 1972.

[5] E.B. DYNKIN - A.A. YUSCHKEVITCH

Theorems and problems in Markov processes - N.Y. Prenum Press 1968.

[6] N.EL. KAROUI

E cole de probabilités de Saint Flour IX - 1979. Lect. Notes in Maths 876, Springer Verlag - 1981.

[7] M.A. MAINGUENEAU

Théorie générale des processus et problèmes d'optimalité. Séminaire de probabilités XII, Lect. Notes in Maths 649, Springer Verlag - 1978.

[8] L. STETTNER

P.H.D. Thesis - Université de Varsovie - 1981.

AN INTRODUCTION TO THE STOCHASTIC CALCULUS OF VARIATIONS

BY

Jean-Michel BISMUT

Université de Paris-Sud

Département de Mathématique

Bâtiment 425

91405 ORSAY CEDEX

+–+–+–+–+

To be presented at the Conference on Stochastic Systems of Bad-Honnef (June 1982).

The purpose of this paper is to give a general introduction to the methods of the stochastic calculus of variations, and to some of its applications. The subject being in full development, we have not tried to cover the whole thing, but we have essentially focused on the aspects of the calculus which may be of special relevance to control and filtering.

The paper is organised around four essential topics.

In section 1, the theory of stochastic flows is reviewed. It has been known for a long time (see Gihman - Skorohod [19]) that the solutions of a stochastic differential equation depend smoothly on the initial conditions in L_2-sense. Blagoveschenskii and Freidlin [12] announced the stronger result that the solutions of such a stochastic differential equation could be modified so as to obtain an a.s. smoothness with respect to the initial conditions. This result was rediscovered by Malliavin [39] who gave a new impetus to the field. The whole subject has been developped in Ventzell [58], Rozovskii [48], Baxendale [1], Elworthy [17], Bismut [5] - [6], Ikeda - Watanabe [26], Kunita [31] - [32]. The basic idea is that in many aspects a stochastic differential equation behaves like deterministic differential equations so that to such an equation, a flow $\varphi_t(\omega,\cdot)$ of diffeomorphisms of R^d onto itself may be associated, so that $\varphi_t(\omega,\cdot)$ is a continuous stationary independent increment process with values in the group of diffeomorphisms of R^d. A number of standard operations may be performed on $\varphi_.(\omega,\cdot)$ like the lifting of $\varphi_t(\omega,\cdot)$ to tensors, the determination of the equation of $\varphi_t^{-1}(\omega,\cdot)$ etc ... The theory of flows extends to general stochastic differential equations the results of Doss [15], Sussmann [56]. It can be applied to the Malliavin calculus of variations and to study conditional diffusions (Bismut - Michel [10] - [11], Kunita [33], [34], [35]).

In section 2, the results of Haussmann [21] - [22] extending Clark [13] are presented. It has been known for a long time that a connection exists between the differentiability properties of functions and their behaviour as random variables. This is clearly illustrated by the well-known decomposition of functions of the Schwartz space S(R) into linear combinations of weighted Hermite polynomials [47].

Clark [13] and Haussmann [21] - [22] interpreted such relations in the setting of diffusions by relating the representation of certain random variables as martingales in terms of their differentials in the sense of Fréchet. In [7], Bismut exhibited the relation between these results and integration by parts on the Wiener space, and used them to recover the integration by parts obtained by Malliavin [39] using the Ornstein - Uhlenbeck process.

In Section 3, we give a brief exposition of the techniques of the Malliavin calculus. The papers of Malliavin [39] - [40] have been the starting point of the development of the whole field. The basic idea of Malliavin was that it was possible to use the construction of the solution of the stochastic differential equation by means of a Brownian motion as an efficient way of obtaining results on partial differential operators. The technique of Malliavin uses an auxiliary infinite dimensional stochastic process to establish a key integration by parts formula. His ideas were simplified by Shigekawa [49], Kusuoka [36] who used functional analysis techniques and the Ornstein Uhlenbeck operator. These aspects are fully developed and much extended in the papers of Stroock [51] - [52] - [53], to which the reader is refered. We have chosen instead to follow the approach taken by us in [7]. Also note that the key estimates were obtained by Malliavin [40], and Ikeda - Watanabe [26] and have been recently considerably improved by Kusuoka and Stroock [37].

In Section 4, the main results of Bismut - Michel [10] - [11] extending Michel [44] on conditional diffusions are developped. In this application, the theory of flows and the Malliavin calculus are used in combination to obtain results which extend the results of Pardoux [46], Krylov - Rozovskii [28] - [29], Davis [14], Eliott - Kohlmann [16]. These results are apparently difficult to obtain using more classical methods. For related developments, we refer to Kunita [33] - [34] - [35]. Another application of the Malliavin calculus is in Holley - Stroock [23], where the obtained results fall out of the reach of classical methods.

We have excluded of this survey some recent developments of the methods of the calculus of variations to jump processes [8] and boundary processes [9].

AN INTRODUCTION TO THE STOCHASTIC

CALSULUS OF VARIATIONS

BY J.M.BISMUT

(1) Stochastic flows

Let Ω be the set $\mathcal{C}(R^+;R^m)$ whose standard element ω is a trajectory

$$\omega = (w_t^1,\dots,w_t^m)$$

Ω is endowed with its canonical filtration

$$F_t = \mathcal{B}(w_s \mid s \leq t)$$

which is eventually regularized on the right.

P is the Brownian measure on Ω, with $P[w_0 = 0] = 1$.

$X_0(x),\dots,X_m(x)$ are $m+1$ vector fields defined on R^d with values in R^d, which are bounded, C^∞, with bounded differentials.

For $x_0 \in R^d$, consider the stochastic differential equation written in Stratonovitch form

$$(1.1) \qquad dx = X_0(x)dt + X_i(x).dw^i$$
$$x(0) = x_0$$

where dw^i is the Stratonovitch differential of w^i. (1.1) can be written in Ito's form

$$(1.1') \qquad dx = \left(X_0(x) + \frac{1}{2}\frac{\partial X_i}{\partial x} X_i(x)\right)dt + X_i(x).\delta w^i$$
$$x(0) = x\ .$$

where δw^i is the Ito differential of w^i (the summation sign $\sum_{i=1}^{m}$ will be generally omitted).

a) Ito calculus and Stratonovitch calculus

In the whole paper, we will use as much as we can Stratonovitch integrals. If H and X_t are continuous semi-martingales, it is known [41] that the Stratonovitch integral $\int_0^t H_s \, d X_s$ can be expressed in terms of the Ito integral $\int_0^t H_s \, \delta X_s$ by the relation

$$(1.2) \qquad \int_0^t H_s \, d X_s = \int_0^t H_s \, \delta X_s + \frac{1}{2} < H,X >_t$$

where $< H,X >_t$ denotes the quadratic variation of H and X.

In classical stochastic differential calculus, it is known [41] that the formal rules of the usual differential calculus are preserved in the Stratonovitch stochastic calculus.

As a consequence, we see that $X_0, X_1 \ldots X_m$ in (1.1) are in fact vector-fields in the sense of differential geometry, i.e. if equation (1.1) is expressed in a new set of coordinates, the new fields $X_0', \ldots, X_m'$ are obtained through the standard rules of transformation of vector-fields. This is of course connected with the fact that the generator $\mathcal{L}$ of the diffusion (1.1) is given by

$$(1.3) \qquad \mathcal{L} = X_0 + \frac{1}{2} \sum_1^m X_i^2$$

where $X_0, X_1, \ldots, X_m$ are considered as first-order differential operators. The invariance of $X_0, \ldots, X_m$ is also connected with the results of Stroock-Varadhan [55] which show that the probability law of $x.$ is the limit of the probability laws of $x^n_.$, where $x^n_.$ is the solution of the differential equation (1.1) where w^i is replaced by its continuous linear dyadic approximation.

The reasons for using the Stratonovitch calculus are even more stringent here.

Since in fact we will deal with infinite dimensional diffusions, even if the Stratonovitch expressions are still easy to write, the corresponding Ito calculus formulas become hairy.

Ito's calculus will of course be the essential tool through which the analysis of the problems which we consider will be carried over, but once the formulas are proved we will come back to Stratonovitch calculus.

b) Construction of the flow

The idea is to associate to equation (1.1) a flow of diffeomorphisms $\varphi_t(\omega,.)$ of R^d onto R^d such that for any $x_0 \in R^d$, $\varphi_t(\omega,x_0)$ is exactly the solution of (1.1).

The differentiability of the solutions of (1.1) on the variable x_0 in the L_2-sense has been known for a long time (Gihman-Skorokhod [19]) and is in fact sufficient to establish "elementary" properties of the diffusion (1.1) like the fact that the associated semi-group T_t maps $C_b^\infty(R^d)$ in $C_b^\infty(R^d)$.

The a.s. differentiability of the solution of (1.1) in the variable x_0 had been announced in Blagoveschenskii and Freidlin [12]. It has been studied by Malliavin [39], Elworthy [17], Baxendale [1], Ikeda-Watanabe [26].

The complete properties of the flow $\varphi_t(\omega,\cdot)$, namely the fact that a.s. it is one-to-one and onto for every $t \in R^+$ were established in Bismut [5], [6] and Kunita [31] who showed these properties extend when w is replaced by a continuous semi-martingale.

Namely, we have :

Theorem 1.1 : *There exists a mapping $\varphi_t(\omega,x)$ defined on $R^+ \times \Omega \times R^d$ with values in R^d such that*

a) *For any $(t,x) \in R^+ \times R^d$, $\omega \to \varphi_t(\omega,x)$ is measurable.*

b) *For any $\omega \in \Omega$, $(t,x) \to \varphi_t(\omega,x)$ is continuous.*

c) *For any* $(\omega,t) \in \Omega \times R^+$, $x \to \varphi_t(\omega,x)$ *is a* C^∞ *diffeomorphism of* R^d *onto* R^d.

d) *For any* $\omega \in \Omega$, *and any multi-index* m, $(t,x) \to \frac{\partial^m \varphi_t}{\partial x^m}(\omega,x)$ *is continuous.*

e) *For any* $x_0 \in R^d$, $t \to \varphi_t(\omega,x_0)$ *is the (essentially) unique solution of (1.1).*

Moreover properties a)-e) determine $\varphi.(\omega,.)$ *uniquely (in the sense of essential uniqueness).*

Proof : We only give here a sketch of the proof. The existence of $\varphi_t(\omega,.)$ having properties a), b), d), e) is the "easy" part. In fact, if $x.^{x_0}$ is the solution of (1.1), it is elementary to show that for any $T > 0$, $p \geqslant 2$, $x_0, y_0 \in R^d$, $s,t \leqslant T$

$$E\left|x_t^{x_0} - x_s^{y_0}\right|^{2p} \leqslant C_T\left[\left|x_0 - y_0\right|^{2p} + \left|t-s\right|^p\right] \tag{1.4}$$

Taking $p > d+1$ in (1.4) Kolmogorov's lemma implies the existence of an a.s. continuous mapping $(t,x_0) \to x_t^{x_0}(\omega)$. Using the differentiability of $x.^{x_0}$ on x_0 in the L_2-sense, it is not hard to prove the differentiability of $x_t^{x_0}$ on x_0 in the sense of distributions. Using the classical form of the L_2 differential of x^{x_0} on x_0 given by the stochastic differential equation

$$dZ^{x_0} = \frac{\partial X_0}{\partial x}(x_t^{x_0})\, Z^{x_0}\, dt + \frac{\partial X_i}{\partial x}(x_t^{x_0})\, Z^{x_0}.\, dw^i \tag{1.5}$$

$$Z^{x_0}(0) = I$$

it is not hard to prove (1.4) for Z^{x_0} and to deduce the a.s. differentiability of order 1 on x_0 for $x.^{x_0}$. The same rules give the C^∞ differentiability. Set $x_t^{x_0} = \varphi_t(\omega,x_0)$. Equation (1.5) permits the easy proof that a.s., for any $(t,x_0) \in R^+ \times R^n$, $\frac{\partial \varphi_t}{\partial x}(\omega,x_0)$ is invertible. The technical difficulty is now to prove the a.s. injectivity and the a.s. onto property of $\varphi_t(\omega,.)$. Note that if R^d is replaced by a compact connected manifold, since φ_0 = identity, a homotopy argument gives precisely the required result (using the invertibility of $\frac{\partial \varphi_t}{\partial x}(\omega,x)$). In the case of R^d, the approximation of the flow $\varphi(\omega,.)$ by the flows $\varphi^n(\omega,.)$ of differential equations and a time reversal argument of Malliavin [39] were used in Bismut [5]-[6] to obtain Property c). In Kunita [31] a priori inequalities on (1.1) give this property. $\square$

For $s \in R^+$, let θ_s be the standard translation operator in Ω, i.e. if $\omega = (w_t)$, $\theta_s\omega = (w_{s+t})$. It is then easy to prove that for any stopping time S, on $(S < +\infty)$, a.s., for any $t \geq 0$.

$$(1.6) \qquad \varphi_{S+t}(\omega,.) = \varphi_t(\theta_S\omega) \circ \varphi_S(\omega,.) .$$

Let G be the group of C^∞ diffeomorphisms of R^d onto R^d, endowed with the C^∞_K topology (which is the topology of uniform convergence of a function and its derivatives on compact sets). Theorem 1.1 and (1.6) say precisely that $\varphi_t(\omega,.)$ is a continuous independent increment process with values in G. The lifting of the diffusion (1.1) to an independent increment process with values in G will be essential in the sequel.

Let $\mathcal{g}$ be the "Lie algebra" of C^∞ vector fields on R^d. $\mathcal{g}$ may be considered (at least formally) as the Lie algebra of the "Lie Group" G. $\mathcal{g}$ is then interpreted as the Lie algebra of the tangent vectors to G which are invariant <u>on the right</u>. Equation (1.1) may be formally rewritten as

$$(1.7) \qquad d\varphi = X_0(\varphi)dt + X_i(\varphi).dw^i$$

$$\varphi(0) = id .$$

$(X_0(\varphi),\dots X_m(\varphi)$ is clearly the image of $X_0(e),\dots,X_m(e)$ through the action of φ on the right).

A case of special interest is the case where the Lie algebra $\mathcal{g}_0$ generated by $X_0,X_1\dots X_m$ is finite dimensional, i.e. there exists a finite number of Lie brackets $Y_1\dots Y_r$ of the vector fields $X_0,\dots,X_m$ such that any Lie bracket of $X_0\dots X_m$ is a linear combination with <u>constant coefficients</u> of $Y_1\dots Y_r$. In this case Theorem 1.1 becomes trivially true, and no proof is even needed. In fact let G_0 be a connected Lie group whose Lie algebra is $\mathcal{G}_0$. The stochastic differential equation (1.7) can be trivially solved on G_0 , which is a finite dimensional manifold (the fact it remains in G_0 follows trivially from Stratonovitch calculus ; infinite life time is trivial by the group structure of G_0 . In fact $\mathcal{G}_0$ and G_0 may be identified to matrices, and everything is trivial). Now G_0 can be identified to a finite dimensional group of diffeomorphisms of R^d, so that the original equation (1.7) in G has in fact been solved.

The case where $X_0, X_1 \dots X_m$ commute is in fact the easiest. If $\varphi_t^0, \varphi_t^1 \dots \varphi_t^m$ are the one parameter groups of diffeomorphisms of R^d associated to $X_0, X_1 \dots X_m$, the previous argument shows that

$$\varphi_t = \varphi_t^o \circ \varphi^1_{w_t^1} \circ \dots \varphi^m_{w_t^m} \quad . \tag{1.8}$$

This is the basis of the papers of Doss [15] and Sussmann [56] (where in fact X_0 is not supposed to commute with $X_1 \dots X_m$).

For other properties of stochastic flows, see Meyer [43].

c) The Ito-Stratonovitch formula on flows

$\varphi_t(\omega,.)$ is now a diffusion with values in G. If z_t is a continuous semi-martingale with values in R^d, a natural question to ask is to know if $\varphi_t(\omega,z_t)$ is a semi-martingale. Of course if φ_t takes its values in a finite dimensional Lie-subgroup G_0, the answer is trivially positive, since $\varphi_t(\omega,z_t)$ is the image of (φ_t,z_t) through the C^∞ mapping $(\varphi,z) \to \varphi(z)$.

In the general case, the problem was given a positive answer (without proof) by Ventzell [58]. The corresponding Ito formula was proved by Rozovskii in [48]. Independently, this result was reproved in Bismut [5], [6] and Kunita [32].

Let z_t be a continuous semi-martingale with values in R^d whose Ito-Meyer decomposition is

$$z_t = z_0 + A_t + \int_0^t H_i \, .\delta w^i \tag{1.9}$$

where $z_0 \in R^n$, A_t is a continuous adapted bounded variation process, such that $A_0 = 0$ and H_i is an adapted process such that $\int_0^t |H_i|^2 \, ds < +\infty$ a.s. Using the convention in (1.2), we have :

<u>*Theorem 1.2 :*</u> *$\varphi_t(\omega, z_t)$ <u>is a continuous semi-martingale, whose Ito decomposition is given by</u>*

$$\varphi_t(\omega,z_t) = z_0 + \int_0^t \Big[X_0\big(\varphi_s(\omega,z_s)\big) + \frac{1}{2}\frac{\partial X_i}{\partial x} X_i\big(\varphi_s(\omega,z_s)\big) + \frac{1}{2}\frac{\partial^2\varphi_s}{\partial x^2}(\omega,z_s)\,(H_{i_s},H_{i_s}) + \frac{\partial X_i}{\partial x}\big(\varphi_s(\omega,z_s)\big)\frac{\partial \varphi_s}{\partial x}(\omega,z_s)H_{i,s}\Big] ds + \int_0^t X_i\big(\varphi_s(\omega,z_s)\big)\delta w^i + \int_0^t \frac{\partial\varphi_s}{\partial x}(\omega,z_s).\delta z_s \tag{1.10}$$

<u>In Stratonovitch form,</u> (1.10) <u>can be written as</u> :

$$\varphi_t(\omega,z_t) = z_0 + \int_0^t X_0\big(\varphi_s(\omega,z_s)\big)ds + \int_0^t X_i\big(\varphi_s(\omega,z_s)\big).\,d w^i + \int_0^t \frac{\partial\varphi_s}{\partial x}(\omega,z_s).\,d z_s . \tag{1.11}$$

<u>Proof</u> : We follow the simple argument of Rozovskii [48]. Let g be a $\geqslant 0$ C^∞ function defined on R^d with compact support such that $\int g(x)dx = 1$. Set $g_\varepsilon(x) = \varepsilon^{-d} g(x/\varepsilon)$. Take $y \in R^n$. By Ito's formula, we have

$$g_\varepsilon(z_t - y)\,\varphi_t(\omega,y) = g_\varepsilon(z_0 - y)\, y + \int_0^t g_\varepsilon(z_s - y)\Big[\Big(X_0 + \frac{1}{2}\frac{\partial X_i}{\partial x}X_i\Big)\big(\varphi_s(\omega,y)\big)ds + X_i\big(\varphi_s(\omega,y)\big).\delta w^i\Big] + \int_0^t \Big[< \frac{\partial g_\varepsilon}{\partial x}(z_s - y), \delta z > + \frac{1}{2}\frac{\partial^2 g_\varepsilon}{\partial x^2}(z_s - y)\,(H_i,H_i)ds\Big]\varphi_s(\omega,y) + \int_0^t < \frac{\partial g_\varepsilon}{\partial x}(z_s - y), H_i > X_i\big(\varphi_s(\omega,y)\big)ds \tag{1.12}$$

Note that since both sides of (1.10) are continuous, we need to prove (1.10) at a fixed time t. Also note that formula (1.10) can be stopped at adequate stopping times. Integrate (1.12) in y. Integration in y and stochastic integration with respect to δw^i, δz can be trivially interchanged on the R.H.S. of (1.12). Integrate by parts in the integrands of the R.H.S. of (1.12) in the variable y, so that only $g_\varepsilon(z_s - y)$ appears and not its differential : this is possible using the smoothness of $\varphi_s(\omega,y)$ in y. Finally make $\varepsilon \to 0$. Convergence of the bounded variations terms is trivial. For the stochastic integrals, stop (1.12) adequately. The proof of the L_2 convergence of these integrals is then elementary.

d) The equation of the inverse flow

In the case where $\varphi.(\omega,.)$ takes its values in a finite dimensional Lie-group G_0 there is nothing to prove. In fact, in this case, let ${}^gX_0, {}^gX_1 \ldots {}^gX_m$ be the corresponding left invariant tangent vectors. The equation of $\psi = \varphi^{-1}$ is then trivially

$$(1.13) \qquad d\psi = - {}^gX_0(\psi)dt - {}^gX_i(\psi).dw^i$$

$$\psi(0) = e.$$

If G_0 is considered as a Lie transformation group of R^d, note that for $j = 0 \ldots m$

$${}^gX_j(\psi) = \frac{\partial\psi}{\partial x} X_j$$

i.e. for any $x \in R^n$

$$\left({}^gX_j(\psi)\right)(x) = \frac{\partial\psi}{\partial x}(x)\, X_j(x) .$$

Now since $\frac{\partial\psi}{\partial x}(x) = \left[\frac{\partial\varphi}{\partial x}\left(\psi(x)\right)\right]^{-1}$, we see that if $y_t = \varphi_t^{-1}(\omega,x)$, then

$$(1.14) \qquad dy_t = -\left[\frac{\partial\varphi_t}{\partial x}(\omega,y_t)\right]^{-1}\left[X_0(x)dt + X_i(x).dw^i\right] .$$

Of course, in the general case, this argument does not make really sense.

In fact the following is proved in Bismut [5]-[6](see also Krylov-Rozovskii [30] for a special case).

Theorem 1.3 : Take z_t as in (1.9). Then $y_t = \varphi_t^{-1}(\omega, z_t)$ is a continuous semi-martingale which is the unique solution of the Ito stochastic differential equation

$$(1.15)\qquad dy_t = \left[\frac{\partial \varphi_t}{\partial x}(\omega, y_t)\right]^{-1}\left[\delta z - X_o(z_t)dt - X_i(z_t)\delta w^i\right]$$

$$+ \left[\frac{\partial \varphi_t}{\partial x}(\omega, y_t)\right]^{-1}\left[-\frac{1}{2}\frac{\partial X_i}{\partial x}(z_t)H_i - \frac{1}{2}\frac{\partial X_i}{\partial x}\left(\varphi_t(\omega, y_t)\right)\right.$$

$$\left(H_i - X_i(z_t)\right) - \frac{1}{2}\frac{\partial^2 \varphi_t}{\partial x^2}(\omega, y_t)\left(\left[\frac{\partial \varphi_t}{\partial x}(\omega, y_t)\right]^{-1}\left(H_i - X_i(z_t)\right)\cdot\right.$$

$$\left.\left.\left[\frac{\partial \varphi_t}{\partial x}(\omega, y_t)\right]^{-1}\left(H_i - X_i(z_t)\right)\right)\right]dt$$

$$y(0) = z_o$$

In Stratonovitch form, (1.15) is written as

$$(1.16)\qquad dy_t = \left[\frac{\partial \varphi_t}{\partial x}(\omega, y_t)\right]^{-1}\left[dz - X_o(z_t)dt - X_i(z_t).dw^i\right].$$

Proof : Since a.s., for $t \geq 0$ $\varphi_t^{-1}(\omega,.)$ exists, we know a priori that $\varphi_t^{-1}(\omega, z_t)$ is a continuous process. Consider equation (1.15). Since its coefficients are C^∞ in y, by Protter, Emery, Doleans - Dade [18], we know it has a solution on a stochastic interval $[0,T[$, where T is a a.s. >0 stopping time. Now it is a simple exercise to check that (1.15) implies (1.16). Apply Theorem 1.2 using (1.11), and check that $z'_t = \varphi_t(\omega, y_t)$ is the solution of an Ito stochastic differential equation, whose unique solution is z_t . Then on $[0,T[$, $y_t = \varphi_t^{-1}(\omega, z_t)$. Deduce from this that $T = +\infty$ a.s. For the complete proof, see [5]-[6]. □

Remark 2 : Except when $X_1 \dots X_m$ have compact support, there is no clear-cut argument proving that the solutions of (1.15) do not explode (except Theorem 1.3 itself !).

When $z_t = x \in R^n$, (1.16) is formally equivalent to (1.14), so that in fact $\psi_t = \varphi_t^{-1}$ is a solution of equation (1.13) on G. Note that in general, (1.13) is a "truly" infinite dimensional equation, which is not at all of the type of the equation (1.7) giving φ.

To see better the difference, let us remark that by Ito's formula, for any $f \in C_b^\infty(R^n)$, $x \in R^n$

$$(1.17) \qquad f\left(\varphi_t(\omega,x)\right) - \int_0^t \left[\left(X_0 + \frac{1}{2} X_i^2\right) f\right] \left(\varphi_s(\omega,x)\right) ds$$

is a martingale given by

$$(1.18) \qquad \int_0^t (X_i f)\left(\varphi_s(\omega,x)\right) . \delta w^i + f(x)$$

while

$$(1.19) \qquad f\left(\psi_t(\omega,x)\right) - \int_0^t \left(- X_0 + \frac{1}{2} X_i^2\right) (x) \left[f\left(\psi_s(\omega,x)\right)\right] ds$$

is a local martingale given by

$$(1.20) \qquad - \int_0^t X_i(x) \left[f\left(\psi_s(\omega,x)\right)\right] \delta w^i + f(x)$$

(in (1.19)-(1.20) the vector fields X_j are acting on $f\left(\psi_s(\omega,x)\right)$ as a function of x).

e) The diffusion of tensors

If $g \in G$, we know that g acts on tensors. Namely, if $X \in T_x(R^d)$, we may define the vector $g_* X \in T_{g(x)}(R^d)$ by

$$g_* X = \frac{\partial g}{\partial x}(x)\, X .$$

If k is a 1-form at g(x), the 1-form $g^* k$ at x is defined by

$$g^* k = \widetilde{\frac{\partial g}{\partial x}}(x)\, k .$$

From these rules, it is obvious how to define the action of g on tensors. However g_* sends vectors at x into vectors at g(x) while g^* sends forms at g(x) into forms at x. We will adopt a unified notation. Namely if K(x) is a tensor field, g^{*-1} K(x) is the tensor field which for each x is obtained as the pull back of K(g(x)) through $\frac{\partial g}{\partial x}$ (x). Namely, if X(x) is a vector field

$$\left(g^{*-1} X\right)(x) = \left(\frac{\partial g}{\partial x}\right)^{-1} (x)\ X(g(x))$$

If k(x) is a 1-form

$$\left(g^{*-1} k\right)(x) = \widetilde{\frac{\partial g}{\partial x}} (x)\ k(g(x)) \ .$$

We now recall the definition of the Lie derivative of a tensor field.

Definition 1.4 : If X(x) is a C^∞ vector field, if g_t is the (local) group of diffeomorphisms associated to X, if K(x) is a C^∞ tensor field on R^n, the tensor field $L_X K$ is defined by

$$L_X K(x) = \left[\frac{d}{dt}\left(g_t^{*-1} K\right)(x)\right]_{t=0} \tag{1.21}$$

Of course if Y(x) is a C^∞ vector field, $L_X Y = [X,Y]$.

Theorem 1.5 : Let K be a C^∞ tensor field on R^n. Then for any $x \in R^n$

$$\begin{aligned} \varphi_t^{*-1} K(x) &= K(x) + \int_0^t \left(\varphi_s^{*-1} L_{X_0} K\right)(x)ds + \int_0^t \left(\varphi_s^{*-1} L_{X_i} K\right)(x).dw^i \\ &= K(x) + \int_0^t \left(\varphi_s^{*-1}\left(L_{X_0} + \frac{1}{2} L_{X_i}^2\right)K\right)(x)ds + \\ &+ \int_0^t \left(\varphi_s^{*-1} L_{X_i} K\right)(x).\delta w^i \ . \end{aligned} \tag{1.22}$$

Proof : The proof is elementary using Theorem 1.1 and the equations of $\frac{\partial \varphi_t}{\partial x}(\omega,x)$,

$\left[\frac{\partial \varphi_t}{\partial x}(\omega,x)\right]^{-1}$. Approximation can also be used (see Bismut [5]). □

Of course if K is a (0,0) tensor f, i.e. a C^∞ function, (1.22) is the classical Ito-Stratonovitch formula. (1.22) lifts this formula to tensors. Note that the Ito decomposition (1.22) has a geometrical sense, since $\varphi_t^{*-1} K(x)$ is a process valued in the vector space of the considered tensors at point x.

The semi-group T_t associated to the diffusion (1.1) which acts on functions can then be lifted to tensors (see Bismut [5], Kunita [32]). Of course this lifting depends explicitly on $X_0,\dots,X_m$ and not only on the generator $\mathcal{L}$. The generator of the lifted semi-group is given by

$$\tilde{\mathcal{L}} = L_{X_0} + \frac{1}{2} L_{X_i}^2$$

② Martingales and integration by parts

Let $f \in C_b^\infty(R^d)$, and $T > 0$. Define $h(t,x)$ by :

$$h(t,x) = E\left[f\left(\varphi_{T-t}(\omega,x)\right)\right] \quad t \leqslant T \tag{2.1}$$

It is easy to see that $h(t,x) \in C_b^\infty([0,T]\times R^d)$, and moreover that

$$\begin{cases} \frac{\partial h}{\partial t} + \frac{1}{2} X_0 h + \frac{1}{2} X_i^2 h = 0 \\ h(T,.) = f \end{cases}$$

Set

$$k(t,x) = d\,h\,(t,x)$$

where d is the (exterior) differentiation operator in the variable x. $k(t,x)$ is then a C^∞ differential form. Use now the known fact that the operator d and the Lie differential L_{X_j} commute so that

$$\frac{\partial k}{\partial t} + \left(L_{X_0} + \frac{1}{2} L_{X_i}^2\right) k = 0 \tag{2.2}$$

$$k(T,.) = d\,f$$

Using (1.22), (2.2) expresses the fact that $(\varphi_t^{*-1} k)(x)$ is a martingale valued in $T_x^*(R^n)$. Since $k(T,.) = df$, we have

$$\left(\varphi_t^{*-1} k\right)(x) = E^{F_t}\left(\varphi_T^{*-1} d\,f\right)(x) \ . \tag{2.3}$$

Note that this martingale property is true although the flow $\varphi_t(\omega,.)$ depends on $X_0, X_1 \dots X_m$ and not only through the generator $\mathcal{L}$.

Of course (2.3) can be obtained directly using (2.1). In fact by (1.6),

$$h\left(t,\varphi_t(\omega,x)\right) = E^{F_t}\left[f\left[\varphi_T(\omega,x)\right]\right]$$

so that by differentiating in x, (2.3) holds trivially.

The representation of k as in (2.3) has been noted by a number of authors like Stroock [50] ; Bismut [2]-[4] used a version of (2.3) in the setting of the general theory of stochastic processes to represent the solution of backward stochastic differential equations appearing in control theory. Haussmann used (2.3) in control theory for the same purpose in [20] to represent the adjoint process in a maximum principle.

In fact using stochastic calculus, we see that

$$h\left(T,\varphi_T(\omega,x)\right) = h(0,x) + \int_0^T (X_i\, h)\left(\varphi_s(\omega,x)\right) \delta w^i$$

Using (2.3), we get

$$h\left(T,\varphi_T(\omega,x)\right) = h(0,x) + \int_0^T < (\varphi_s^{*-1} X_i)(x), E^{F_s}(\varphi_T^{*-1} d\,f)(x) > . \,\delta w^i \tag{2.4}$$

i.e.

$$f\left(\varphi_T(\omega,x)\right) = h(0,x) + \int_0^T < (\varphi_s^{*-1} X_i)(x), E^{F_s}(\varphi_T^{*-1} d\,f)(x) > \delta w^i \ . \tag{2.5}$$

The interpretation of (2.5) was that it was related with a sort of "multidimensional Girsanov transformation" (this was clear in [3]) connected with the martingale representation of certain random variables. Haussmann was then led in [21]-[22] to extend the result of Clark [13] expressing the representation of functionals of Brownian motion in terms of their differentials.

Namely let g be a bounded function defined on $\mathcal{C}([0,T]\,;R^d)$ with values in R, which is continuous and differentiable. For every $y \in \mathcal{C}([0,T]\,;R^d)$ $dg(y)$ is an element of the dual of $\mathcal{C}([0,T]\,;R^d)$, i.e. is given by a bounded measure $d\mu^y(t)$ on [0,T] with values in R^d, so that for $z \in \mathcal{C}([0,T]\,;R^d)$

$$(2.6) \qquad < dg(y),z > = \int_{[0,T]} < z_t, d\mu^y(t) > .$$

From the point of view of differential geometry, since z_t is a variation of y_t, z_t is in $T_{y_t} R^n$, so that $d\mu^y(t)$ can be identified to a generalized element of $T^*_{y_t}(R^d)$. If S_t is a continuous function defined on [0,T] with values in $T_x R^d$, it is feasible to set

$$\int_{[0,T]} < S_t\, , \varphi_t^{*-1}\, d\mu^{\varphi.(\omega,x)}(t) > = \int_{[0,T]} < \varphi_t^*(x)\; S_t\, , d\mu^{\varphi.(\omega,x)}(t) > .$$

We now have the main result of Haussmann [21]-[22] extending Clark [13].

Theorem 2.1 : If $u = (u^1,\dots,u^m)$ is a bounded predictable process defined on $\Omega \times R^+$ with values in R^m, then for any $x \in R^d$

$$(2.7)\; \mathbf{E}\left[g\big(\varphi.(\omega,x)\big)\int_0^T u^i\, \delta w^i\right] = E\int_0^T u^i\, ds < \varphi_s^{*-1} X_i(x), \int_{[s,T]} \varphi_v^{*-1}\, d\mu^{\varphi.(\omega,x)}(v) > \Big]$$

Proof : The proof in Haussmann [21] is a beautiful example of the use of the Girsanov transformation. We follow our presentation [7] of Haussmann's proof, as simplified by Williams [59].

For $\ell \in R$, consider the martingale

(2.8) $$Z_t^\ell = \exp\left\{\int_0^t - \ell u^i \delta w^i - \frac{1}{2}\int_0^t |\ell u^i|^2 ds\right\} .$$

Let Q^ℓ be the probability measure on Ω whose density on each F_t relative to P is Z_t^ℓ, i.e.

(2.9) $$\frac{dQ^\ell}{dP}\Big| F_t = Z_t^\ell$$

Now by a classical property of the Girsanov transformation [54], under Q^ℓ, $w_t^\ell = w_t + \int_0^t \ell u\ ds$ is a Brownian martingale.

Consider the stochastic differential equation

(2.10) $$dx^\ell = X_0(x^\ell)dt + X_i(x^\ell)\ (dw^i + \ell u^i\ ds)$$
$$x^\ell(0) = x .$$

By Theorem 1.3, we know that $x_t^\ell = \varphi_t(\omega, y_t^\ell)$ where y_t^ℓ is the solution of the differential equation

(2.11) $$dy^\ell = \left(\varphi_t^{*-1} X_i\right)(y^\ell)\ \ell u^i\ ds$$
$$y^\ell(0) = x .$$

Note that Theorem 1.3 only says that (2.11) has an a.s. non exploding solution, i.e. the a.s. can depend on ℓ. To avoid difficulties, it is feasible to assume that $X_1, \ldots, X_m$ have compact support so that (2.11) is really a "standard" differential equation, and later to remove this assumption. Now under Q^ℓ, $x_.^\ell$ has the same law as $x_. = \varphi_.(\omega, x)$ under P. This implies that

(2.12) $$E\left[Z_T^\ell\, g\,(x^\ell)\right] = E\left[g(x)\right]$$

(expectations are taken with respect to P).

The differential of (2.12) in the variable P is then 0. Now

$$\left.\frac{dZ_t^\ell}{d\ell}\right|_{\ell=0} = -\int_0^T u^i\,\delta w^i \; . \tag{2.13}$$

Moreover using (2.11), y^ℓ is seen to be differentiable and moreover

$$\left.\frac{dy_t^\ell}{d\ell}\right|_{\ell=0} = \int_0^t \left(\varphi_s^{*-1} X_i\right)(x)u^i\,ds \tag{2.14}$$

For any $\omega \in \Omega$ $\ell \to x^\ell$ is differentiable as a function from R into $\mathcal{C}([0,T]\,;R^d)$ and then

$$\left.\frac{dx_t^\ell}{d\ell}\right|_{\ell=0} = \varphi_t^* \int_0^t \left(\varphi_s^{*-1} X_i\right)(x)u^i\,ds \quad . \tag{2.15}$$

After checking that differentiation under E is feasible in the L.H.S. of (2.12), (2.7) follows. □

We now give the main result of Haussmann [21]-[22].

Corollary : The following equality holds a.s. on Ω :

$$g\left(\varphi_\cdot(\omega,x)\right) = E\left[g\left(\varphi_\cdot(\omega,x)\right)\right] + \int_0^T < \varphi_s^{*-1} X_i(x),\ H_s > .\ \delta w^i \tag{2.16}$$

where H is the predictable projection of U_s given by

$$U_s = \int_{[s,T]} \varphi_v^{*-1}\, d\mu^{\varphi_\cdot(\omega,x)}(v) \tag{2.17}$$

Proof : We know by a result of Ito [41] that any square integrable F_T-measurable random variable M_T can be uniquely represented as

$$M_T = a + \int_0^T u^i\,\delta w^i$$

where $a \in R$, and $u = (u^1 \ldots u^m)$ is adapted and such that $E\int_0^T |u|^2\,ds < +\infty$. Using (2.7), it is not hard to deduce (2.16). □

Observe that the Corollary of Theorem 2.1 is the natural extension of (2.3). In fact Davis in [60] gave a proof of this result using a representation of the type (2.3). Also note that the theory of flows has been used to reduce differentiation in ℓ of (2.10) to a standard calculus of variations on a differential equation. When differentiation of tensors will be later needed, this will help us to keep track of the various computations.

However, a superficial look at formula (2.7) shows that g appears on the L.H.S., and dg on the R.H.S., so that formula (2.7), has all the features of an <u>integration by parts</u> formula. Of course this fact was certainly obscured by the fact that the purpose of control theorists was to obtain a martingale representation result.

In fact let $g(x)\,dx$ be the standard gaussian law on the euclidean space R^k, and $f \in C_c^\infty(R^k)$. If $a \in R^k$, we have

$$(2.18) \qquad \int_{R^k} < df(x),a > g(x)dx = \int_{R^k} f(x) < a,x > g(x)dx$$

Of course (2.18) is trivial because we can use the translation invariance of the Lebesgue measure and standard differential calculus on R^k. An unnatural way of obtaining (2.18) would be to observe that under the probability law $\exp(-<\ell a ,x > - \frac{1}{2}|\ell a|^2)$ $g(x)dx$, the law of $x+\ell a$ is equal to $g(x)dx$ so that

$$(2.19) \qquad \int_{R^k} f(x+\ell a) \exp\left[- < \ell a,x > - \frac{1}{2}|\ell a|^2\right] g(x)dx = \int_{R^k} f(x)\; g(x)dx$$

and so (2.18) follows by differentiation in the variable ℓ.

If $Y(x)$ is a C^∞ vector field on R^k with compact support, (2.18) extends to

$$(2.20) \qquad \int_{R^k} (Yf)(x)\; g(x)dx = \int_{R^k} f(x)\left[< Y(x) ,x > - (\operatorname{div} Y)(x)\right] g(x)dx \;.$$

To obtain (2.20) from (2.18), it suffices to apply (2.18) to $a = e_i$, replacing f par fY^i $(1 \leq i \leq k)$ and to sum the formulas in $1 \leq i \leq k$.

The analogy of (2.7) and (2.20) can now be described. P may be considered as the gaussian cylindrical measure on the Hilbert space $(L_2(R^+ ; R))^m$, which is in fact of 0 measure for P . Now an elementary property of gaussians shows P is quasi-invariant under constant (i.e. non random) translations in $(L_2(R^+ ; R))^m$ so that (2.19), and then (2.18) makes sense. However in (2.7), we have used the full strength of the Girsanov transformation i.e. used the quasi-invariance of P under non anticipating random translations in $(L_2(R^+ ; R))^m$, and so we obtain a formula very similar to (2.20) without having to do a necessarily infinite summation to obtain it.

③ The Malliavin calculus

a) Finite dimensional calculus

Consider the space R^k endowed with a probability law $g(x)dx$ where $g \in C^\infty(R^k)$. Let Φ be a C^∞ mapping from R^k into R^d, and let μ be the probability law of Φ on R^d. A natural question to ask is to know if μ is absolutely continuous with respect to the Lebesgue measure, and if the corresponding density is smooth.

Natural assumptions are that $k \geq d$, and that at each $x \in R^k$, the rank of $\Phi'(x) = \frac{\partial \Phi}{\partial x}(x)$ is maximal i.e. equal to d.

a) If Φ is proper, i.e. if for any bounded set B in R^d, $\Phi^{-1}(B)$ is bounded, then it is trivial to see that μ has a smooth density with respect to the Lebesgue measure using localization and the implicit function theorem.

b) If Φ is not proper, the existence of a density for μ is still trivial, but smoothness is not guaranteed. (take $k = 1$, $g = \dfrac{1}{\pi(1+x^2)}$, $\Phi(x) = \text{Arctg}\, x$,

$d\mu(x) = 1_{|x| \leq \frac{\pi}{2}} \dfrac{dx}{\pi}$). In case b), a direct study is necessary.

Take $f \in C_b^\infty(R^d)$. Let $h(x)$ be a C^∞ vector field on R^k with compact support. Now using integration by parts as in (2.20) (where g was supposed to be gaussian)

we get

$$(3.1) \qquad \int_{R^k} < f'(\Phi(x)), \Phi'(x)\ h(x) > g(x)dx +$$

$$\int_{R^k} f(\Phi(x)) \left(\frac{\operatorname{div} h g}{g}\right)(x)\ g(x)dx = 0 \quad .$$

To prove that μ is given by a C^∞ density, it suffices to show that its differentials in distribution sense are bounded measures. In fact this implies that if $\hat{\mu}(\alpha)$ is the Fourier transform of μ, for any m, $|\alpha|^m\ \hat{\mu}(\alpha)$ is bounded, and so the result follows trivially.

If we want to show that the first order differentials of μ are bounded measures, it suffices to prove that formula (3.1) still makes sense if h can be chosen in (2.21) so that $\Phi'(x)\ h(x) = \frac{\partial}{\partial y^\ell}$ $(1 \leqslant \ell \leqslant d)$ (of course h will not have compact support !).

If R^k is endowed with a natural Hilbert space structure (which is the case when g(x)dx is the gaussian law), a natural choice of h(x) is to take the element in $\left\{H \in R^k\ ;\ \Phi'(x)\ H = \frac{\partial}{\partial y^\ell}\right\}$ (which is not empty) of minimal norm namely

$$(3.2) \qquad h(x) = \Phi'^*(x) \left[\Phi'\Phi'^*\right]^{-1}(x)\ \frac{\partial}{\partial y^\ell}$$

where $\Phi'^*(x)$ is the transpose of $\Phi'(x)$ (which maps $T^*_{\Phi(x)}R^k$ into the dual space of R^k, identified to R^k).

Of course it must still be proved that this choice of h is feasible, i.e. that the integral in the second term in the L.H.S. of (3.1) makes sense (which is not the case in the one dimensional counterexample given before).

When applying this procedure repeatedly, i.e. by doing as many integrations by parts as needed, we can then prove that the derivatives of μ are bounded measures and then that μ has a C^∞ density.

In his seminal papers [39], [40] Malliavin has shown this procedure could be applied to the solutions of stochastic differential equations.

b) The Malliavin calculus

Namely for $x_0 \in R^d$, consider the solution $\varphi_t(\omega, x_0)$ of (1.1). For a given $T > 0$, consider the mapping

$$\omega \to \varphi_T(\omega, x_0) \ . \tag{3.3}$$

In the previous argument R^k is replaced by Ω, $g(x)dx$ by P and Φ by the mapping (3.3). Of course, to have a complete analogy with the argument of a), P should be regarded as the gaussian cylindrical measure on the Hilbert space $\left[L^2(R^+ ; R)\right]^m$.

In [39], Malliavin has developped a technique of integration by parts based on the use of the Ornstein-Uhlenbeck operator A which is an unbounded self-adjoint operator operating on $L_2(\Omega, P)$. This technique has been completely clarified and extended by Stroock [51]-[53], and Shigekawa [49].

In [40], Malliavin showed that the extension to this situation of the choice of h given by (3.2) was feasible in the sense made precise in a), in an argument later completed by Ikeda-Watanabe [26].

In [7], we suggested a different route to the integration by parts result, which was precisely what has been described in section 2.

Since the approach taken by Malliavin [39]-[40], Stroock [51],[53], Shigekawa [49], Ikeda-Watanabe [26] to the integration by parts is completely described in these papers, we will focus now on our paper [7].

Before going into details, we feel that the essential contribution of Malliavin has been to show that the classical differential and integral calculus can be successfully extended on the probability space of Brownian motion, by using the classical Ito calculus on diffusions to express the various quantities appearing in the integration by parts procedure.

We now develop as in [7] the first application of the Malliavin calculus.

c) An application : Hörmander's theorem

In his celebrated paper [24], Hörmander proved that if 0 is an open set in R^d such that at each $x \in 0$, the vector space spanned by $X_0(x),\dots,X_m(x)$ and their Lie brackets at x is equal to R^d, the operator $\mathcal{L}$ is hypoelliptic, i.e. if u is a distribution such that $\mathcal{L}u \in C^\infty(0)$, then $u \in C^\infty(0)$.

If we apply this result to the differential operator $\bar{\mathcal{L}}$ operating on $C^\infty(R \times R^d)$

$$\bar{\mathcal{L}} = \frac{\partial}{\partial t} + \mathcal{L}$$

we see that if 0 is an open set in R^d such that at any $x \in 0$, the vector space T'_x in R^d spanned by $X_1(x),\dots,X_m(x)$ and all the brackets of length ≥ 2 of $X_0, X_1 \dots X_m$ at x span the whole space R^d, $\bar{\mathcal{L}}$ is hypoelliptic on $R \times 0$.

Now if $p_t(x,dy)$ are the transition probabilities of the diffusion (1.1), we know that if $\mathcal{L}^*$ is the adjoint operator of $\mathcal{L}$ (with respect to the Lebesgue measure), then for any $x \in R^n$,

$$\left(\frac{\partial}{\partial t} - \mathcal{L}^*_y\right) p_t(x,y) = 0 \qquad \text{on }]0,+\infty[\times R^d . \tag{3.4}$$

By applying Hörmander's theorem, we see that under the previous conditions, for any $x \in R^d$, $p_t(x,dy) = p_t(x,y)\,dy$, where $p_t(x,y)$ is C^∞ in the variables $(t,y) \in]0,+\infty[\times 0$. In fact Hörmander's theorem has been used by Ichihara and Kunita [25] to prove the corresponding results on the transition probabilities.

We will now try to show how the Malliavin calculus can be used to prove directly that the $p_t(x,dy)$ are smooth in (t,y). Note that using pseudo-differential operators, Kohn [27] has simplified Hörmander's original proof (for a systematic exposition, see Trèves [57]).

We will not show how the hypoellipticity of $\bar{\mathcal{L}}$ can be recovered from the smoothness of $p_t(x,dy)$ but concentrate on a probabilistic proof of the smoothness of $p_t(x,dy)$.

Let H be a continuous function defined on $\mathscr{C}([0,T];R^d)$ with values in $T_x^*(R^d)$ which is bounded, differentiable, with a uniformly bounded differential. For $y \in \mathscr{C}([0,T];R^d)$, dH(y) can be identified to a finite measure $dv^y(t)$ on $[0,T]$ with values in $R^d \otimes R^d$, so that

(3.5)
$$z \in \mathscr{C}([0,T];R^d) \to \; <dH(y),z> \; = \int_{[0,T]} dv^y(t)\, z_t$$

$dv^y(t)$ can be identified to a generalized linear mapping from $T_{y_t}(R^d)$ into $T_x^*(R^d)$. If $\ell \in T_x(R^d)$ we define the action of $\int_0^T \varphi_t^{*-1}\, dv^y(t)$ on ℓ by

(3.6)
$$\int_0^T \varphi_t^{*-1}\, dv^y(t)\,(\ell) = \int dv^y(t)\, \varphi_t^*(\ell)\;.$$

We then have the following result in Bismut [7].

Theorem 3.1 :   _Let_ $f \in C_b^\infty(R^d)$. _Then_ _the_ _following_ _equality_ _holds_

(3.7)
$$E\left[f\left(\varphi_T(\omega,x)\right) < H\left(\varphi.(\omega,x)\right), \int_0^T \left(\varphi_s^{*-1} X_i\right)(x)\delta w^i > \right] =$$

$$E < H\left(\varphi.(\omega,x)\right), \int_0^T \left(\varphi_s^{*-1} X_i\right)(x) > <\left(\varphi_s^{*-1} X_i\right)(x)ds, \left(\varphi_T^{*-1}\, df\right)(x) > +$$

$$+ E\left[f\left(\varphi_T(\omega,x)\right) \int_0^T < \left(\varphi_s^{*-1} X_i\right)(x), \int_{[s,T]} [\varphi_h^{*-1}(\omega,x)dv^{\varphi.(\omega,x)}(h)] \right.$$

$$\left. \left(\varphi_s^{*-1} X_i\right)(x) > ds \right].$$

Proof : Write $h = \sum_1^d h_k\, e^k$. Use then formula (2.7), where g is $h_k(x)\, f(x_T)$, and $u^i = \left(\varphi_s^{*-1} X_i\right)^k(x)$ (which is _adapted_ !). Summing the corresponding formulas in k, we get (3.7). □

Of course we have obtained (3.7) by analogy with (3.1). In fact using (2.15),

$\Phi'(\omega)$ can be interpreted as the mapping.

$$u \in L_2([0,T]\,;R^m) \to \varphi_T^* \int_0^T \varphi_s^{*-1} X_i\; u_i\; ds \in T_{\varphi_T(\omega,x)}\; R^d$$

The adjoint mapping $\Phi'^*(\omega)$ is clearly

$$p \in T^*_{\varphi_T(\omega,x)}\; R^d \to < \varphi_T^* \varphi_s^{*-1} X_i\,(x), p > \in L_2([0,T]\,;R^m)$$

so that $\Phi'\;\Phi'^*(\omega)$ is the mapping

$$(3.8)\qquad p \in T^*_{\varphi_T(\omega,x)}\; R^d \to \varphi_T^* \int_0^T \varphi_s^{*-1} X_i\,(x) < \varphi_s^{*-1} X_i\,,\, \varphi_T^{*-1}\; p > ds$$

Formula (3.7) is then the analogous of formula (3.1) with

$$(3.9)\qquad h = \Phi'^*(\omega)\; \varphi_T^*\, H$$

The choice of h in (3.8) is of course justified by the argument leading to (3.2).

Consider now the following assumption H1 : $x \in R^d$ is such that the vector space spanned by $X_1(x),\dots X_m(x)$, and the brackets of $X_0, X_1 \dots X_m$ of length $\geqslant 2$ at x generate R^d.

Definition 3.2 : *For $T > 0$, $C_T(\omega)$ is the linear mapping from $T_x^* R^d$ into $T_x R^d$ given by*

$$(3.10)\qquad p \to \sum_{i=1}^{m} \int_0^T < \varphi_s^{*-1}\; X_i\,(x), p > (\varphi_s^{*-1}\; X_i\,)(x) ds\;.$$

$C_T(\omega)$ defines a symmetric positive form on $T_x^*\,(R^d)$, and by (3.8) is clearly related to $\Phi'\;\Phi'^*(\omega)$.

Recall that in a), Φ' was supposed to be of maximal rank d (here $k = +\infty$!).

The first spectacular result of Malliavin [39] was that precisely H1 implies that $\Phi' \Phi'^*$ is a.s. invertible. Namely

Proposition 3.3 : Under H1, a.s., for any $T > 0$, $C_T(\omega)$ is invertible.

Proof : We sketch the proof in [39]-[7]. If $f \neq 0$ is such that $C_T(\omega)\, f = 0$, clearly $< \varphi_s^{*-1} X_i(x), f > = 0$ for $s \leqslant T$. Using (1.22), it follows that $< \varphi_s^{*-1}[X_j, X_i](x), f > = 0$ $(1 \leqslant i,j \leqslant m \ldots, s \leqslant T)$ and so $< \varphi_s^{*-1} [X_j,[X_j,X_i]](x), f > = 0$ $(s \leqslant T)$, so that cance-ling the drift term in $< \varphi_s^{*-1} X_i(x), f >$, $< (\varphi_s^{*-1} X_0)(x), f > = 0$ $(s \leqslant T)$. Iterating the procedure, we find that in particular f is orthogonal to $X_1(x) \ldots X_m(x)$ and to the brackets considered in H1, so that $f = 0$. Of course negligible sets must be worked out [7]. □

As noted in Malliavin [39] (see Stroock [53]) Proposition 3.3 is enough to show that the law of $\varphi_T(\omega, x)$ is given by a density $p_t(x,y)dy$.

Recall that in a), it was underlined that a feasibility assumption was to be checked on the choice (3.2).

Malliavin [40] and Ikeda-Watanabe [26] were able to prove that when H1 is still verified excluding X_0 - i.e. by considering only $X_1 \ldots X_m$ and their brackets - this is the case. More recently, Kusuoka and Stroock [37]-[52] were able to prove that this is still the case under H1.

Namely

Theorem 3.4 : If $x \in R^d$ is such that H1 is verified, then for any $t > 0$, $\| C_t^{-1} \|$ belongs to all the $L_p (1 \leqslant p < +\infty)$.

Proof : It is out of question to give here the whole proof of Malliavin [40], Ikeda-Watanabe [26], Kusuoka and Stroock [37]-[52], so that we only give a brief sketch.

Take $u \in R$, and let b_t be a one-dimensional Brownian motion such that $b_0 = 0$. We have the easy equality

$$(3.11) \qquad E\left[\exp - \frac{\alpha^2}{2} \int_0^T |x + b_s|^2 \, ds\right] = (\operatorname{ch} \alpha T)^{-1/2} \exp\left[- \frac{\alpha x^2}{2} \, t\,h(\alpha T)\right] .$$

Using (3.11) and Čebyšev inequality, it can be shown that if X_t is a continuous semi-martingale on a filtered probability space $(\Omega, \{F_t\}_{t \geq 0}, P)$ whose Ito-Meyer decomposition is

(3.12) $$X_t = X_0 + \int_0^t A_s \, ds + N_t$$

where

- X_0 is F_0-measurable
- A is predictable and such that $|A| \leq M$
- N_t is a Brownian martingale such that $N_0 = 0$

then for any $\varepsilon > 0$, $T > 0$

(3.13) $$P\left[\int_0^T |X_s|^2 ds < \varepsilon\right] \leq \sqrt{2} \exp - \left[16\left(\frac{\varepsilon}{T^2} + \frac{M^2 T}{3}\right)\right]^{-1}$$

Now under the restricted assumptions of Malliavin [40], Ikeda-Watanabe [26], for $f \in R^d$, $\|f\| = 1$, there is one bracket of $X_1 \ldots X_m$, which will be written $X_{[I]}$ such that $|<f, X_I(x)>| \geq \eta > 0$. If $I = i$ $(1 \leq i \leq m)$, $< C_t f, f >$ will be large enough. If $X_{[I]} = [X_i, X_{[J]}]$, by using (1.22), we see that $< f, \varphi_s^{*-1} X_{[J]} >$ is a semi-martingale such that its quadratic variation is given by

(3.14) $$\sum_{k=1}^m \int_0^t < f, \varphi_s^{*-1} [X_k, X_{[J]}] >^2 ds$$

By doing a time change on $< f, \varphi_s^{*-1} X_{[J]} >$, we can then go back to the situation studied in (3.12)-(3.13) and find that the probability that $\int_0^t < f, \varphi_s^{*-1} X_{[J]} >^2 ds$ is "small" is itself small enough. By induction, we can prove that for any $k \in N$

(3.15) $$P\left[\int_0^t < f, \varphi_s^{*-1} X_i >^2 ds < \varepsilon\right] = o(\varepsilon^k) \quad \varepsilon \to 0$$

The theorem follows easily from (3.15) as in [40]-[26]. In fact in [40]-[26], instead of (3.11), an estimate on the variance of b_s on [0,1] was obtained using the Fourier series representation of the Brownian motion.

In the case where X_0 is necessary to fulfill H1, Kusuoka and Stroock [37]-[52] use the fact that the drift term in the Ito-Meyer decomposition of (1.22) is a.s. a Hölder function. Combining this fact with the previous estimates, they obtain the Theorem in full generality. □

From Theorem 3.5, it is then possible to obtain :

Theorem 3.5 : If $x \in R^d$ is such that H1 is verified at x, for any $t > 0$, the law of $\varphi_t(\omega,x)$ is given by $p_t(x,y)dy$, where $p_t(x,.) \in C_b^\infty(R^d)$.

Remark : Using the techniques of Stroock [51], the previous results can be localized, so that the smoothness of $p_t(x,dy)$ depends only on the behaviour of $X_0, X_1 \ldots X_m$ on a neighbourhood of y.

Moreover the techniques of Bismut [7] used in combination with Kusuoka-Stroock [37]-[52] give the existence of a smooth resolvent operator $V^\lambda(x,y)$ ($\lambda > 0$, $y \neq x$) associated to $\mathcal{L}$, when $\mathcal{L}$ verifies Hörmander's conditions (the resolvent can be smooth while the semi-group is not !).

④ Application to filtering

The Malliavin calculus has been applied to situations where standard analytic techniques do not work.

In [23], Holley and Stroock have obtained results concerning the existence and smoothness of the finite-dimensional distributions of an infinite system of interacting diffusions. In this case, while infinite dimensional analogues of Partial Differential equations appear to be difficult to use, the Malliavin calculus is an efficient tool to prove the indicated smoothness results.

Here, we will concentrate on the results of Bismut-Michel [11] extending Michel [44] on conditional diffusions.

a) A finite dimensional analog

We start again with the situation studied in Section 3 a). Besides we assume that $x = (x',x")$ where $x' \in R^{k'}$, $x" \in R^{k"}$ (of course $k' + k" = k$).

We want to study the smoothness of the conditional law of $\Phi(x)$ given x". A natural assumption is that $k' \geqslant d$, and the partial differential $\Phi'_{x'}(x',x")$ is of maximal rank d.

Following the ideas of 3 a), it is natural to try to obtain a conditional formula of integration by parts. Namely let h(x) be a C^∞ function defined on R^k with valus in $R^{k'}$ which has compact support. Let U(x") be a measurable bounded function on $R^{k"}$. For $f \in C_b^\infty(R^d)$, we have

(4.1) $$\int_{R^k} U(x") < f'(\Phi(x)), \Phi'_{x'}(x)\, h(x) > g(x)dx$$

$$+ \int_{R^k} U(x")\, f(\Phi(x))\, \frac{\mathrm{div}_{x'}(h\, g)}{g}(x)\, g(x)dx = 0$$

(in (4.1) the divergence is taken with respect to the variable x'). Of course (4.1) is a consequence of (3.1), where instead of taking a general h, we chose h with values in $R^{k'}$ (the extension of (4.1) to a measurable U is trivial).

Assuming that the choice

$$h(x) = \Phi'^*_{x'}\,(\Phi'_{x'}\,\Phi'^*_{x'})^{-1}(x)\,\frac{\partial}{\partial y^\ell}$$

is feasible in (4.1), we see that

(4.2) $$\int_{R^k} U(x")\,(\frac{\partial}{\partial y^\ell} f)\,(\Phi(x))\,g(x)dx + \int_{R^k} U(x")\, f(\Phi(x))\, K_\ell(x)\, g(x)dx = 0$$

If $P_{x"}(dx)$ is the conditional law of x given x", we get from (4.2)

$$(4.3) \qquad \int_{R^k} \frac{\partial}{\partial y^\ell} f(\Phi(x))\, P_{x''}(dx) = -\int_{R^k} f(\Phi(x))\, K_\ell(x)\, P_{x''}(dx) \quad \text{a.s.}$$

By eliminating negligible sets adequately, it is clear that a.s., the differential of the law of $\Phi(x)$ given x'' is a bounded measure.

By iterating the procedure, we may obtain the smoothness of the conditional law of $\Phi(x)$ given x''.

b) Application to diffusions

From (2.7), we may obtain the analog of (4.1). Assume that $w = (w', w'')$ where $w' = (w^1 \ldots w^{m'})$, $w'' = (w^{m'+1} \ldots w^m)$. Assume that in (2.7), g is replaced by $U(w'')\, g(\varphi.(\omega,x))$, where we suppose (at the beginning !) that U verifies the same type of assumptions as g.

If, in (2.7), we assume that $(u^{m'+1}, \ldots, u^m)$ are all 0, it is clear that on the r.h.s. of (2.7), no differential of U will ever appear, so that (2.7) can be extended by assuming that U is only bounded and measurable.

Conditional expectations with respect to w'' may then be taken on both sides of (2.7).

c) The main results on conditional diffusions

Consider the system of stochastic differential equations

$$(4.4) \qquad \begin{aligned} dx &= X_0(x,z)dt + \sum_{i=1}^{m} X_i(x,z).dw^i + \sum_{j=1}^{d} \tilde{X}_j(x,z)(d\tilde{w}^j + \ell^j(x,z)dt) \\ x(0) &= x_0 \\ dz &= Z_0(z)dt \qquad\qquad\qquad\qquad + \sum_{j=1}^{d} Z_j(z)\,(d\tilde{w}^j + \ell^j(x,z)dt) \\ z(0) &= z_0 \end{aligned}$$

where $w = (w^1 \ldots w^m)$, $\tilde{w} = (\tilde{w}^1 \ldots \tilde{w}^d)$ are independent Brownian motions. Of course the vector fields and the functions appearing in (4.4) are supposed to be smooth with bounded differentials.

$x_t \in R^n$ is supposed to be the state of the system and $z_t \in R^p$ the observation on the system.

Set

$$F_t^z = \mathcal{B}(z_s | s \leqslant t)$$

Using the results of Schwartz in the theory of prediction [42], it is easy to see that there exists a continuous process τ_t^z with values in the set Π of probability measures on $\mathcal{C}(R^+ ; R^n)$ such that for every $\{F_t^z\}_{t \geqslant 0}$-stopping time T, on $(T < +\infty)$, τ_T^z is the conditional law of the process x given F_T^z .

In Bismut-Michel [11], the techniques of the calculus of variations have been applied to study in detail the process $\{\tau_T^z\}$.

Note that the technique briefly summarized in b) allows us not to make any distinction between smoothing, filtering, and prediction, since the process $\tau_{\mathbb{T}}^z$ can be analyzed globally.

The main results in Bismut-Michel [11] are

Theorem 4.1 : Assume that $y_0 = (x_0, z_0)$ is such that the vector subspace of $R^n \times \{0\}$ spanned by $((X_1(y_0),0)\dots((X_m(y_0),0))$ and all the Lie brackets at y_0 of the vectors $((X_1,0),\dots,(X_m,0), (\tilde{X}_1,Z_1),\dots,(\tilde{X}_m,Z_m))$ in which $(X_1,0),\dots,(X_m,0)$ appear at least once is equal to $R^n \times \{0\}$. Then a.s., for every $T \geqslant 0$, $t > 0$, the law of x_t for τ_T^z is equal to $q_{t,T}^z(y)dy$, where $q_{t,T}^z(y) \in C_b^\infty(R^n)$.

This result extends the results of Pardoux [45] where x_0 was assumed to have itself a probability law given by a density - while here we assume that x_0 is fixed - and moreover in [45] (4.4) was supposed to be partially elliptic in the sense that $X_1 \dots X_m$ had to span R^n . Theorem 4.1 is an extension of Hörmander's theorem, but is still a probabilistic result (i.e. there is a a.s. in it).

Theorem 4.2 : Assume that $p = d$ and that $Z_1 \ldots Z_d$ span R^d. Then if $X_1 \ldots X_m$, $\ell_1 \ldots \ell_d$ have compact support, a.s., for every $T \geqslant 0$, τ_T^z is the law of a non homogeneous Feller process.

The assumption on $Z_1 \ldots Z_d$ is standard. The compactness assumption is technical, and can be weakened.

Since in general for τ_T^z, x_t is not a semi-martingale, a change of variables is done in [11] with the help of a flow depending only on z so as to get a standard diffusion process (under τ_T^z).

Under the same assumptions as Theorem 4.2, the unnormalized filtering equation is reduced to a standard - i.e. with no diffusion term - partial differential equation, with coefficients very irregular in time. This is done in general by using the theory of stochastic flows inside the Girsanov density so as to perform an integration by parts inside the Girsanov density (which is trivial in the standard filtering problem [38]).The results of Eliott and Kohlmann [16] and Davis [14] are then extended in full generality.

Note that except in the partially elliptic case where the filtering equation can be solved "pointwise", under the assumptions of Theorem 4.1, it is not clear at all how could classical analysis techniques be applied, because of the non differentiability of the coefficients of the equation in time, in order to obtain analytically the same result. In the partially elliptic case, the results of Pardoux [45] (on filtering) can be reobtained using the filtering equation of [11].

REFERENCES

[1] P. Baxendale :

Wiener processes on manifolds of maps ,
Proceedings Royal Soc. Edinburgh, 87A, 127-152 (1980).

[2] J.M. Bismut :

Linear-quadratic optimal stochastic control with random coefficients ,
SIAM J. of Control 14, 419-444 (1976)

[3] J.M. Bismut :

An introductory approach to duality in optimal stochastic control ,
SIAM Review 20, 62-78 (1978).

[4] J.M. Bismut :

Contrôle des systèmes linéaires-quadratiques ,
Séminaire de Probabilités n° XII, p. 180-264, Lecture Notes in Math. 649
Berlin-Springer 1978.

[5] J.M. Bismut :

Mécanique aléatoire ,
Lecture Notes in Math. n° 866, Berlin Springer 1981.

[6] J.M. Bismut :

A generalized formula of Ito and some other properties of stochastic flows ,
Z. Wahrsch, 55, 331-350 (1981).

[7] J.M. Bismut :

Martingales, the Malliavin calculus and hypoellipticity under general Hörmander's conditions ,
Z. Wahrsch, 56, 469-505 (1981).

[8] J.M. Bismut :

Calcul des variations stochastique et processus de sauts ,
A paraître.

[9] J.M. Bismut :

The calculus of boundary processes ,
To be presented at the Kyoto Conference in Probability (1982).

[10] J.M. Bismut - D. Michel :
Structure des diffusions conditionnelles et calcul des variations ,
C.R.A.S. 292, Série I, 731-734 (1981).

[11] J.M. Bismut - D. Michel :
Diffusions conditionnelles ,
J. of Funct. Anal. Part I 44, 174-211 (1981), Part II 45, 274-292 (1982).

[12] Y.N. Blagoveschenskii - M.I. Freidlin :
Certain properties of processes depending on parameters ,
Sov. Math. Dokl. 2, 633-636 (1961).

[13] J.M.C. Clark :
The representation of functionals of Brownian motion by stochastic integrals ,
Ann. Math. Stat 41, 1282-1295 (1970), 42, 1778 (1971).

[14] M.H.A. Davis :
Pathwise non linear filtering ,
"Stochastic systems"(M. Hazewinkel ed.) Dordrecht : Reidel. To appear.

[15] H. Doss :
Liens entre équations différentielles stochastiques et ordinaires ,
Ann. Inst. H. Poincaré 13, 99-125 (1977).

[16] R.J. Eliott - M. Kohlmann :
Robust filtering for correlated multidimensional observation ,
Preprint 1981.

[17] K.D. Elworthy :
Stochastic dynamical systems and their flows ,
In "Stochastic Analysis" A. Friedman and M. Pinsky ed. p 79-95 ,
New-York : Acad. Press 1978.

[18] M. Emery :
Equations différentielles stochastiques lipschitziennes : étude de la stabilité ,
Séminaire de Probabilités n° 13 , p. 281-293, Lecture Notes in Math 721,
Berlin : Springer 1979.

[19] I.I. Gihman - A.V. Skorohod :
Stochastic differential equations ,
Erg. der Math. und ihrer Grenzgeb, Band 72, Berlin : Springer 1972.

[20] U. Haussmann :
On the stochastic maximum principle ,
SIAM J. of Control and Opt. 16, 236-251 (1978).

[21] U. Haussmann :
Functionals of Ito processes as stochastic integrals ,
SIAM J. of Control and Opt. 16, 252-269 (1978).

[22] U. Haussmann :
On the integral representation of Ito processes ,
Stochastics 3, 17-27 (1979).

[23] R. Holley - D. Stroock :
Diffusions on an infinite dimensional torus ,
J. of Funct. Anal. 42, 29-63 (1981).

[24] L. Hörmander :
Hypoelliptic second order differential equations ,
Acta Math. 119, 147-171 (1967).

[25] K. Ichihara - H. Kunita :
A classification of second order degenerate elliptic operators and its probabilistic characterization ,
Z. Wahrsch, 30, 235-254 (1974).

[26] N. Ikeda - S. Watanabe :
Stochastic differential equations and diffusion processes ,
Amsterdam, North-Holland 1981.

[27] J.J. Kohn :
Pseudo-differential operators and hypoellipticity ,
Proc. Symp. Pure Math. 23, 61-69 (1973).

[28] N.V. Krylov - B.L. Rozovskii :
On the Cauchy problem for linear stochastic partial differential equations ,
Math. USSR Izv. 11, 1267-1284 (1977).

[29] N.V. Krylov - B.L. Rozovskii :

On conditional distributions of diffusion processes ,

Math. USSR Izv. 12, 336-356 (1978).

[30] N.V. Krylov - B.L. Rozovskii :

On the first integral and Liouville equations for diffusion processes ,

In "Stochastic differential systems", M. Arato and D. Vermes ed. p. 117-125,Lecture Notes in Control and Inf. Sci. 36, Berlin, Springer 1981.

[31] H. Kunita :

On the decomposition of solutions of stochastic differential equations ,

In "Stochastic Integrals", D. Williams ed, p. 213-255, Lecture Notes in Math. n° 851, Berlin, Springer 1981.

[32] H. Kunita :

Some extensions of Ito's formula ,

Séminaire de Probabilités n° XV, p. 118-141, Lecture Notes in Math. n° 850, Berlin, Springer 1981.

[33] H. Kunita :

Stochastic partial differential equations connected with non linear filtering ,

Proceedings of CIME ; To appear.

[34] H. Kunita :

Cauchy problems for stochastic partial differential equations arising in non linear filtering theory ,

Systems and Control Letters 1, 37-41 (1981).

[35] H. Kunita :

Densities of a measure valued process governed by a stochastic differential equation ,

Systems and Control Letters 1, 100-104 (1981).

[36] S. Kusuoka :

Analytic functionals of Wiener processes and absolute continuity ,

In "Functional Analysis in Markov processes" p. 1-46, Lecture Notes in Math. n° 923, Berlin, Springer 1982.

[37] S. Kusuoka - D. Stroock :

To appear.

[38] R.S. Liptser - A.N. Shiryayev :

Statistics of random processes ,

Vol. I, II, Berlin : Springer 1977-1978.

[39] P. Malliavin :

Stochastic calculus of variations and hypoelliptic operators ,

Proceedings of the Conference on Stochastic differential equations of Kyoto (1976), p. 195-263, Tokyo : Kinokuniya and New-York : Wiley 1978.

[40] P. Malliavin :

C^k *hypoellipticity with degeneracy* ,

In "Stochastic Analysis", A. Friedman and M. Pinsky ed, p. 199-214, New-York : Acad. Press 1978.

[41] P.A. Meyer :

Un cours sur les intégrales stochastiques ,

Séminaire de Probabilités n° X, p. 245-400, Lecture Notes in Math n° 511, Berlin : Springer 1973.

[42] P.A. Meyer :

Sur les désintégrations régulières de L. Schwartz ,

Séminaire de Probabilités n° 7, p. 217-222, Lecture Notes in Math n° 321, Berlin : Springer 1973.

[43] P.A. Meyer :

Flot d'une équation différentielle stochastique ,

Séminaire de Probabilités n° 15, p. 103-117, Lecture Notes in Math n° 850, Berlin : Springer 1981.

[44] D. Michel :

Régularité des lois conditionnelles en théorie du filtrage non linéaire et calcul des variations stochastique ,

J. of Funct. Anal. 41, 8-36 (1981).

[45] E. Pardoux :

Stochastic partial differential equations and filtering of diffusion processes ,

Stochastics 3, 127-167 (1979).

[46] E. Pardoux :

Equations du filtrage non linéaire, de la prédiction et du lissage ,

Preprint Université de Provence 1980.

[47] M. Reed - B. Simon :
Methods of modern mathematical Physics ,
Vol I, New-York : Acad. Press 1972.

[48] B.L. Rozovskii :
A formula of Ito-Ventcell,
Vestnik Moskou Univ. Ser 1 Math Meh 26-32 (1973).

[49] I. Shigekawa :
Derivatives of Wiener functionals and absolute continuity of induced measures ,
J. Math. Kyoto Univ. 20, 263-289 (1980).

[50] D. Stroock :
On certain systems of parabolic equations ,
Comm. Pure and Applied Math. 23, 447-457 (1970).

[51] D. Stroock :
The Malliavin calculus and its applications to second order parabolic differential equations ,
Math Systems Theory, Part I 14, 25-65 (1981), Part II 141-171 (1981).

[52] D. Stroock :
Some applications of stochastic calculus to partial differential equations ,
Ecole d'été de Probabilités de Saint Flour. A paraître.

[53] D. Stroock :
The Malliavin calculus : a functional analytic approach ,
J. Funct. Anal. 44, 212-257 (1981).

[54] D. Stroock - S.R.S. Varadhan :
Multidimensional diffusion processes ,
Grundlehren Math Wissenschaften 233, Berlin : Springer 1979.

[55] D. Stroock - S.R.S. Varadhan :
On the support of diffusions with applications to the strong maximum principle ,
6 th Berkeley Symposium in Probability and Statistics, Vol III, 333-359, Berkeley : University of California Press 1972.

[56] H. Sussmann :

On the gap between deterministic and stochastic ordinary differential equations ,

Annals of Prob. 6, 19-41 (1978).

[57] F. Trèves :

Introduction to Pseudodifferential and Fourier Integral operators ,

Vol I, New-York : Plenum Press 1981.

[58] A.D. Ventzell :

On equations of the theory of conditional processes ,

Theory Prob. and Appl. 10, 357-361 (1965).

[59] D. Williams :

" To begin at the beginning " ,

In "Stochastic Integrals", D. Williams ed. p. 1-55, Lecture Notes in Math. n° 851, Berlin : Springer 1981.

[60] M.H.A. Davis :

Functionals of Ito processes as stochastic integrals ,

Proc. Cambridge Philos. Soc. 87, 157-166 (1980).

On one-dimensional Markov SDEs

M.P. Ershov, Kl. Gooßen
University of Essen-GHS
4300 Essen 1 (Germany-West)

In [1] the existence of strong solutions of one-dimensional SDEs with coefficients depending only on the present was established. The proof was based on the Ito change-of-variables formula applied to a specially chosen function. However that function was not sufficiently smooth to enable the application of the Ito formula immediately.

The aim of this paper is (1) to get Ito's formula for "bad" functions including the function of [1] (similar results were obtained in [3] and also communicated to us by M. Yor) and (2) to show that the result of [1] mentioned can also be obtained without the Ito formula from purely deterministic lemmas. In fact, it is seen from those lemmas that the specific properties of the Wiener process are more or less unessential in the proof.

1. Let (Ω,F,P) be a complete probability space with a right continuous increasing family $(F_t)_{t\geq 0}$ of σ-sub-fields of F each containing all P-null sets in F. Let $\xi = ((\xi_1(t),\dots,\xi_d(t)), F_t)_{t\geq 0}$ be a continuous d-dimensional semimartingale having the following representation

$$\xi_i(t) = \eta_i + A_i(t) + M_i(t) \quad , \ t \geq 0, \ i=1,\dots,d,$$

where η_i are F_o-measurable random variables, A_i (F_t)-adapted processes with $A_i(0) = 0$ a.s. and continuous paths of bounded variation on each finite interval, M_i continuous local (F_t)-martingales with $M_i(0) = 0$ a.s., $i=1,\dots,d$.

Proposition 1. Let $f:\mathbb{R}^d \to \mathbb{R}$ be a continuous function such that

(i) $f|U$ is twice continuously differentiable,

(ii) $f_i'|UK$ and $ff_{ij}''|UK$ are bounded for all compact $K \subset \mathbb{R}^d$ and $i,j=1,\ldots,d$ where $U = f^{-1}(\mathbb{R}\setminus\{0\})$.

Then, for all $t \geq 0$,

$$(1.1)\quad f(\xi(t))-f(\xi(0)) = \sum_{i=1}^{d} \int_0^t 1_U(\xi(s))f_i'(\xi(s))d\xi_i(s) +$$

$$\frac{1}{2}\,\mathrm{P\text{-}lim}_{\varepsilon\to 0} \sum_{i,j=1}^{d} \int_0^t \{1_{[\varepsilon,\infty)}(|f(\xi(s))|)f_{ij}''(\xi(s)) +$$

$$\frac{1}{\varepsilon}1_{(0,\varepsilon)}(|f(\xi(s))|)[(\mathrm{sgn} f\, f_i'f_j' + |f|f_{ij}'')(\xi(s))]\}d<M_i,M_j>(s).$$

Proof. Consider, for $\varepsilon > 0$ and $n \in \mathbb{N}$,

$$g_{n\varepsilon}(y) := \frac{1}{2n-1}\{\tfrac{2}{3}n^2\varepsilon^{-2}y^3\, 1_{(0,\frac{\varepsilon}{2n})}(|y|) + [\tfrac{n}{\varepsilon}y|y| - \tfrac{y}{2} + \tfrac{\varepsilon}{12n}\mathrm{sgn}\, y]1_{[\frac{\varepsilon}{2n},\frac{(2n-1)\varepsilon}{2n}]}(|y|)$$

$$+[-\tfrac{2}{3}n^2\varepsilon^{-2}y^3 + \tfrac{2}{\varepsilon}n^2 y|y| - (2n^2-2n+1)y + \tfrac{\varepsilon}{6}(4n^2-6n+3)\mathrm{sgn}\, y]1_{(\frac{(2n-1)\varepsilon}{2n},\varepsilon)}(|y|)\}$$

$$+\ (y - \tfrac{\varepsilon}{2}\mathrm{sgn}\, y)\, 1_{[\varepsilon,\infty)}(|y|).$$

It is easy to check that $g_{n\varepsilon}$ is twice continuously differentiable and

$$(1.2)\quad g_{n\varepsilon}'(y) = \frac{1}{2n-1}\{2n^2\varepsilon^{-2}y^2 1_{(0,\frac{\varepsilon}{2n})}(|y|) + [\tfrac{2n}{\varepsilon}|y| - \tfrac{1}{2}]\, 1_{[\frac{\varepsilon}{2n},\frac{(2n-1)\varepsilon}{2n}]}(|y|)$$

$$-[2n^2\varepsilon^{-2}y^2 - \tfrac{4}{\varepsilon}n^2|y| + 2n^2-2n+1]\, 1_{(\frac{(2n-1)\varepsilon}{2n},\varepsilon)}(|y|)\} + 1_{[\varepsilon,\infty)}(|y|),$$

$$(1.3)\quad g_{n\varepsilon}''(y) = \frac{1}{2n-1}\{4n^2\varepsilon^{-2}y 1_{(0,\frac{\varepsilon}{2n})}(|y|) + \frac{2n}{\varepsilon}\mathrm{sgn}\, y\ \ 1_{[\frac{\varepsilon}{2n},\frac{(2n-1)\varepsilon}{2n}]}(|y|)$$

$$-\ \frac{4n^2}{\varepsilon}[\frac{y}{\varepsilon} - \mathrm{sgn}\, y]\ 1_{(\frac{(2n-1)\varepsilon}{2n},\varepsilon)}(|y|).$$

Under the assumptions on f, $g_{n\varepsilon}\circ f$ is continuous and $g_{n\varepsilon}\circ f|U$ is twice continuously differentiable with

$$(1.4)\qquad (g_{n\varepsilon}\circ f)'_i|U = g'_{n\varepsilon}\circ f\cdot f'_i|U$$

$$(g_{n\varepsilon}\circ f)''_{ij}|U = g''_{n\varepsilon}\circ f\cdot f'_i f'_j + g'_{n\varepsilon}\circ f\cdot f''_{ij}|U \qquad (i,j=1,\dots,d).$$

We shall show that $g_{n\varepsilon}\circ f$ is twice continuously differentiable everywhere with

$$(1.5)\qquad (g_{n\varepsilon}\circ f)'_i|U^c = (g_{n\varepsilon}\circ f)''_{ij}|U^c = 0 \qquad (i,j=1,\dots,d).$$

Let $x = (x_1,\dots,x_d)\in U^c$, $i\in\{1,\dots,d\}$ and $\alpha > 0$. Under the assumptions on f, the function

$$h:[x_i-\alpha,x_i)\cup(x_i,x_i+\alpha]\to\mathbb{R},\ h(z):=\frac{1}{z-x_i}\, f(x_i,\dots,x_{i-1},z,x_{i+1},\dots,x_d)$$

is bounded. Hence $f^3(x_1,\dots,x_{i-1},\cdot,x_{i+1},\dots,x_d)$ is differentiable in x_i and $(f^3)'_i(x) = 0$.

By (1.4) and the assumptions on f, it is now easy to deduce that f^3 is twice continuously differentiable with

$$(f^3)'_i|U^c = (f^3)''_{ij}|U^c = 0.$$

By the definition of $g_{n\varepsilon}$, this implies (1.5). Ito's formula, in connection with (1.4), (1,5), yields

$$(1.6)\ g_{n\varepsilon}\circ f(\xi(t))-g_{n\varepsilon}\circ f(\xi(0)) = \sum_{i=1}^{d}\int_0^t 1_U(\xi(s))g'_{n\varepsilon}(f(\xi(s)))f'_i(\xi(s))d\xi_i(s)$$

$$+\frac{1}{2}\sum_{i,j=1}^{d}\int_0^t 1_U(\xi(s))[g''_{n\varepsilon}\circ f\cdot f'_i f'_j + g'_{n\varepsilon}\circ f\, f''_{ij}](\xi(s))d\langle M_i,M_j\rangle(s).$$

Applying the boundedness conditions for f to the compact sets $\xi([0,t])$ and (1.2), (1.3), it is not difficult to show that the integrands on the right side of (1.6) are bounded, independently of $s \in [0,t]$ and $n \in \mathbb{N}$. Now A_i and $<M_i,M_j>$ are differences of two monotone and continuous processes and therefore define bounded signed Borel measures on $[0,t]$ $(i,j=1,\dots,d)$. Thus the limit on the right side of (1.6) as $n \to \infty$ can be calculated by Lebesgue's dominated convergence theorem, and we obviously get

$$g_\varepsilon \circ f(\xi(t)) - g_\varepsilon \circ f(\xi(0)) =$$

$$\sum_{i=1}^{d} \int_0^t \{\tfrac{1}{\varepsilon} 1_{(0,\varepsilon)}(|f(\xi(s))|)|f(\xi(s))| + 1_{[\varepsilon,\infty)}(|f(\xi(s))|)\} f_i'(\xi(s))\, d\xi_i(s)$$

$$+\frac{1}{2} \sum_{i,j=1}^{d} \int_0^t \{\tfrac{1}{\varepsilon} 1_{(0,\varepsilon)}(|f(\xi(s))|)[\operatorname{sgn} f\, f_i' f_j' + |f| f_{ij}''](\xi(s)) +$$

$$1_{[\varepsilon,\infty)}(|f(\xi(s))|)\, f_{ij}''(\xi(s))\} d<M_i,M_j>(s)$$

where $g_\varepsilon : \mathbb{R} \to \mathbb{R}$ is defined by

$$g_\varepsilon(y) := \frac{1}{2\varepsilon} y|y|\, 1_{(0,\varepsilon)}(|y|) + (y - \tfrac{\varepsilon}{2} \operatorname{sgn} y)\, 1_{[\varepsilon,\infty)}(|y|).$$

Letting $\varepsilon \downarrow 0$, we arrive at (1.1) by Lebesgue's theorem and the boundedness of f_i'.

<u>Remarks.</u> 1) If instead of $ff_{ij}''|UK$ even $f_{ij}''|UK$ is bounded for all compact $K \subset \mathbb{R}^d$, the P-limit in (1.1) evidently equals to

$$(1.7) \qquad \sum_{i,j=1}^{d} \int_0^t 1_U(\xi(s))\, f_{ij}''(\xi(s))\, d<M_i,M_j>(s) +$$

$$\text{P-}\lim_{\varepsilon \to 0} \frac{1}{\varepsilon} \sum_{i,j=1}^{d} \int_0^t 1_{(0,\varepsilon)}(|f(\xi(s))|)[\operatorname{sgn} f\, f_i' f_j'](\xi(s))\, d<M_i,M_j>(s).$$

2) If $f \in C^2(\mathbb{R}^d)$ a comparison of (1.1), (1.7) and Ito's formula leads to

$$\frac{1}{2}\ P\text{-}\lim_{\varepsilon\to 0} \frac{1}{\varepsilon} \sum_{i,j=1}^{d} \int_0^t 1_{(0,\varepsilon)}(|f(\xi(s))|)[\operatorname{sgn} f\, f_i' f_j'](\xi(s))\ d\langle M_i, M_j\rangle(s)$$

$$= \sum_{i=1}^{d} \int_0^t 1_{\{0\}}(f(\xi(s)))f_i'(\xi(s))\ d\xi_i(s) +$$

$$\frac{1}{2} \sum_{i,j=1}^{d} \int_0^t 1_{\{0\}}(f(\xi(s)))f_{ij}''(\xi(s))\ d\langle M_i, M_j\rangle(s).$$

Example. Let $p \in [2,\infty)$ and

$$f(x) := \|x\|_p := \Big(\sum_{k=1}^{d} |x_k|^p\Big)^{1/p},\quad x \in \mathbb{R}^d .$$

Then (1.1) yields

$$\|\xi(t)\|_p - \|\xi(0)\|_p = \sum_{i=1}^{d} \int_0^t 1_{\mathbb{R}^d-\{0\}}(\xi(s))\ \|\xi(s)\|_p^{1-p}|\xi_i(s)|^{p-2}\xi_i(s)d\xi_i(s)$$

$$+ \frac{1}{2}\ P\text{-}\lim_{\varepsilon\to 0} \sum_{i,j=1}^{d} \int_0^t \{\tfrac{1}{\varepsilon} 1_{(0,\varepsilon)}(\|\xi(s)\|_p)[(p-1)\|\xi(s)\|_p^{2-p}|\xi_i(s)|^{p-2}\ \delta_{ij}$$

$$+ (2-p)\ \|\xi(s)\|_p^{2-2p}|\xi_i(s)|^{p-2}\xi_i(s)|\xi_j(s)|^{p-2}\xi_j(s)]$$

$$+(p-1)1_{[\varepsilon,\infty)}(\|\xi(s)\|_p)\ \|\xi(s)\|_p^{1-p}|\xi_i(s)|^{p-2}(\delta_{ij}-$$

$$- \|\xi(s)\|_p^{-p}|\xi_j(s)|^{p-2}\xi_i(s)\xi_j(s))\}d\langle M_i, M_j\rangle(s)$$

where δ_{ij} is Kronecker's symbol.

2. Let C_o be the set of continuous functions $x=(x_t)_{t\in[0,1]}$ with $x_o=0$. We consider C_o as a Polish space with the usual uniform metric.

Let a be a measurable function on $[0,1]\times\mathbb{R}$. Denote by S the subset of C_o which consists of all $x\in C_o$ such that the functions

$$t \longrightarrow a(t,x_t)$$

are Lebesgue-integrable on [0,1] and

$$\forall t \quad x_t = \int_0^t a(s,x_s)ds$$

holds.

<u>Lemma 1.</u> S is closed under the finite minimum and maximum operations.

(The idea to prove this lemma in order to apply it to the problem of the existence of strong solutions was also pointed out by N.V. Krylov. M. Yor informed us that he too proved this simple lemma in a slightly different way.)

<u>Proof.</u> Obviously it is sufficient to prove that S is closed with respect to one of the two operations, $\wedge$, say.

Since, for any four real numbers $\alpha,\beta,\gamma,\delta$, we have

$$|\alpha\wedge\beta - \gamma\wedge\delta| \leq |\alpha-\gamma| \vee |\beta-\delta|,$$

the minimum of two absolutely continuous functions is also an absolutely continuous function. Therefore $x^1\wedge x^2$ is a.e. differentiable, just as x^i themselves, and x^i, $x^1\wedge x^2$ are the integrals of their derivatives.

On the open set

$$\{t : x_t^1 < x_t^2\}$$

we have obviously (a.e.)

$$\frac{d(x_t^1 \wedge x_t^2)}{dt} = \frac{dx_t^1}{dt} = a(t,x_t^1)$$

$$= a(t,x_t^1 \wedge x_t^2).$$

The same is of course true on

$$\{t : x_t^1 > x_t^2\}.$$

The remaining closed set

$$\{t : x_t^1 = x_t^2\}$$

may consist of countably many isolated points (which may be neglected) and the dense-in-itself part D. On D we also have (a.e.)

$$\frac{d(x_t^1 \wedge x_t^2)}{dt} = \frac{dx_t^i}{dt} = a(t,x_t^i)$$

$$= a(t,x_t^1 \wedge x_t^2). \quad \square$$

<u>Lemma 2.</u> Assume that $S \neq \emptyset$ and a satisfies the following additional conditions:

(i) for almost all t, the functions

$$\mathbb{R} \ni y \longrightarrow a(t,y)$$

are continuous from the right (left);

(ii) the set of functions

$$\{t \longrightarrow a(t,x_t) \; : \; x \in S\}$$

is uniformly integrable;

(iii) $$\sup_{x \in S} \left| \int_s^t a(u,x_u)du \right| \longrightarrow 0 \quad \text{as } |s-t| \longrightarrow 0.$$

Then S is closed with respect to the infimum (supremum) operations.

<u>Proof.</u> Let $S' \subset S$ and

$$\hat{x} = \inf S',$$

i.e.

$$\forall t \qquad \hat{x}_t = \inf \{x_t \; : \; x \in S'\} \qquad (\hat{x} = (\hat{x}_t)_{t \in [0,1]}).$$

It follows from (ii) (or also from (iii)) that

$$\forall t \quad |\hat{x}_t| < \infty .$$

By using Lemma 1, we can construct a monotone sequence $(x^n) \subset S$ converging to $\hat{x}$ at each rational point.

From (iii), it follows that (x^n) converges to $\hat{x}$ everywhere. Furthermore: $\hat{x}$ ist continuous and, by (i) and (ii), belongs to S.

The assertion for the supremum operation can be proved in the same way. □

Remark 1. From the proof of Lemma 2, it is easy to see, that, under conditions (i) - (iii), there always exists a countable $S' \subset S$ such that

$$\inf S' = \inf S, \ \sup S' = \sup S.$$

We introduce the mapping F from C_o into itself by

$$(z_t)_{t \in [0,1]} = z = F(x) \ : \forall t \ z_t = x_t - \int_o^t a(s,x_s)ds.$$

Lemma 3. Let a satisfy the conditions of Lemma 2, the latter two with C_o instead of S.

Then the mapping f:

$$z \longrightarrow f(z) = \inf F^{-1}(z) \qquad (\sup F^{-1}(z))$$

is an analytic-measurable selector for F, i.e.

$$F \circ f = id_{C_o}$$

and

$$f^{-1}(\mathcal{C}) \subset \mathcal{A}$$

where $\mathcal{C}$ is the Borel σ-algebra in C_o and $\mathcal{A}$ the σ-algebra in C_o generated by all analytic sets in C_o.

Proof. First of all, F is surjective. This can be proved in the same way as in [2].

For each fixed $z \in C_o$, the function a_z on $[0,1] \times \mathbb{R}$:

$$a_z(t,y) = a(t, z_t + y)$$

satisfies the conditions of Lemma 2 with

$$S = S_z = \{x : \forall t \; x_t = \int_0^t a_z(s, x_s) ds\}.$$

Therefore f is a selector for F, i.e.

$$f : C_o \longrightarrow C_o \; , \; F \circ f = id_{C_o} .$$

For each fixed $x \in C_o$, in the case f = inf, say, we have

$$\{z \in C_o : f(z) \leq x\} = \{z \in C_o : \inf F^{-1}(z) \leq x\}$$

$$(\inf F^{-1}(z) \in F^{-1}(z) \text{ by Lemma 2})$$

$$= \{z \in C_o : \exists x' \in F^{-1}(z) \text{ with } x' \leq x\}$$

$$= \{F(x') : x' \leq x\} = F(\{x' \in C_o : x' \leq x\}).$$

Since all sets of the form

$$\{x' \in C_o : x' \leq x\}$$

generate $\mathcal{C}$ and the F-images of Borel sets are analytic in C_o, it completes the proof. □

Remark 2. This lemma can be considered as a deterministic sharpening of the corresponding result of [1] (Theorem 2.3).

Remark 3. Since F is causal, i.e.

$$\forall s \leq t \quad x_s^1 = x_s^2 \Rightarrow F(x^1)|_t = F(x^2)|_t,$$

and onto, the "pieces" of functions from $F^{-1}(z)$ up to time t depend only on z_s, $s \leq t$. In other words, f is causal too:

$$\forall s \leq t \quad z_s^1 = z_s^2 \Rightarrow f(z^1)|_t = f(z^2)|_t.$$

Corollary 1. Assume that a process $\xi = (\xi_t)_{t \in [0,1]}$ satisfies the equation

$$\xi_t = \int_0^t a(s,\xi_s)ds + W_s$$

with a as at the beginning of Section 2 and a Wiener process $W=(W_t)_{t\in[0,1]}$ with respect to the natural filtration of ξ.

If this equation (with ξ and W on an arbitrary probability space) uniquely determines the function

$$t \longrightarrow E\, f(\xi_t),$$

where f is some strictly increasing function with $E|f(\xi_t)| < \infty$, then ξ is a strong (causal) solution, i.e. ξ_t is measurable with respect to $\sigma(W_s;\ s \leq t)$.

Proof is based on Lemma 1 applied to a_W and the Yamada-Watanabe theorem [4] , just as in [2] .

Corollary 2. Let a satisfy conditions of Lemma 3.

Then the equation

$$\xi_t = \int_0^t a(s,\xi_s)ds + W_t$$

has a strong (causal) solution.

Proof. Put $\xi = f(W)$, where f is the selector from Lemma 3. Then ξ depends on W causally, and hence, for each t, ξ_t is $\sigma(W_s;\ s \leq t)$-measurable a.s., and satisfies the equation. □

Remark 4. Note that the last assertion holds true for any process W with sample paths in C_o and not only for the Wiener process.

References

[1] R.J. Chitashvili; T.A. Toronjadze, On one-dimensional SDEs with unit diffusion coefficient. Structure of solutions, Stochastics, 4 (1981), 281-315.

[2] M.P. Ershov, On non-anticipating solutions of stochastic equations, Proc. Third USSR-Japan Sympos on Probability Theory, August 1975, Lecture Notes in Math., Vol 550.

[3] A.T. Wang, Generalized Ito's formula and additive functionals of Brownian motion, Z.Wahrscheinlichkeitstheorie verw.Gebiete, 41 (1977), 153-159.

[4] T. Yamada; Sh. Watanabe, On the uniqueness of solutions of SDEs, J. Math. Kyoto Univ., 11 (1971), 155-167.

SOME PROBLEMS IN SEQUENTIAL ANALYSIS

Avner Friedman
Northwestern University
Evanston, IL 60201/USA

§1. Simple hypotheses for a diffusion process

A process $\xi(t)$ has the structure $w(t)+\lambda t$ where $w(t)$ is a one-dimensional Brownian motion and λ is a real parameter which is not known; it is considered as a random variable independent of the process $w(t)$. We observe the process $\xi(t)$ for some time τ, at cost c per unit time, and then make a decision (an "educated guess") about the true value of λ; there is a cost incurred for an incorrect decision.

To be specific, suppose λ is known to take one of $n+1$ real values $\lambda_0,\lambda_1,\dots,\lambda_n$ with

$$P(\lambda=\lambda_i)=\pi_i \qquad (\pi_i\geq 0,\ \sum_{i=0}^{n}\pi_i=1).$$

Denote by $\mathcal{F}_t$ the σ-field generated by $\xi(s)$, $s\leq t$. Set

$$\begin{aligned} W(\lambda,d) &= 0 \quad \text{if } d=i,\ \lambda=\lambda_i, \\ &= a_i \quad \text{if } d=i,\ \lambda\neq\lambda_i, \qquad (a_i>0). \end{aligned}$$

W is called the risk function and a_i is the cost for making decision $d=i$ when $\lambda\neq\lambda_i$. The total cost is then

$$J_\pi(\tau,d)=E^\pi\Big[\int_0^\tau f(\lambda(\omega))dt+W(\lambda(\omega),d(\omega))\Big] \tag{1.1}$$

where $f(\lambda)=c$. Let

$\mathcal{A}$ = all stopping times τ with respect to the σ-fields $\mathcal{F}_t$, $t\geq 0$;

$\mathcal{D}_\tau$ = all $\mathcal{F}_\tau$ measurable functions d with values in $\{0,1,\dots,n\}$.

Problem: Find $\tilde{\tau}\in\mathcal{A}$, $\tilde{d}\in\mathcal{D}_{\tilde{\tau}}$ such that

$$J_\pi(\tilde{\tau},\tilde{d})=\inf_{\tau\in\mathcal{A},\,d\in\mathcal{D}_\tau} J_\pi(\tau,d). \tag{1.2}$$

This is a problem in sequential testing with $n+1$ simple hypotheses.

Notice that $\xi(t)$ is not a Markov process and therefore (1.2) is not a standard stopping time problem; the process

$$(\xi(t),\lambda) \qquad (\xi(t)=w(t)+\sum_{i=0}^{n}\lambda_i\, t\, I_{\{\lambda=\lambda_i\}})$$

is a Markov process. In order to reduce problem (1.2) into a Markovian problem we need to introduce the filter

Partially supported by National Science Foundation Grant MCS 791 5171.

$$\pi_j(t) = P^\pi[\lambda = \lambda_j | \mathcal{F}_t].$$

The process $\pi(t) = (\pi_0(t),\ldots,\pi_n(t))$ lies in

$$\Pi_{n+1} = \{\pi = (\pi_0,\pi_1,\ldots,\pi_n),\ \ \pi_i \geq 0,\ \ \sum_{i=0}^{n} \pi_i = 1\}$$

and it is a Markov process given by ([2])

$$\pi_j(t) = \pi_j\{\sum_{k=0}^{n} \pi_k z_{j,k}(t)\}^{-1},$$

(1.3)

$$z_{j,k}(t) = \exp\{(\lambda_k - \lambda_j)\cdot\xi(t) - \tfrac{1}{2}(|\lambda_k|^2 - |\lambda_j|^2)t\}$$

with generator

$$\text{(1.4)}\qquad Mu(\pi) = \frac{1}{2}\sum_{i,j=0}^{n} \pi_i\pi_j\Big(\lambda_i - \sum_{k=0}^{n}\lambda_k\pi_k\Big)\Big(\lambda_j - \sum_{\ell=0}^{n}\lambda_\ell\pi_\ell\Big)\frac{\partial^2 u(\pi)}{\partial\pi_i\partial\pi_j}\ .$$

We can rewrite (1.1) in the form

$$J_\pi(\tau,d) = E^\pi\Big[\int_0^\tau h(\pi(t))dt + \sum_{i=0}^{n}(1 - \pi_i(\tau))a_i\, I_{\{d(\omega)=i\}}\Big]$$

where $h(\pi) = c$. Given τ it is clear that the optimal decision $d(\omega)$ is that which minimizes $(1 - \pi_i(\tau))a_i$. Consequently,

$$\text{(1.5)}\qquad J_\pi(\tau) \equiv \inf_{d\in\mathcal{D}_\tau} J_\pi(\tau,d) = E^\pi\Big[\int_0^\tau h(\pi(t))dt + g(\pi(\tau))\Big]$$

where

$$\text{(1.6)}\qquad g(\pi) = \min_{0\leq i\leq n}\{a_i(1 - \pi_i)\}.$$

Problem (1.2) then reduces to:

$$\text{(1.7)}\qquad \text{Find } \tilde{\tau}\in\mathcal{A} \text{ such that } J_\pi(\tilde{\tau}) = \inf_{\tau\in\mathcal{A}} J_\pi(\tau) \equiv u(\pi).$$

This is a standard stopping time problem (see [5],[11]). Since, however, the generator M is a degenerate elliptic operator, we cannot use elliptic estimates as in the general theory of stopping time problems. Thus we shall have to use more probabilistic methods in studying problem (1.7).

Denote by G_i the points of Π_{n+1} nearest to the vertex e_i; here e_j = $(0,0,\ldots,0,1,0,\ldots,0)$ with 1 in the $(j+1)^{th}$ place. Denote by N_ε the ε-neighborhood of the set of vertices $\{e_0,e_1,\ldots,e_n\}$. Let

$C = \{\pi;\ u(\pi) < g(\pi)\}$ the continuation set,

$S = \{\pi;\ u(\pi) = g(\pi)\}$ the stopping set,

$\Gamma = \mathring{\Pi} \cap \partial C$ the free boundary,

where $\mathring{\Pi} = \text{int}\ \Pi_{n+1}$.

<u>Theorem 1</u>. The set $S_i = S \cap G_i$ is convex and the function $u(\pi)$ is a concave function.

<u>Theorem 2</u>. There exist positive constants δ_1, δ_2 such that

$$(1.8) \qquad S \subset N_{\delta_2 c}, \qquad C \supset \Pi_{n+1} \setminus N_{\delta_1 c}.$$

Thus the free boundary Γ lies in $N_{\delta_2 c} \setminus N_{\delta_1 c}$.

Theorem 1 is proved directly from (1.1). Theorem 2 is proved by comparison. Notice that formally $\hat{u} = u - g$ satisfies

$$M\hat{u} + k \geq 0, \quad \hat{u} \leq 0, \quad (M\hat{u} + k)\hat{u} = 0 \quad \text{where } k = h + Mg.$$

If we can find a smooth function $\tilde{u}$ such that

$$M\tilde{u} + \tilde{k} \geq 0, \quad \tilde{u} \leq 0, \quad (M\tilde{u} + \tilde{k})\tilde{u} = 0 \quad \text{where } \tilde{k} \leq k,$$

then $\tilde{u} \leq \hat{u}$ (this follows by representing $\hat{u}$ and $\tilde{u}$ stochastically, using Ito's formula), which gives a lower bound on the set $\{u = g\}$. Choosing an appropriate $\tilde{u}$ we find that $S \supset N_{\delta_1 c}$. The second assertion in (1.8) is proved similarly; for details see [8].

The behavior of the solution as $c \to 0$ is of particular interest in sequential analysis. Here we can state:

<u>Theorem 3</u>. As $c \to 0$

$$(1.9) \qquad u(\pi) = \Big(\sum_{i=0}^{n} \gamma_i \pi_i\Big) c \log \frac{1}{c} + 0\Big(c\big(\log \frac{1}{c}\big)^{\frac{1}{2}}\Big)$$

where

$$\frac{1}{2}\gamma_i = \{\min_{k \neq i} |\lambda_k - \lambda_i|\}^{-1}.$$

The proof follows from the estimate

$$(1.10) \qquad E^{\pi}\tau_{\varepsilon} = \Big(\sum_{i=0}^{n} \gamma_i \pi_i\Big) \log \frac{1}{\varepsilon} + 0\Big(\big(\log \frac{1}{\varepsilon}\big)^{\frac{1}{2}}\Big)$$

where τ_ε = exit time from N_ε. In proving this estimate one uses the representation (1.3).

One can also show that as $c \to 0$

$$\frac{u(cy)}{c} \to u_0(y) \qquad \Big(y_i = \frac{\pi_i}{\pi_0}, \quad 1 \leq i \leq n\Big)$$

where u_0 is a solution of a stopping time problem.

The above results extend to the case where $w(t)$ and λ_i are m-dimensional. Theorem 2 extends also to the case where $\lambda(\omega)$ is replaced by a Markov process $\lambda(t,\omega)$ with states $\lambda_0, \lambda_1, \ldots, \lambda_n$. If $m \geq n$ and M is non-degenerate in $\mathring{\Pi}$, then one can show that the free boundary is analytic and that the ridge $\big(\bigcup_{i \neq j} \partial G_i \cap \partial G_j\big)$ is in C. All these results are proved in [8]. We finally mention a procedure for implementing the optimal strategy. For simplicity we take $n = 2$ and

$$\lambda_0 = 0, \quad \lambda_1 = -1, \quad \lambda_2 = 1,$$

and set $p_i = \pi_i$, $\gamma = p_0^2/p_1p_2$. Set

$$\eta(s) = \frac{1}{2}\log\frac{p_1\pi_1(t)}{p_2\pi_2(t)}, \qquad s = \log\frac{\pi_0^2(t)}{\gamma\pi_1(t)\pi_2(t)} .$$

The mapping σ from $(\eta(s),s)$ into $\pi(t)$ is 1-1 from the half-plane $\{s > 0\}$ onto the subset in $\mathring{\Pi}$ with $\pi_0^2 > \gamma\pi_1\pi_2$. The stopping sets S_i are mapped onto Σ_i (see the accompanying figure), and

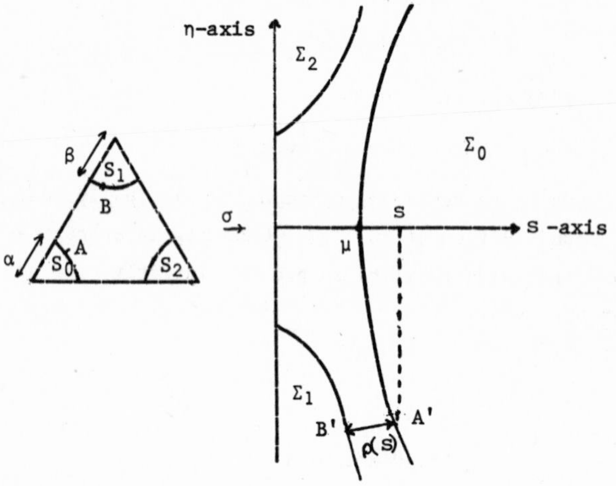

$$\rho(s) \to \frac{2}{\sqrt{5}}\log\frac{\alpha(1-\beta)}{\beta(1-\alpha)} \qquad \text{as } s \to \infty,$$

$$\mu \sim \log\frac{1}{\delta^2c^2} \qquad \text{as } c \to 0 \qquad (\delta \text{ constant} \neq 0);$$

α and β can be computed by solving a problem with two simple hypotheses.

§2. Simple hypotheses for a Poisson process

Here $\xi(t)$ is a Poisson process with parameter λ which is a random variable. Otherwise the setting is the same as before. The filter $\pi(t)$ is given by ([1],[12])

$$(2.1) \qquad \pi_j(t) = \pi_j\Big\{\sum_{k=0}^{n}\pi_k z_{j,k}(t)\Big\}^{-1}$$

where

$$(2.2) \qquad z_{j,k}(t) = \exp\Big\{\log\frac{\lambda_k}{\lambda_j}\cdot\xi(t) - (\lambda_k - \lambda_j)t\Big\}$$

and the generator is

$$(2.3) \qquad Au(\pi) = \sum_{i,j=0}^{n}\pi_i\pi_j(\lambda_i - \lambda_j)\frac{\partial u}{\partial\pi_j} + \Big(\sum_{i=0}^{n}\lambda_i\pi_i\Big)(u(\Lambda\pi) - u(\pi)),$$

where

$$\Lambda\pi = \left[\frac{\lambda_k\pi_k}{\sum_{\ell=0}^{n}\lambda_\ell\pi_\ell}\right]_{k=0,1,\dots,n} .$$

We again obtain (1.5)-(1.7) and Theorems 1, 2 are valid. In proving Theorem 2 we can actually use the same comparison functions.

Theorem 4. As $c \to 0$

$$u(\pi) = \Big(\sum_{i=0}^{n} \gamma_i\pi_i\Big)c \log\frac{1}{c} + O(c) \tag{2.4}$$

where

$$\gamma_i = \{\lambda_i \min_{k\neq i} \psi\Big(\frac{\lambda_k}{\lambda_i}\Big)\}^{-1}, \qquad \psi(x) = x - 1 - \log x. \tag{2.5}$$

Since $\psi(x) > 0$ if $0 < x < \infty$, $x \neq 1$, $\gamma_i \in (0,\infty)$.

The proof of Theorem 4 is based on the method of proof of Theorem 3 coupled with the estimate

$$E^\pi\tau_\varepsilon \leq A \log\frac{1}{\varepsilon} .$$

To establish this estimate we check that the function

$$v(\pi) = \sum_{i=0}^{n} \log\frac{\pi_i}{\varepsilon} \quad \text{in } \Pi_{n+1}\setminus N_\varepsilon$$

satisfies $v \geq 0$, $Mv \geq -C$, and that

$$v = 1 \quad \text{on } \partial N_\varepsilon \cap \mathring{\Pi},$$

and then apply Ito's formula.

§3. Two-armed bandit problem

An arm L of a gambling machine (a "bandit") has expectation λ; λ is a random variable taking values $\lambda_0,\lambda_1,\dots,\lambda_n$ with probabilities π_i for $\lambda = \lambda_i$. The observed process is, say, $\xi(t) = w(t) + \lambda t$. The total duration of the game is T, that is, we may stop at any time τ subject to $\tau \leq T$. The net profit is then

$$\int_0^\tau f(\lambda(\omega))dt \qquad (f(\lambda_i) = c_i)$$

where $c_i = \lambda_i - \beta$, β = cost per unit time of playing the machine. In terms of the filter, the cost is

$$J_\pi(\tau) = E^\pi\Big[\int_0^\tau h(\pi(t))dt\Big]$$

where $h(\pi) = \sum_{i=0}^{n} c_i\pi_i$. This leads, formally, to the parabolic variational inequality

$$\frac{\partial u}{\partial t} + Mu + h \leq 0, \quad u \geq 0, \quad \Big(\frac{\partial u}{\partial t} + Mu + h\Big)u = 0 \quad \text{a.e. in } \Pi\times(0,T), \tag{3.1}$$

and to the terminal condition $u(\pi,T) = 0$. Since M is degenerate, u may not be differentiable and thus (3.1) need not make sense.

The case of a two-armed bandit has been considered in the statistics literature (see [6],[7],[9] and the references given there). Here there is another independent arm R with expectation ρ which is random. One may switch from one arm to another any number of times. The total duration T is prescribed. Even when R is known (i.e., when ρ is a given number) this problem has not been reduced, up to now, to a variational or quasi-variational inequality. In order to accomplish such a reduction one must first find the structure of processes of the form

$$\tilde{\pi}_j(t) = P^{\pi}[\lambda = \lambda_j \mid \mathcal{F}_{\tau_0}, \mathcal{F}_t^{\tau_0+\sigma_0}]$$

where τ_0 is a stopping time with respect to the σ-fields $\mathcal{F}_t$, σ_0 is $\mathcal{F}_{\tau_0}$ measurable and

$$\mathcal{F}_t^{\tau_0+\sigma_0} = \sigma(\xi(\tau_0 + \sigma_0 + s),\ 0 \leq s \leq t).$$

Consider a simplified version of the two-armed bandit problem whereby ρ is a random variable taking values $\rho_0, \rho_1, \ldots, \rho_n$ and both arms are <u>continuously observed</u>; however, a penalty k is imposed for any switch in the arms. Then (cf. [3],[4]) the problem reduces to a system of quasi-variational inequalities

$$\begin{aligned}
&\frac{\partial u}{\partial t} + Mu + h \leq 0, \quad u \geq v - k,\\
&\left(\frac{\partial u}{\partial t} + Mu + h\right)(u - v + k) = 0,\\
(3.2)\qquad &\frac{\partial v}{\partial t} + M_1 v + h_1 \leq 0, \quad v \geq u - k,\\
&\left(\frac{\partial v}{\partial t} + M_1 v + h_1\right)(v - u + k) = 0
\end{aligned}$$

with the terminal conditions $u(\pi,T) = v(\pi,T) = 0$; here h_1, M_1 are defined as h, M with respect to ρ. As $k \to 0$ the functions u, v both converge to the same function w (see [10]) satisfying:

$$(3.3)\qquad \frac{\partial w}{\partial t} + \max\{Mw + h, M_1 w + h_1\} = 0 \ \text{ in } \Pi_{n+1} \times (0,T), \qquad w(\pi,T) = 0.$$

If instead of finite horizon T we have an infinite horizon with a discount factor α, (3.2) and (3.3) become elliptic with M, M_1 replaced by $M - \alpha$, $M_1 - \alpha$ respectively.

It would be interesting to study the systems (3.1), (3.2), (3.3) and their elliptic counterparts, and to establish results analogous to Theorems 1 - 3.

§4. Composite hypotheses

Consider a one-dimensional process z(t):

$$dz(t) = y\,dt + \sigma\,dw(t) \qquad (z(0) = 0)$$

where y is random. Unlike the situation in §§1-3 we now let y take continuum of values, say

$$y = \mu_0 + \xi \qquad (\mu_0 \text{ a real number})$$

where ξ is a normal variable $N(0,\sigma_0^2)$. The hypotheses are

H_1: accept that $y > 0$,

H_2: accept that $y < 0$.

We take the risk function

$$\begin{aligned} W(y,d) &= k|y| \quad \text{if } d = 1,\ y < 0 \quad \text{or } d = 2,\ y > 0, \\ &= 0 \qquad \text{in all other cases.} \end{aligned}$$

Here $d = i$ means accepting the hypothesis H_i.

The total cost is

$$J = E[c\tau + W(y(\omega),d(\omega))] \qquad (c > 0)$$

where τ is any stopping time with respect to $\mathcal{F}_t = \sigma(z(s),\ s \le t)$. Again we introduce the filter

$$\hat{y}(t) = E[y \mid \mathcal{F}_t]$$

which, by the Kalman-Bucy theory, satisfies

$$d\hat{y}(t) = \frac{p(t)}{\sigma}\, d\hat{w}(t), \qquad \hat{y}(0) = \mu_0,$$

and

$$p(t) = \frac{1}{\frac{t}{\sigma^2} + \frac{1}{\sigma_0^2}} .$$

Setting $s = p(t)$, $\tilde{y}(s) = \hat{y}(t)$, $\inf J$ becomes $u(\mu_0,\sigma_0^2)$ where $u(x,s)$ satisfies ([13])

$$u_s - \frac{1}{2} u_{xx} \le \frac{c\sigma^2}{s^2},$$

(4.1)
$$u \le k\Psi$$

$$(u_s - \frac{1}{2} u_{xx} - \frac{c\sigma^2}{s^2})(u - k\Psi) = 0$$

and

$$u(x,0) = 0,$$

where

$$\Psi(x,s) = \sqrt{s}\, \psi\left(\frac{x}{\sqrt{s}}\right),$$

$$\psi(u) = \phi(u) + u\Phi(u), \qquad u > 0,$$

$$\psi(u) = \psi(-u), \qquad u < 0,$$

$$\phi(v) = \frac{1}{\sqrt{2\pi}}\, e^{-v^2/2},$$

$$\Phi(u) = \int_{-\infty}^{u} \phi(v)dv.$$

<u>Theorem 5</u>. The free boundary in $\{x > 0\}$ is given by $x = \zeta(s)$ where, as $s \to 0$,

$$\zeta(s) = \frac{s^2}{4\alpha} - \frac{s^5}{48\alpha^2} + O(s^8) \qquad \left(\alpha = \frac{c\sigma^2}{k}\right), \tag{4.2}$$

$$\zeta'(s) = \frac{s}{2\alpha} + O(s^4). \tag{4.3}$$

Chernoff [7] obtained, in fact, an asymptotic series

$$\zeta(s) = \frac{s^2}{4\alpha}\left(1 + \sum_{j=1}^{\infty} \gamma_j \, s^{3j}\right).$$

The method for obtaining both (4.2) and (4.3) is given in [14] and it applies to more general problems; see for instance (4.4), (4.5) below.

Consider the case of three hypotheses:

H_1: accept that $y > a$,

H_2: accept that $-a < y < a$,

H_3: accept that $y < -a$.

This leads to the variational inequality (4.1) with

$$\Psi = \min(\Psi_1, \Psi_2, \Psi_3)$$

where Ψ_i can be explicitly computed (see [13],[14]). Now two free boundaries $x = \zeta_i(s)$, $(i = 1,2)$, initiate at $x = a$ and symmetrically two free boundaries initiate at $x = -a$. The asymptotic behavior of the curves $x = \zeta_i(s)$ near $x = a$ is

$$\begin{aligned} \zeta_1(s) &= a + \frac{s^2}{4\alpha} - \frac{s^5}{48\alpha^2} + O(s^8), \\ \zeta_2(s) &= a - \frac{s^2}{4\alpha} + \frac{s^5}{48\alpha^2} + O(s^8), \end{aligned} \tag{4.4}$$

and

$$\zeta_1'(s) = \frac{s}{2\alpha} + O(s^4), \qquad \zeta_2'(s) = -\frac{s}{2\alpha} + O(s^4). \tag{4.5}$$

Although $\zeta_1(s)$, $\zeta_2(s)$ tend to be symmetric with respect to the line $\{x = a\}$ as $s \to 0$, they are definitely not symmetric with respect to $\{x = a\}$ as the figure below indicates.

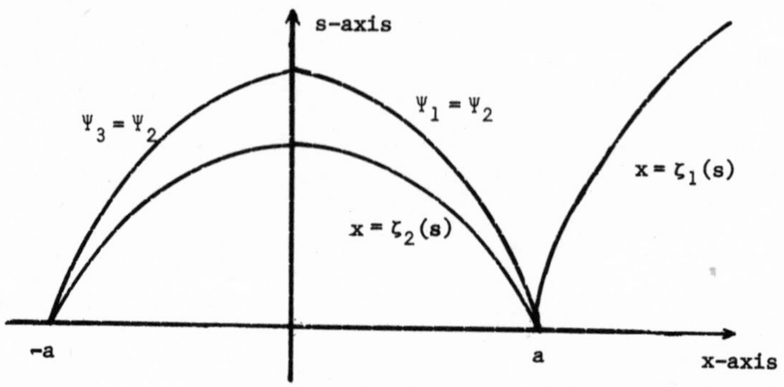

References

[1] R. F. Anderson and A. Friedman, A quality control problem and quasi-variational inequalities, Archive Rat. Mech. Anal., 63 (1977), 205-252.

[2] R. F. Anderson and A. Friedman, Multi-dimensional quality control problems and quasi-variational inequalities, Trans. Amer. Math. Soc., 246 (1978), 31-76.

[3] A. Bensoussan and A. Friedman, On the support of the solution of a system of quasi-variational inequalities, J. Math. Anal. Appl., 65 (1978), 660-674.

[4] A. Bensoussan and J. L. Lions, Contrôle impulsionnel et systèmes d'inéquations quasi variationnelles, C. R. Acad. Sci. Paris, 278 (1974), 747-751.

[5] A. Bensoussan and J. L. Lions, Applications des inégalités variationnelles en contrôle stochastique, Dunod, Paris, 1978.

[6] D. A. Berry and B. Fristedt, Two-armed bandits with a goal, I. One arm known, Adv. Appl. Prob., 12 (1980), 775-798.

[7] D. A. Berry and B. Fristedt, Two-armed bandits with a goal, II. Dependent arms, Adv. Appl. Prob., 12 (1980), 958-971.

[8] L. A. Caffarelli and A. Friedman, Sequential testing of several simple hypotheses for a diffusion process and the corresponding free boundary problem, Pacific J. Math., 93 (1981), 49-94.

[9] H. Chernoff, Sequential Analysis and Optimal Design, Regional Conference Series in Applied Mathematics, #8, SIAM, Philadelphia, 1972.

[10] L. C. Evans and A. Friedman, Optimal stochastic switching and the Dirichlet problem for the Bellman equation, Trans. Amer. Math. Soc., 253 (1979), 365-389.

[11] A. Friedman, Stochastic Differential Equations and Applications, vol. 2, Academic Press, New York, 1976.

[12] A. Friedman, Optimal stopping for random evolution of multi-dimensional Poisson processes with partial observation, Stochastic Analysis, Academic Press, 1978, pp. 109-126.

[13] A. Friedman, Variational inequalities in sequential analysis, SIAM J. Math. Anal., 12 (1981), 385-397.

[14] A. Friedman, Asymptotic behavior for the free boundary of parabolic variational inequalities and applications to sequential analysis, Ill. J. Math., to appear.

A STOCHASTIC DIFFERENTIAL EQUATION FOR FELLER'S ONE-DIMENSIONAL DIFFUSIONS

Jürgen Groh
Friedrich-Schiller-Universität Jena
DDR-6900 Jena, UHH

INTRODUCTION

In this note we are concerned with a special class of one-dimensional diffusion processes X generated by Feller's generalized second order differential operator $D_m D_s^+$ [5], [15]. Here m stands for the so-called speed measure, charging all non-empty open sets in $\mathbb{R}$, and s denotes the road map, or natural scale, for the process X. To omit difficulties with boundary conditions both boundaries of the state space $\mathbb{R}$ are assumed to be inaccessible. Assuming additionally an identical road map $s(x) = x$ and even natural boundaries [4], the diffusion X is a martingal, see Arbib [1]. Moreover, we suppose that the speed measure m has the special form

$$(1) \qquad dm(x) = \sigma(x)^{-2} 2dx + d\mu(x), \qquad x \in \mathbb{R}$$

with a measurable bounded uniformly positive function σ and a (with respect to the Lebesgue measure) singular measure μ. Then there exist two disjoint Borel sets Λ and Γ whose union is $\mathbb{R}$ and $\mu(\Lambda) = \int_\Gamma dx = 0$.

In a first part of this paper we give a representation of the diffusion X as solution to some stochastic differential equation. For the case of classical diffusion processes with absolutely continuous speed measure m (i.e. $\mu = 0$) such a representation is the heart of Itô's theory, but for a non vanishing singular component μ of the speed m it seems to be new. Essentially, we use the well-known fact, that a diffusion X with generator $D_m D_x^+$ can be constructed via a time change procedure from a suitable Brownian motion process B^0, see Itô/McKean [12] or Freedman [7]. There is the crucial point that the time change process is constructed by the help of local times (Trotter [19]) of the underlying Brownian motion process. Using this Brownian motion and the time change we construct a new Brownian motion process, defined over an enlarged probability space, which leads to the desired stochastic equation for X. Our method is adapted to the approach of Gihman/Skorohod [8], describing classical one-dimensional diffusion processes with sticky boundaries, and to that of S.Watanabe [21] and Y.Oshima [17], [18] concerning the several dimensional case.

The second part of this note is devoted to some applications of our differential equation. Firstly, it is possible now to give an explicit version of Itô's change of variable formula for the diffusion X. Secondly, following Orey [16], we consider absolutely continuous diffusion processes. Here it is possible to get an explicit expression for the corresponding Radon-Nikodym derivatives and a stochastic differential equation for all diffusions which are absolutely continuous with respect to the given process X. Whereas the methods in the first part are presented in detail, the assertions of the second part are given without proofs. For the full story we refer the reader to forthcoming papers, see [10], [11].

PRELIMINARIES

Let $X = (x_t, \underline{\underline{F}}^X_t, \mathbb{P}^o_x)$ be a diffusion process generated by the differential operator $D_m D_x^+$. We regard this diffusion as constructed via a time change T from some Brownian motion process B^o. This is not a serious restriction because under the present conditions on the speed measure m the time change process is invertible. Thus let us agree that $B^o = (B^o_t, \underline{\underline{F}}^o_t, \mathbb{P}^o_x)$ is the "right" Brownian motion over the sample space $(\Omega^o, \underline{\underline{F}}^o)$ starting $\mathbb{P}^o_x$-a.s. from $x \in \mathbb{R}$ and with the local times $\{l^o(t,y);\ y \in \mathbb{R}\}$. The time change T, defined for all t = 0 by the relation

$$(2) \qquad t = \int_{\mathbb{R}} l^o(T_t, y)\, dm(y)$$

leads by means of the substitutions

$$x_t = B^o_{T(t)}, \qquad \underline{\underline{F}}^X_t = \underline{\underline{F}}^o_{T(t)}, \qquad t \geq 0$$

to the diffusion $X = (x_t, \underline{\underline{F}}^X_t, \mathbb{P}^o_x)$. As it is well-known, even the process X has local times, which can be expressed by

$$l(t,y) = l^o(T_t, y), \quad t \geq 0,\ y \in \mathbb{R}.$$

Moreover, for each bounded measurable function f it holds

$$(3) \qquad \int_0^t f(x_u)\, du = \int_{\mathbb{R}} l(t,y) f(y)\, dm(y), \quad t \geq 0.$$

1. THE QUADRATIC VARIATION PROCESS

First we are interested in an explicit representation of the time change T in terms of the diffusion X. Using properties of local times and the basic assumption (1) for the speed m we can write

$$\int_0^t I_\Lambda(x_u)du = \int_{\mathbb{R}} l(t,y)I_\Lambda(y)dm(y) = \int_{\mathbb{R}} l^0(T_t,y)\sigma^{-2}(y)2dy =$$

$$= \int_0^{T(t)} \sigma^{-2}(B_u^0)du = \int_0^t \sigma^{-2}(B^o_{T(u)})dT_u = \int_0^t \sigma^{-2}(x_u)dT_u,$$

remember $\mu(\Lambda) = 0$ and the fact that the Brownian motion B^0 has the speed measure 2dx. By means of a formal calculation with the corresponding differentials and observing $T_0 = 0$ we get the formula

$$T_t = \int_0^t \sigma^2(x_u)I_\Lambda(x_u)du, \quad t \geq 0. \tag{4}$$

Now we apply a result of Kazamaki [13] concerning quadratic variations for time changed processes. Since the time change T is continuous, it holds

$$\langle X\rangle_t = \langle B^o_T\rangle_t = \langle B^0\rangle_{T(t)} = T_t, \quad t \geq 0.$$

Thus, we have derived an explicit expression for the quadratic variation process of X, which is nothing else but Meyer's natural increasing process for the locally square integrable martingale X (see Fisk [6], Wong [22]).

$$\langle X\rangle_t = \int_0^t \sigma^2(x_u)I_\Lambda(x_u)du, \quad t \geq 0. \tag{5}$$

Observe that the quadratic variation $\langle X\rangle$ depends on the behaviour of the process within the "nonsingular" set Λ only. Furthermore, $\langle X\rangle$ is absolutely continuous with respect to the Lebesgue measure. Following Doob [2; p.449], the diffusion X can be represented as a stochastic integral with respect to some Brownian motion process, where it may be necessary to enlarge the underlying probability space by the adjunction of another Brownian motion. This procedure will be crucial in constructing a stochastic differential equation for X in the next section.

2. THE STOCHASTIC DIFFERENTIAL EQUATION

Let us explain what we will understand by a solution to the stochastic equation

$$(6)\qquad \begin{cases} x_t = x + \int_0^t \sigma(x_u) I_\Lambda(x_u)\, dB_u, & t \geq 0 \\ \int_0^t g(x_u) I_\Gamma(x_u)\, du = \int_{\mathbb{R}} l(t,y) g(y)\, d\mu(y), & \\ \qquad t \geq 0,\ g \text{ bounded measurable}, & \end{cases}$$

where x is an arbitrary initial point of the state space $\mathbb{R}$.

Definition. A solution of equation (6) is a family of stochastic processes $\mathbb{X} = (x_t, B_t, L_t)$ defined on a filtered probability space $(\Omega, \underline{\underline{F}}, \underline{\underline{F}}_t, \mathbb{P}_x)$ satifying the following conditions.

(i) B_t is a Brownian motion starting from zero.

(ii) $L_t = \{l(t,y);\ y \in \mathbb{R}\}$ is a family of local times for the continuous process x_t, i.e. $l(t,y)$ is ($\mathbb{P}_x$-a.s.) non-negative, (t,y)-continuous, increasing in t with

$$(7)\qquad \int_0^t I_{\{u: x_u = y\}}(r)\, l(dr,y) = l(t,y).$$

(iii) The processes x_t, B_t and L_t satisfy $\mathbb{P}_x$-a.s. equation (6).

Now we are in position to formulate our assertion concerning the existence of a solution to the stochastic equation (6).

THEOREM 1. Let X be a diffusion process with infinitesimal generator $D_m D_x^+$. Then, for each initial value $x \in \mathbb{R}$, we can find processes B_t and L_t over some filtered probability space $(\Omega, \underline{\underline{F}}, \underline{\underline{F}}_t, \mathbb{P}_x)$ such that $\mathbb{X} = (x_t, B_t, L_t)$ is a solution to equation (6).

P r o o f. For an arbitrary starting point $x \in \mathbb{R}$ we introduce along with $(\Omega^0, \underline{\underline{F}}^0, \mathbb{P}_x^0)$ a probability space $(\hat{\Omega}, \hat{\underline{\underline{F}}}, \hat{\mathbb{P}})$ and a Brownian motion $\{B_t, \underline{\underline{F}}_t;\ t \geq 0\}$ from zero over this space. We lift X and B to two independent stochastic processes over the product space

$$(\Omega, \underline{\underline{F}}, \mathbb{P}_x) = (\Omega^0 \times \hat{\Omega}, \underline{\underline{F}}^0 \otimes \hat{\underline{\underline{F}}}, \mathbb{P}_x^0 \otimes \hat{\mathbb{P}})$$

and construct a new process B by setting

$$\underline{\underline{F}}_t = \underline{\underline{F}}^0_{T(t)} \otimes \hat{\underline{\underline{F}}}_t, \qquad t \geq 0,$$

$$B_t = \int_0^t \sigma(x_u)^{-1} I_\Lambda(x_u)\, dx_u + \int_0^t I_\Gamma(x_u)\, d\hat{B}_u, \qquad t \geq 0.$$

Here the last stochastic integral is really defined over the filtered product space $(\Omega, \underline{\underline{F}}, \underline{\underline{F}}_t, \mathbb{P}_x)$.

We show that the process $B = \{B_t, \underline{\underline{F}}_t;\ t \geq 0\}$ is a Brownian motion over

$(\Omega, \underline{\underline{F}}, \mathbb{P}_x)$. Indeed, for all $t \geq 0$, $h > 0$ it holds

$$\mathbb{E}_x\{B_{t+h} - B_t \mid \underline{\underline{F}}_t\} = 0, \qquad \mathbb{E}_x\{(B_{t+h} - B_t)^2 \mid \underline{\underline{F}}_t\} =$$

$$= \mathbb{E}_x\{\int_t^{t+h} \sigma(x_u)^{-2} I_\Lambda(x_u) d\langle X\rangle_u + 2\int_t^{t+h} \sigma(x_u)^{-1} I_\Lambda(x_u) I_\Gamma(x_u) d\langle X, \hat{B}\rangle_u +$$

$$+ \int_t^{t+h} I_\Gamma(x_u) du \mid \underline{\underline{F}}_t\} =$$

$$= \mathbb{E}_x\{\int_t^{t+h} I_\Lambda(x_u) du + \int_t^{t+h} I_\Gamma(x_u) du \mid \underline{\underline{F}}_t\} = h,$$

where we have used formula (5) concerning the quadratic variation process $\langle X\rangle$. In other words, the process $B = \{B_t, \underline{\underline{F}}_t;\ t \geq 0\}$ is a continuous local martingale with quadratic variation $\langle B\rangle_t = t$.

Furthermore,

$$\langle \int_0^\cdot I_\Gamma(x_u) dx_u\rangle_t = \int_0^t I_\Gamma(x_u) d\langle X\rangle_u = \int_0^t I_\Gamma(x_u) I_\Lambda(x_u) \sigma^2(x_u) du = 0,$$

and, cosequently

$$\int_0^t I_\Gamma(x_u) dx_u = 0, \qquad \mathbb{P}_x\text{-a.s.}$$

Now we deduce immediately the first part of equation (6)

$$\int_0^t \sigma(x_u) I_\Lambda(x_u) dB_u = \int_0^t \sigma(x_u) I_\Lambda(x_u) \sigma(x_u)^{-1} I_\Lambda(x_u) dx_u +$$

$$+ \int_0^t \sigma(x_u) I_\Lambda(x_u) I_\Gamma(x_u) d\hat{B}_u =$$

$$= \int_0^t I_\Lambda(x_u) dx_u = \int_0^t dx_u - \int_0^t I_\Gamma(x_u) dx_u = x_t - x, \qquad \mathbb{P}_x\text{-a.s.}$$

The second part of equation (6) is a consequence of relation (3) concerning the local times of the diffusion X and of assumption (1) for the speed m. Namely, for every $t \geq 0$ and any bounded measurable function g it holds

$$\int_0^t g(x_u) I_\Gamma(x_u) du = \int_{\mathbb{R}} l(t,y) g(y) I_\Gamma(y) dm(y) = \int_{\mathbb{R}} l(t,y) g(y) d\mu(y).$$

Consequently, the triplet $\mathbb{X} = (x_t, B_t, L_t)$ defined over $(\Omega, \underline{\underline{F}}, \underline{\underline{F}}_t, \mathbb{P}_x)$ is the solution to equation (6).

One can learn from equation (6) that the dynamic of the process X within the set Λ is totally determined through the stochastic differential equation, which forms the first part of (6). On the other hand, the points from the set Γ are sticky for X, the Lebesgue measure of the sojourn time of the process at a point in Γ is, according to the second part of (6), a.s. positive.

3. ABOUT ITO'S FORMULA

The knowledge of the natural increasing process $\langle X\rangle$ implies an explicit version of Itô's change of variable formula for the diffusion X, compare Kunita/Watanabe [14].

THEOREM 2. *Let* f *be a twice continuously differentiable function on* $\mathbb{R}$. *Then, for every* $t \geq 0$ *and* $x \in \mathbb{R}$, *it holds the equation*

$$f(x_t) - f(x_0) = \frac{1}{2}\int_0^t f''(x_u)\sigma^2(x_u)I_\Lambda(x_u)du + $$

$$+ \int_0^t f'(x_u)\sigma(x_u)I_\Lambda(x_u)dB_u, \qquad \mathbb{P}_x\text{-a.s.}$$

We remark that this formula can be used to construct a stochastic differential equation for diffusions with a non identical, but smooth road map s, see [10]. On the other hand, the above Itô formula contains some essential gaps. Observe that for all test functions $f \in C^2$ the equation do not contain explicit information about the behaviour of the process in the "singular" set Γ, because it is independent of the measure μ. This is connected with the fact, that in general a function $f \in C^2$ does not belong to the domaine of definition of the infinitesimal generator $D_m D_x^+$, which is the second gap for the formula above. But this problem can be overcomed by the following

THEOREM 3. *Let* f *be a function from the domain of definition of the generalized differential operator* $D_m D_x^+$. *Then it holds for all* $t \geq 0$ *and* $x \in \mathbb{R}$

$$f(x_t) - f(x_0) = \int_0^t (D_m D_x^+ f)(x_u)du + \int_0^t (D_x^+ f)(x_u)dx_u, \qquad \mathbb{P}_x\text{-a.s.}$$

The proof of this formula is related to Wang [22], where we do not need condition (1).

4. CHANGE OF LAW

In this section we are concerned with diffusions, absolutely continuous with respect to the given process $X = (x_t, \underline{\underline{F}}_t, \mathbb{P}_x)$, determined by the generator $D_m D_x^+$. In [16] S.Orey gave an analytical condition for absolute continuity in terms of the corresponding speed measures and scale functions. Let $(x_t, \underline{\underline{F}}_t, Q_x)$ be another diffusion process with generator $D_n D_s^+$ (inaccessible boundaries are assumed tacitly). Roughly speaking, the Q_x are absolutely continuous with respect to $\mathbb{P}_x$ for all $x \in \mathbb{R}$, if and only if $(dn/dm)(ds/dx) = 1$ and s" exists almost everywhere.

Using the stochastic differential equation for X one obtains an explicit formula for the corresponding Radon-Nikodym derivatives, for all $x \in \mathbb{R}$ it holds

$$\frac{dQ_x \mid \underline{\underline{F}}_t}{d\mathbb{P}_x \mid \underline{\underline{F}}_t} = \exp\left\{\int_0^t b(x_u)\sigma(x_u)I_\Lambda(x_u)dB_u - \frac{1}{2}\int_0^t b^2(x_u)\sigma^2(x_u)I_\Lambda(x_u)du\right\}, \quad t \geq 0,$$

where $b = -s''/2s'$. Moreover, the process

$$C_t = B_t - \int_0^t b(x_u)\sigma(x_u)I_\Lambda(x_u)du, \quad t \geq 0$$

forms a Brownian motion over the filtered probability space $(\Omega, \underline{\underline{F}}, \underline{\underline{F}}_t, Q_x)$. Finally, x_t is a solution to the stochastic differential equation

$$x_t = x + \int_0^t b(x_u)\sigma^2(x_u)I_\Lambda(x_u)du + \int_0^t \sigma(x_u)I_\Lambda(x_u)dC_u, \quad t \geq 0; \; Q_x\text{-a.s.}$$

This establishes another way getting diffusions with non identical road maps, which might be useful in nonlinear filtering and optimal control of such processes.

REFERENCES

[1] M.A.Arbib, Hitting and martingale characterizations of one-dimensional diffusions, Z. Wahrscheinlichkeitstheorie und Verw. Gebiete 4 (1965), 232-247.

[2] J.L.Doob, Stochastic Processes, John Wiley & Sons, New York 1953.

[3] E.B.Dynkin, Markov Processes, Springer, Berlin 1965.

[4] W.Feller, The parabolic differential equations and the associated semi-groups of transformations, Ann. Math. 55 (1952), 468-519.

[5] ——, On the intrinsic form for second order differential operators, Illinois J. Math. 2 (1958), 1-18.

[6] D.L.Fisk, Sample quadratic variation of sample continuous, second order martingales, Z. Wahrscheinlichkeitstheorie und Verw. Gebiete 6 (1966), 273-278.

[7] D.Freedman, Brownian Motion and Diffusion, Holden-Day, San Francisco 1971.

[8] I.I.Gihman and A.V.Skorohod, Stochastic Differential Equations, Springer, Berlin 1972.

[9] J.Groh, A stochastic differential equation for Feller's one-dimensional diffusions, preprint N/81/72, FSU Jena.

[10] ——, On a stochastic calculus for Feller's one-dimensional diffusions, preprint N/82/10, Fiedrich-Schiller-Universität Jena.

[11] ——, On absolute continuity of one-dimensional diffusions, preprint N/82/38, Friedrich-Schiller-Universität Jena.

[12] K.Itô and H.P.McKean, Jr., Diffusion Processes and their Sample Paths, Springer, Berlin 1965.

[13] N.Kazamaki, Changes of time, stochastic integrals, and weak martingales, Z. Wahrscheinlichkeitstheorie und Verw. Gebiete 22 (1972), 25-32.

[14] H.Kunita and S.Watanabe, On square integrable martingales, Nagoya Math. J. 30 (1967), 209-245.

[15] P.Mandl, Analytical Treatment of One-dimensional Markov Processes, Academia, Prague and Springer, Berlin 1968.

[16] S.Orey, Conditions for the absolute continuity of two diffusions, Trans. Amer. Math. Soc. 193 (1974), 413-426.

[17] Y.Oshima, On stochastic differential equations characterizing some singular diffusion processes, Proc. Japan Acad. 57, Ser.A (1981), 151-154.

[18] ——, Some singular diffusion processes and their associated stochastic differential equations, Z. Wahrscheinlichkeitstheorie und Verw. Gebiete 59 (1982), 249-276.

[19] H.F.Trotter, A property of Brownian motion paths, Illinois J. Math. 2 (1958), 425-433.

[20] A.T.Wang, Generalized Itô's formula and additive functionals of Brownian motion, Z. Wahrscheinlichkeitstheorie und Verw. Gebiete 41 (1977), 153-159.

[21] S.Watanabe, On stochastic differential equations for multi-dimensional diffusion processes with boundary conditions I, II, J. Math. Kyoto Univ. 11 (1971), 169-180, 545-551.

[22] E.Wong, Representation of martingales, quadratic variation and applications, SIAM J. Control 9 (1971), 621-633.

A RESULT OF THE ITERATED LOGARITHM TYPE FOR A CERTAIN CLASS OF STOCHASTIC PROCESSES

K. Helmes

Institut für Angewandte Mathematik
Universität Bonn

ABSTRACT

Let $(W_t) = (W_t^1, W_t^1, \cdots, W_t^d)$, $d \geq 2$, $t \geq 0$, *denote a (standard) d-dimensional Brownian motion and let* $x(t)$, $A(t)$ *be measurable functions from* $[0,\infty)$ *into* $\mathbb{R}^d$ *and* s^d, s^d *the space of dxd skew-symmetric matrices, respectively which are bounded on every interval* $[0,T]$, $T>0$. *Define the stochastic process* $(L_t^{A,x})$ *by*

$$L_t^{A,x} := \int_0^t \langle A(s)(W_s + x(s)), dW_s \rangle,$$

the stochastic integral being defined in the sense of K. Itô. Based on a formula for the joint characteristic function of W_t *and* $L_t^{A,x}$ *we give an iterated logarithm type result for those processes where* $A(t) \equiv A$ *and* $x(t) \equiv x$.

1. INTRODUCTION

Let (W_t), $t \geq 0$, be a d-dimensional (standard) Brownian motion, $A(t)$ a measurable function from $[0,\infty)$ into s^d, s^d the space of dxd real valued skew-symmetric matrices, and $x(t)$ a measurable function from $[0,\infty)$ into $\mathbb{R}^d$. Assume that the functions $A(t)$ and $x(t)$ are bounded on every interval $[0,T]$, $T>0$. Then for every $t>0$ the (Itô) stochastic integral

$$L_t^{A,x} := \int_0^t \langle A(s)(W_s + x(s)), dW_s \rangle, \tag{1}$$

$\langle \cdot,\cdot \rangle$ Euclidean scalar product in $\mathbb{R}^d$, is well defined. In this note we are concerned with asymptotic fine properties of those processes $L_t^{A,x}$, where A and x are constants. A prominent example within this class of stochastic processes is P. Lévy's 'stochastic area'.

$$(2) \qquad L_t := \frac{1}{2} \int_0^t (W_s^1 dW_s^2 - W_s^2 dW_s^1)$$

which is given by (1) when choosing $x(t) \equiv 0$ and $A(t) \equiv \frac{1}{2} J$,

$$J = \begin{pmatrix} 0 & -1 \\ 1 & 0 \end{pmatrix}.$$

The 'stochastic area' is related to several, seemingly uncorrelated, objects in different areas of mathematics, e.g. Brownian motion on some Lie groups , fundamental solutions of certain hypoelliptic operators, the Schrödinger operator for a particle in a constant magnetic field and parameter estimation of two-dimensional Gaussian Markov processes; cf. Ref. 5 and the literature cited therein for the first two examples, Ref.10, p.168, together with Ref. 6 for the third one and Ref. 8, section 17.2, for the estimation problem.

A law of the iterated logarithm for (L_t) was proved by Berthuet in [2]. In Ref. 3 Crepel and Roynette have shown a log log law for random walks on the (first) Heisenberg group, cf. [3], i.e. discrete time processes which correspond to (L_t) the way sums of i.i.d. random variables correspond to Brownian motion. Here we extend Berthuet's result to higher dimensions. The proof mimics the demonstration of Khintchine's law of the iterated logarithm as given by Mc Kean in [9] section 1.5. For this kind of proof a formula for the joint characteristic function of W_t and $L_t^{A,x}$, Ref. 6, Corollary 2, is instrumental.

Since Brownian motion obeys the law of large numbers we may restrict ourselves to the case x = 0 from the beginning (see Theorem 1, below) and we shall write L_t^A instead of $L_t^{A,0}$; we call (L_t^A) the 'generalized stochastic area' of the Wiener Process (W_t).

2. THE LAW OF THE ITERATED LOGARITHM

In [6] the following result is derived as a Corollary to a general formula but it could also be derived from Lévy's formula for the conditional characteristic function of L_t given W_t, cf. [5]. To formulate the result let us recall that for any skew-symmetric matrix A there is an orthogonal matrix O such that O'AO is a skew-symmetric matrix formed from [d/2] blocs

$$\begin{pmatrix} 0 & -a_k \\ a_k & 0 \end{pmatrix} , \quad 1 \leq k \leq [d/2],$$

where the numbers a_k are simple algebraic functions of the entries of the matrix A. If d is odd the last row- and column-vector of O'AO consists of zeros. Put n: = [d/2] and let, from now on,

(3) $$a: = \max\{|a_k|;\ 1 \leq k \leq n\}.$$

Lemma 1. *Let $\gamma \in \mathbb{R}^d$ and $\Lambda \in \mathbb{R}$, $T>0$ such that $|\Lambda T a| < \pi/2$. Then for every $t \in [0,T]$,*

(4) $$E[\exp\{\langle\gamma, W_t\rangle + \langle\Lambda, L_t^A\rangle\}]$$

$$= \prod_{k=1}^{n} \frac{1}{\cos(\Lambda t a_k)} \exp\{\tfrac{1}{2}[(O\gamma)^2_{2k-1} + (O\gamma)^2_{2k}] \frac{tg(\Lambda t a_k)}{\Lambda a_k}\}.$$

The log log law for the 'generalized stochastic area' will now be proved using Doob's submartingale inequality and two complemments to Lemma 1.

Lemma 2. *Let $\delta>0$ be given and fix Θ, $1<\Theta<2$, such that*

(5) $$\frac{tg[(\Theta-1)\pi/2]}{(\Theta-1)\pi/2} < 1.$$

Put $\Lambda_m = \frac{\pi\Theta^{-m}}{2a(1+\delta)}$, $m \in \mathbb{N}$. Then, for every $\rho>0$,

(6) $$P[\sup_{\theta^m \leq t \leq \Theta^{m+1}} (\exp[\Lambda_m L_t^A]) > \rho] \leq \frac{C}{\rho}$$

for all $m \geq m_0(\delta,\theta)$ and the constant C only depends on δ and θ; in particular, it is independent of $m \geq m_0$.

Proof. Put $T_m: = \Theta^m(\Theta-1)$ and define $\tilde{W}_u: = W_{u+\theta^m} - W_{\theta^m}$, $0 \leq u \leq T_m$. Denote by $(\tilde{L}_u^A)$ the 'generalized stochastic area' associated with $(\tilde{W}_u)$. Since

$$L_t^A = \int_{\theta^m}^{t} \langle A(W_r - W_{\theta^m}), dW_r\rangle + \langle AW_{\theta^m}, W_t - W_{\theta^m}\rangle + L_{\theta^m}^A$$

$$= \tilde{L}^A_u + \langle AW_{\theta^m}, \tilde{W}_u \rangle + L^A_{\theta^m}, \qquad u=t-\theta^m,$$

and since $(\tilde{W}_u)$ is independent of F_{θ^m} Doob's submartingale inequality applied to $s \to \exp[\Lambda_m \tilde{L}^A_s + \langle \Lambda_m \gamma, \tilde{W}_s \rangle + \gamma]$, $\gamma \in \mathbb{R}^d$, $0 \leq s \leq T_m$, implies

$$P[\sup_{\theta^m \leq t \leq \theta^{m+1}} (\exp[\Lambda_m L^A_t]) > \rho]$$

$$\leq \frac{1}{\rho} E[E[\exp\{\Lambda_m \tilde{L}^A_{T_m} + \Lambda_m \langle AW_{\theta^m}, \tilde{W}_{T_m} \rangle + \Lambda_m L^A_{\theta^m}\} | F_{\theta^m}]]$$

and, by Lemma 1,

$$\leq \frac{C_1(\delta)}{\rho} E[\exp\{\Lambda_m L^A_{\theta^m} + \frac{\theta^{-m}}{2} C_2(\delta,\theta) |W_{\theta^m}|^2\}].$$

Since the conditional distribution of L^A_s given W_t is a function of $|W_t|^2$, see [6] or [7], p. 388, and since the constants θ and Λ_m are chosen according to (5) there is a constant C such that for $m \geq m_0(\delta,\theta)$ we find the probability to be

$$\leq \frac{C}{\rho} .$$

∎

Part of the notation which will be used in the proof of the Theorem will be introduced in the following lemma which will allow us to show the 'lower class result' of the iterated logarithm.

Lemma 3. Assume $\theta > 1$ *and define* $g(\theta^m) = (\frac{2}{\pi} \log \log (\theta^m))^{1/2}$, $\Lambda_m = (2/\pi)^{-1/2} \theta^m \sqrt{\theta - 1} (a g(\theta^m)))^{-1}$. *Let* $(\gamma_m) \in \mathbb{R}^d$ *such that* $\overline{\lim} \|\gamma_m\| / (2\theta^m \log[\log \theta^m]^{1/2}) = 1$; *define* $Y_m = \Lambda_m \tilde{L}_{T_m} + \Lambda_m \langle A\gamma_m, \tilde{W}_{T_m} \rangle$. *Then for every* $\beta > 0$

$$P[Y_m > x] \geq \exp(-\frac{\pi(\theta-1)}{2} \frac{1}{2} (1+\beta) x^2)$$

for all x, m *sufficiently large.*

Proof. The bilateral Laplace transform of Y_m exists in a neighbourhood of zero, $|z| \leq \rho_0$ (uniformly for m large) and, by Lemma 1, can be estimated from above and from below as follows: For every $\varepsilon > 0$ and m sufficiently large

$$\underline{C}_m(z) \exp(\frac{\pi}{2}(1-\varepsilon) z^2) \leq E[e^{zY_m}] \leq \overline{C}_m(z) \exp(\frac{\pi}{2} z^2 D_m(z))$$

where $\bar{C}_m(z) = (\cos(z\kappa_m))^{-n}$, $D_m(z) = tg(z\kappa_m)/(z\kappa_m)$, $\underline{C}_m(z) = \cos^{-1}(z\kappa_m \cdot a_1/a)\cdot\ldots\cdot\cos^{-1}(z\kappa_m a_k/a)$ and $\kappa_m := (\pi(\theta-1)/2\log\log\theta^m)^{1/2}$. Note that the function $\bar{C}_m(z)$ and $D_m(z)$ tend (monotonically) to one. Now switch from the Laplace transform of Y_m to the logarithm of the transform and expand it in a power series. Compare the coefficients of the expansion with those of the upper and lower bound. Taking the well known estimate for the tail probability of the normal distribution into account the assertion follows, cf. [4], XVI.7. ■

Theorem

$$P[\overline{\lim_{t\to\infty}} \frac{L_t^{A,x}}{t \log\log(t)} = \frac{2a}{\pi}] = 1$$

and

$$P[\underline{\lim_{t\to\infty}} \frac{L_t^{A,x}}{t \log\log(t)} = -\frac{2a}{\pi}] = 1.$$

Proof. The second assertion follows from the first one if there we put -A and -x. Since $L_t^{A,x} = L_t^A + \langle x, W_t\rangle$ and $|W_t|/t \to 0$, $t\to\infty$, we may assume x=0. First we show the upper class result

(7) $$\overline{\lim_{t\to\infty}}\ L_t^A/t \log\log(t) \le 2a/\pi.$$

Let $\delta>0$, $h(t) = \log\log(t)$ and fix $\theta>1$ such that (5) holds. Consider the events $B_m := \{L_t^A > (1+\delta)^2 2ath(t)/\pi$ for at least one $t\in[\theta^m, \theta^{m+1}]\}$. since $th(t)\uparrow$ if $t\uparrow$,

$$B_m \subset \{\sup_{\theta^m \le t \le \theta^{m+1}} (L_t^A) > \frac{2}{\pi}(1+\delta)^2 a\theta^m h(\theta^m)\}.$$

Define Λ_m as in Lemma 2 and put $\rho = \exp((1+\delta)h(\theta^m))$. By Lemma 2 we get

$$P[B_m] \le \frac{C(\delta,\theta)}{m^{(1+\delta)}(\log(\theta))^{1+\delta}}, \qquad m\ge m_0(\delta,\theta).$$

Hence $\Sigma P[B_m] < \infty$, so that, by the Borel-Cantelli lemma $P[B_m \text{ i.o.}] = 0$ (where 'i.o.' stands for infinitely often). But this means that

$$\overline{\lim_{t\to\infty}}\ (L_t^A/t \log\log(t)) \le 2a(1+\delta)^2/\pi.$$

Since δ is arbitrary, (7) follows. We shall next prove the 'lower class result'

(8) $$\overline{\lim_{t\to\infty}}(L_t^A/t \log\log(t)) \geq 2a/\pi.$$

Put $\phi(t) := (2a/\pi)t \log\log(t)$. For $\theta>1$ and $\eta>0$, to be specified later, consider the events

$$\overline{B}_m := \{L^A(\theta^{m+1}) - L^A(\theta^m) > (1-\eta)\phi(\theta^{m+1})\}.$$

Now adopt the notation introduced in Lemma 3 and put $\gamma_m := W_{\theta^m}$. By Lemma 3, for $\beta>0$ (chosen later) and m sufficiently large

$$P[\overline{B}_m/F_{\theta^m}] \geq ((m+1)^{\alpha/2}\log^{\alpha/2}(\phi))^{-1} \qquad \text{a.e.}$$

where $\alpha = (1+\beta)(1-\eta)^2\theta^2h^2(\theta^{m+1})/h^2(\theta^m)$. Consequently, choosing β and θ sufficiently small so that $\alpha<2$ (note that $h(\theta^{m+1})/h(\theta^m)\to 1$ if $m\to\infty$), $\Sigma P[\overline{B}_m | F_{\theta^m}] = \infty$ a.e. Hence, the extended Borel-Cantelli lemma, [1], p.96, implies

(9) $$P[L^A(\theta^{m+1}) > (1-\eta)\phi(\theta^{m+1}) + L(\theta^m) \quad \text{i.o.}] = 1.$$

By the proof of (7) applied to L_t^{-A},

$$P[L^A(\theta^m) < -(1+\eta)\phi(\theta^m) \quad \text{i.o.}] = 0.$$

Putting it together with (9) we deduce that a.s.

$$L(\theta^{m+1}) \geq \phi(\theta^{m+1})[1-\eta) - (1+\eta)\phi(\theta^m)/\phi(\theta^{m+1})].$$

Now, for any given $\delta>0$, if we choose η and θ so small that

$$(1-\eta) - (1+\eta)/\theta > 1-\delta,$$

we get that

$$P[L(\theta^{m+1}) > (1-\delta)\phi(\theta^{m+1}) \quad \text{i.o.}] = 1.$$

Since δ is arbitrary, this completes the proof of the Theorem.

∎

REFERENCES

[1] BREIMAN, L.: *Probability*, Addison Wesley, New York, 1968.

[2] BERTHUET, R.: *Loi du logarithme itéré pour certaines intégrales stochastiques*, C. R. Acad. Sc. Paris, t. 289 (1979), pp. 813-815.

[3] CREPEL, P. and ROYNETTE, B.: *Une loi du logarithme itéré pour le groupe d'Heisenberg*, Z. Wahrscheinlichkeitstheorie verw. Gebiete 39 (1977), pp. 217-229

[4] FELLER,W.: *An introduction to probability theory and its applications*, vol. 2, 2. ed., John Wiley, New York, 1971.

[5] GAVEAU, B.: *Principe de moindre action, propagation de la chaleur et estimée sous elliptiques sur certains groupes nilpotents*, Acta Math. 139 (1977), pp. 95-153.

[6] HELMES, K. and SCHWANE, A.: *Lévy's stochastic area in higher dimensions*, to appear in Proc. on 'Recent Advances in Filtering and Optimization', 1982, eds. W. H. Fleming and L. G. Gorostiza.

[7] IKEDA, N. and WATANABE, S.: *Stochastic differential equations and diffusions processes*, North-Holland Publ., Amsterdam, 1981.

[8] LIPTSER, R. and SHIRYAYEV, A: *Statistics of random processes*, vol.2, Springer-Verlag, New York, 1977.

[9] McKEAN, H. P.: *Stochastic integrals*, Academic Press, New York, 1969.

[10] SIMON, B.: *Functional integration and quantum physics*, Academic Press, New York, 1979.

APPROXIMATION OF LARGE DEVIATIONS ESTIMATES AND ESCAPE TIMES AND APPLICATIONS TO SYSTEMS WITH SMALL NOISE EFFECTS

Harold J. Kushner

Divisions of Applied Mathematics and Engineering
Lefschetz Center for Dynamical Systems
Brown University
Providence, Rhode Island 02912

1. Introduction.

For the purpose of estimating the escape time from a given set, or other statistical properties of systems with small noise effects, it is generally assumed in applications that the system noise is white Gaussian. The Gaussian assumption greatly simplifies the computation, but is not adequate for many important classes of applications to control and communication theory. For example, when the noise is small, the mean escape time from a set can be quite sensitive to the underlying statistics even though in the study of the effects of the noise over any fixed finite time interval, the Gaussian approximation might be a good one. We are concerned with the sensitivity of these statistical quantitites to the underlying statistical structure, when the noise effects are small, and also with the question of when the Gaussian assumption makes sense. Consider a sequence of systems with small noise effects whose statistics converge in some sense to those of a "limit" system. The techniques developed involve approximation and limit theorems for a sequence of variational problems associated with the minimization of the action functionals which arise when the theory of large deviations is applied to the above mentioned systems. Degenerate and non-degenerate cases with both bounded and Gaussian noise are considered. Several examples appear in Section 4.

We will be concerned with robustness, approximation and applications of large deviations methods [1] - [7] for processes of the type (1.1) - (1.2), where σ might be zero. The $\xi(\cdot)$, $\{\xi_n\}$ are bounded and stationary, $w(\cdot)$ is a standard Wiener process, $\{\rho_n\}$ is i.i.d. Gaussian, and $\{\rho_n\}$ and $w(.)$ are independent of $\{\xi_n\}$ and $\xi(\cdot)$, and $E\rho_n = 0$, covar $\rho_n = I$. The functions $\sigma(\cdot)$, $\bar{b}(\cdot)$ and $b(\cdot,\xi)$ are Lipschitz continuous (uniformly in ξ). In all cases, $x \in R^k$, Euclidean k-space. For concreteness, we specialize to (1.2), although similar results hold for (1.1).

$$dx^\gamma = b(x^\gamma,\xi(t/\gamma))dt + \sqrt{\gamma}\sigma(x^\gamma)dw \tag{1.1}$$

$$x^\gamma_{k+1} = x^\gamma_k + \gamma b(x^\gamma_k,\xi_k) + \gamma\sigma(x^\gamma_k)\rho_k. \tag{1.2}$$

We suppose that there is a $\bar{b}(\cdot)$ such that $\bar{b}(x) = \lim_N \frac{1}{N} E \sum_0^N b(x,\xi)$. Only a sketch of the results and a few proofs are presented here. See [17] for a fuller development.

The various assumptions introduced below are not always used together.

Let G be a bounded open set with piecewise differentiable boundary, define $N_{\varepsilon_1}(G) = G_1$, an ε_1-neighborhood of G (henceforth fixed), and assume

(A1.1) $\dot{x} = \bar{b}(x)$ has a unique stable point $\bar{x}_0$ in G_1 and all trajectories originating in G_1 tend to $\bar{x}_0$. Also, these trajectories are never tangent to ∂G.

Define the H-functionals

(1.3) $$H(x,\alpha) = \lim_{\substack{N\to\infty \\ M\to\infty}} \frac{1}{(N-M)} \log E \exp \sum_{M}^{N-1} \alpha'(b(x,\xi_n) + \sigma(x)\rho_n)$$

When we wish to emphasize the Gaussian component, an affix σ will be used and we write $H^{\sigma}(x,\alpha) = H^{0}(x,\alpha) + \tilde{H}^{\sigma}(x,\alpha)$.

(A1.2) In (1.3), let the convergence be uniform for $x \in G_1$ and also in the initial data, if E is replaced by the expectation given the $\xi(\cdot)$, or $\{\xi_n\}$ data up to time T_0/γ.

The limits in (1.3) and assumption (A1.2) are phrased as they are because we wish to treat the escape time problem when the noise is not necessarily Markov.

Define the Cramer transformation $L(x,\beta) = \sup_{\alpha}[\beta'\alpha - H(x,\alpha)]$, and the set $U(x) = \{\beta : L(x,\beta) < \infty\}$. Define $S(T,\phi) = \int_0^T L(\phi(s),\dot{\phi}(s))ds$, if $\phi(\cdot)$ is absolutely continuous, and set it equals to ∞ otherwise. For $T(\phi) = \inf\{t : \phi(t) \notin G\}$, define $S(\phi) = S(T(\phi),\phi)$, $S_0(x) = \inf\{S(\phi) : \phi(0) = x\}$, $S_0 = S_0(\bar{x}_0)$ and set $\tau_G^{\gamma} = \min\{t : x^{\gamma}(t) \notin G\}$.

Under broad conditions

(1.4) $$\lim_{\gamma\to 0} \gamma \log E\tau_G^{\gamma} = S_0 .$$

See [3,5]. In [5], $\sigma = 0$ and (A1.2) was implicitly assumed. With $\sigma = 0$, the proof in [5] (Theorem 5.1) is valid for more general $\xi(\cdot)$, provided (A1.2), (A1.3) hold, and using the convergence of (1.3) uniformly in $x \in G_1$ and in the initial data. It can also be extended to cover σ = constant (see appendix of [17]). The case where $\sigma \neq 0$, but $\xi(\cdot)$ does not appear, is in [3]. Criteria for (A1.3) appear in theorems 3.8 and 3.9 of [17] and in [5].

(A1.3) For $x \in G_1$, $H(\cdot,\cdot)$ is continuous and $H(x,\cdot)$ is differentiable.

In Theorem 3.7, the "continuity" condition A1.4 will be used. For open Q

containing $\bar{x}_0$ define $S(Q) = \inf\{S(T,\phi) : \phi(0) = \bar{x}_0, \phi(T) \in \partial Q\}$. Then $S(Q)$ is lower semi-continuous in Q, in that if $Q_\alpha \downarrow Q$, then $\underline{\lim}_{\alpha} S(Q_\alpha) \geq S(Q)$. But it is continuous at 'most' Q in the following sense. Let $Q_\rho = N_\rho(Q)$ and $\rho < \rho_1$ with $S(Q_{\rho_1}) < \infty$. Then for all but a countable number of $\rho_0 < \rho_1$, $S(Q_\rho) \to S(Q_{\rho_0})$ as $\rho \downarrow \rho_0$.

(A1.4) $S(G_\rho)$ is continuous at $\rho = 0$.

The quantity (1.4) is of considerable importance in numerous applications in control and communication theory, and in various applications to stochastic approximation, particularly in estimating escape times from regions in which an algorithm or process has a 'stability' property. Normally such estimates are hard to get unless ε is small. Except for the purely Gaussian case, it is now almost impossible to calculate $H(\cdot,\cdot)$, $L(\cdot,\cdot)$ or S_0, and so the purely Gaussian model is used almost exclusively. However, the value of S_0 can be quite sensitive to the underlying statistical assumptions, and it is not normally satisfactory to use a 'local diffusion of Gaussian' approximation [15]. We study the problem of robustness and approximatability for such problems. Section 2 contains some background results, and Section 3 gives the main approximation theorems, and some examples are in Section 4. Section 6 of [17] discusses an application to a phase locked loop, and various problems which arise in connection with that application. This class of applications seems to be both natural and of increasing popularity for the applications of large deviations or singular perturbation type (partial differential equation based) methods. Since the physical noise in such systems is not white Gaussian, or even Gaussian at all (strictly speaking), that application provides a good example of the role of our results.

Let A denote any set of continuous functions on $[0,T]$. Approximation results analogous to Theorems 3.4 - 3.8, but for the inequality (1.5), appear in [17, Theorem 4.9].

$$-\inf_{\phi\in A^0} S(T,\phi) \leq \underline{\lim}_{\gamma}\, \gamma \log P\{x^\gamma(\cdot) \in A\}$$

$$\leq \overline{\lim}_{\gamma}\, \gamma \log P\{x^\gamma(\cdot) \in A\} \leq -\inf_{\phi\in \bar{A}} S(T,\phi). \tag{1.5}$$

2. Preliminary Results.

It is well known that $H(x,\cdot)$, $L(x,\cdot)$ and the set $U(\cdot)$ are convex, and that $L(\cdot,\cdot)$ and $U(\cdot)$ (in the Hausdorff topology) are continuous. When minimizing $S(\phi)$ or $S(T,\phi)$, we have $\dot{\phi}(t) \in U(\phi(t))$, and the questions of finiteness and approximatability of S_0 and $\inf_{\phi\in A} S(T,\phi)$ are related to properties of $U(x)$. Here we state only one simple result (see [17]).

Theorem 2.1: *Let $\{\xi_n\}$ be a finite state Markov chain with state apace D, and transition probabilities $\{p_{ij}\}$, and with all states communicating with each other. Then $U(x)$ is the set of β_0 such that there are $N_n \to \infty$ and z_i such that $p_{z_i z_{i+1}} > 0$ all i and*

$$\beta_0 = \lim_n \frac{1}{N_n} \sum_{i=0}^{N_n-1} b(x,z_i). \tag{2.1}$$

Proof. By the communication, such β_0 form a closed convex set. Let $\{z_i\}$, satisfy the hypothesis and define β_0 by (2.1). There is a $q > 0$ such that $p_{z_i z_{i+1}} \geq q$, all i. Then (the limit below exists by the discrete parameter version of Theorem 2.2 of [5]; see also Theorem 3.8 of [17]).

$$\sup_\alpha[\alpha'\beta_0 - \lim_N \frac{1}{N} \log E \exp \alpha' \sum_0^{N-1} b(x,\xi_i)]$$

$$\leq \sup_\alpha[\alpha'\beta_0 - \lim_n \frac{1}{N_n} \log E \exp \alpha' \sum_0^{N_n-1} b(x,\xi_i)]$$

$$\leq \sup_\alpha[\alpha'\beta_0 - \underline{\lim}_n \frac{1}{N_n} \log(\exp \alpha' \sum_0^{N_n-1} b(x,z_i)) \prod_{i=1}^{N_n-1} p_{z_{i-1}z_i}]$$

$$\leq - \log q.$$

Thus $\beta_0 \in U(x)$. We omit the proof of the reverse case.

The continuity of $L(\cdot,\cdot)$.

Theorem 2.2: *Let $H(\cdot,\cdot)$ be continuous and let $x_0,\beta_0,N_\varepsilon(\beta_0)$ satisfy $L(x_0,\beta) < \infty$ for $\beta \in N_\varepsilon(\beta_0)$. Then $L(\cdot,\cdot)$ is continuous at (x_0,β_0).*

Proof. $H(x,\cdot)$ is convex. By the hypothesis and the concavity (in α) of $\alpha'\beta - H(x,\alpha)$, the set of maximizing α (at x_0,β_0) is bounded. Otherwise, for an appropriate arbitrarily small $\delta\beta$ we would get $L(x_0,\beta_0 + \delta\beta_0) = \infty$. Also, $\alpha'\beta - H(x,\alpha) \to \alpha'\beta_0 - H(x_0,\alpha_0)$ uniformly bounded α-sets as $(x,\beta) \to (x_0,\beta_0)$. The concavity and the last three sentences imply that the set of maximizing α in $\sup_\alpha[\alpha'\beta - H(x,\alpha)]$ must also converge to the set of maximizing α for x_0,β_0. Thus $L(x,\beta) \to L(x_0,\beta_0)$.

Q.E.D.

A remark on the degenerate case.

Suppose that (1.2) has the degenerate form $(x = (x_1,x_2), \alpha = (\alpha_1,\alpha_2), \beta = (\beta_1,\beta_2))$

$$\begin{aligned} x^\gamma_{1,k+1} &= x^\gamma_{1,k} + \gamma \bar{b}_1(x^\gamma_k), \quad x_1 \in R^{k-\ell}, \quad x_2 \in R^\ell, \\ x^\gamma_{2,k+1} &= x^\gamma_{2,k} + \gamma b_2(x^\gamma_k,\xi_k) + \gamma\sigma_2(x^\gamma_k)\rho_k. \end{aligned} \tag{2.2}$$

Then

(2.3) $$H(x,\alpha) = \alpha_1'\bar{b}_1(x) + \alpha_2'\sigma_2(x)\sigma_2'(x)\alpha_2/2 + \lim_N \frac{1}{N}\log E\exp \alpha_2' \sum_0^{N-1} b_2(x,\xi_n),$$

and $L(x,\beta) = \infty$ if $\beta_1 \neq \bar{b}_1(x)$. But Theorem 2.2 can be used to study continuity with respect to β_2 when $\beta_1 = \bar{b}_1(x)$. If $\sigma_2(x)\sigma_2'(x)$ is uniformly positive definite and $H(\cdot,\cdot)$ is continuous, then $L(x,\beta)$ is continuous in (x,β_2) when $\beta_1 = \bar{b}_1(x)$. Define $U_2(x) = \{\beta_2 : L(x;\bar{b}_1(x),\beta_2) < \infty\}$. When we refer to the degenerate case, the form (2.2) is always intended. In the non-degenerate case, we assume that $U(x)$ has a non-empty interior, and in the degenerate case that $U_2(x)$ has a non-empty interior.

In the non-degenerate case, define $\vec{U}^\delta(x)$ by: $\beta \in \vec{U}^\delta(x)$ if $\beta \in U(x)$ and $d(\beta,\partial U(x)) \geq \delta$, where d = Euclidean distance. The set $\vec{U}^\delta(x)$ is called a 'δ-interior' set. Let $\vec{U}_2^\delta(x)$ denote the δ-interior set for $U_2(x)$, and in the degenerate case, define $\vec{U}^\delta(x)$ by $\beta \in \vec{U}^\delta(x)$ if $\beta = (\beta_1,\beta_2)$, $\beta_1 = \bar{b}_1(x)$ and $\beta_2 \in \vec{U}_2^\delta(x)$. The continuity of a set valued function is always in the Hausdorff topology.

An argument similar to that of Theorem 2.2 proves the following.

Corollary 2.3: Let $U(\cdot)$ and $H(\cdot,\cdot)$ be continuous and let $H_n(x,\alpha) \to H(x,\alpha)$ uniformly on bounded (x,α) sets, where H_n and H are H-functionals. Then, in the non-degenerate case and for any compact set K and $\delta > 0$, $L_n(x,\beta) \to L(x,\beta)$ uniformly on $\{x,\beta : x \in K, \beta \in \vec{U}^\delta(x)\} \equiv \vec{U}^\delta$. In the degenerate case, let $\bar{b}_n(\cdot)$ denote the 'mean' dynamics for the system yielding H_n. Then $L_n(\cdot;\bar{b}_{1n}(\cdot),\cdot) \to L(\cdot;\bar{b}_1(\cdot),\cdot)$ uniformly on $\{x,\beta_2 : x \in K, \beta_2 \in \vec{U}_2^\delta(x)\}$. (If $U(\cdot)$ is not continuous, the convergence holds but might not be uniform.)

The following non-degeneracy result is useful in the proofs of the results of Section 3.

Theorem 2.4: For each x, let the H-functional $H(x,\cdot)$ be differentiable at $\alpha = 0$, and let K be compact. For each $\delta > 0$, there is an $\varepsilon > 0$ such that $L(x,\beta) \geq \varepsilon$ for $|\beta - \bar{b}(x)| \geq \delta$, $x \in K$.

3. Approximating $U(x)$, $S(T,A)$ and S_0.

Lemma 3.1 and Theorems 3.2,3 show that if $H_n \to H$, $\phi_n \to \phi$, then $\underline{\lim}_n S_n(\phi_n) \geq S(\phi)$, a basic result for the general approximation results. The remaining theorems show that $S_n \to S$ if $H_n \to H$ and some other conditions hold.

One or more of the following conditions will be used throughout the section, and will occasionally be weakened. The variable x is always assumed to be in G_1.

(A3.1) The H-functionals $H_n(\cdot,\cdot)$ converge to H uniformly on bounded (x,α) sets.

(A3.2) $U(\cdot)$ is continuous in the Hausdorff topology.

(A3.3) $U(x)$ and $\{\xi_n\}$ or $\xi(\cdot)$ are uniformly bounded. (We will also treat the unbounded case.)

For simplicity we consider 2 cases,the non-degenerate and the degenerate of (2.2).

(A3.4) There is an $\varepsilon_0 > 0$ such that for all x either (non-degenerate case) $N_{\varepsilon_0}(\bar{b}(x)) \in U(x)$ or (degenerate case) $N_{\varepsilon_0}(\bar{b}_2(x)) \in U_2(x)$.

Lemma 3.1: Under (A3.1), $\underline{\lim}_n L_n(x_n,\beta_n) \geq L(x,\beta)$, if $x_n \to x$, $\beta_n \to \beta$.

Proof: Let $R_N = \{\alpha : |\alpha| \leq N\}$ and define $L^N(x,\beta) = \sup_{\alpha\in R_N} (\alpha'\beta - H(x,\alpha))$. Then $L^N(x,\beta) \uparrow L(x,\beta)$ as $N \to \infty$. Also, $L_n(x_n,\beta_n) \geq \sup_{\alpha\in R_N} (\alpha'\beta_n - H_n(x_n,\alpha)) \to L^N(x,\beta)$. As $n \to \infty$. The assertion follows from this, and the arbitrariness of N. Q.E.D.

Let $S_n(T,\phi)$, $S_N(\phi)$ denote the action functionals corresponding to the H-functional H_n. The next theorem is basic for the subsequent approximation results.

Theorem 3.2: Let $\phi_n(\cdot) \to \phi(\cdot)$ uniformly and $\overline{\lim}_n T(\phi_n) = T < \infty$. Then, under (A3.1) - (A3.4) and (A1.3),

$$\underline{\lim}_n S_n(\phi_n) \geq S(\phi).$$

Proof: Assume w.l.o.g. (choose a subsequence if necessary), that $\overline{\lim}_n S_n(\phi_n) < \infty$, and that $T(\phi_n) \to T \geq T(\phi)$, $T < \infty$, and let $m(\cdot)$ denote Lebesgue measure. For any $\varepsilon > 0$, $m\{t : \phi_n(t) \notin N_\varepsilon(U(\phi_n(t))),\ t \leq T_n\} \to 0$ as $n \to \infty$, since $L_n(x,\beta) \to \infty$ uniformly in (β,x) in any compact subset of $\{\beta,\ x : \beta \notin N_\varepsilon(U(x))\}$. To see this, suppose that there are $\{x_n,\beta_n\}$ and $K < \infty$ such that $L_n(x_n,\beta_n) \leq K$, where $\beta_n \notin N_\varepsilon(U(x_n))$ and $x_n \to x$, $\beta_n \to \beta$. But $\beta \notin N_\varepsilon(U(x))$ and $\underline{\lim}_n L_n(x_n,\beta_n) \geq L(x,\beta) = \infty$ by Lemma 3.1, a contradiction. By this, the convexity of $U(x)$, continuity of $U(\cdot)$, and weak convergence of $\dot{\phi}_n(\cdot)$ to $\dot{\phi}(\cdot)$, we have $m\{t : \dot{\phi}(t) \notin \overline{U(\phi(t))},\ t \leq T\} = 0$; in fact it can be shown that $\overline{U(\phi(t))}$ can be replaced by $U(\phi(t))$ there.

Now recall the definition of the δ-interior set $\overline{U}^\delta(x)$ and define $U_\varepsilon(x) = N_\varepsilon(U(x))$ and let $I_\varepsilon^n(\cdot)$ be the indicator of the set on which $\dot{\phi}_n(s) \in U_\varepsilon(\phi_n(s))$. We have

$$\text{(3.1)}\quad B_1 \equiv \underline{\lim}_n \int_0^{T_n} L_n(\phi_n(s),\ \dot{\phi}_n(s))ds \geq \underline{\lim}_n \int_0^{T_n} L_n(\phi_n(s),\ \dot{\phi}_n(s))I_\varepsilon^n(s)ds \equiv B_2.$$

Let $\delta > \varepsilon$. For large n and small δ and ε, there is a measurable function (see [17], Theorem 4.2) $\Delta_n(\cdot)$ with $|\dot{\Delta}_n(t)| \leq 2\delta$ and such that $\dot{\phi}_n(t) + \dot{\Delta}_n(t) \in \overline{U}^\delta(\phi_n(t))$ for all t such that $\dot{\phi}_n(t) \in U_\varepsilon(\phi_n(t))$ and such that for these t and small δ and ε,

$$(*) \qquad L_n(\phi_n(t), \dot\phi_n(t) + \dot\Delta_n(t)) \le L_n(\phi_n(t), \dot\phi_n(t)) + \delta_1,$$

where $\delta_1 \to 0$ as $\delta \to 0$.

By (3.1) and (*)

$$B_2 \ge \underline{\lim}_n \int_0^{T_n} L_n(\phi_n(t), \dot\phi_n(t) + \dot\Delta_n(t)) I_\varepsilon^n(s)ds - \delta_1 T.$$

By Corollary 2.3,

$$\overline{\lim_\varepsilon}\ \overline{\lim_n} \sup_{\substack{|x-y|\le\varepsilon \\ \beta\in\overline{U}^\delta(x)}} |L_n(y,\beta) - L_n(x,\beta))| = 0, \ y,x \in \text{compact in } G_1.$$

Thus, by the uniform convergence $\phi_n(\cdot) \to \phi(\cdot)$, for each $\delta_0 > 0$ there is an $\varepsilon_0 > 0$ such that $|t - \tau| \le \varepsilon_0$ implies that for large n

$$|L_n(\phi_n(\tau),\beta) - L_n(\phi_n(t),\beta)| \le \delta_0, \ \beta \in \overline{U}^\delta(\phi_n(t)).$$

Define a finite sequence $\{t_i, i = 1, \ldots, q\}$ such that $t_{i+1} > t_i$, $t_{i+1} - t_i \le \varepsilon_0$, $t_0 = 0$, $t_q = T + \varepsilon_0$, and set $L_n(\phi_n(t), \dot\phi_n(t)) = 0$ for $t \ge T_n$. Then (the last inequality below uses Jensen's inequality and the convexity of $L_n(x,\cdot)$)

$$\begin{aligned} B_2 &\ge -\delta_1 T - \delta_0 T \\ (3.2) \qquad &+ \underline{\lim}_n \sum_i \int_{t_i}^{t_{i+1}} L_n(\phi_n(t_i), \dot\phi_n(s) + \dot\Delta_n(s)) I_\varepsilon^n(s)ds \\ &\ge -(\delta_1 + \delta_0)T + \underline{\lim}_n \sum_i (t_{i+1} - t_i) L_n(\phi_n(t_i), f_i^{n,\varepsilon}), \end{aligned}$$

where

$$f_i^{n,\varepsilon} \equiv \frac{1}{(t_{i+1} - t_i)} \int_{t_i}^{t_{i+1}} (\dot\phi_n(s) + \dot\Delta_n(s)) I_\varepsilon^n(s)ds.$$

Assume (or take a suitable subsequence) that $\Delta_n(\cdot)$ converges to a function $\Delta(\cdot)$. Define

$$f_i = \left[\frac{\phi(t_{i+1}) - \phi(t_i)}{t_{i+1} - t_i} + \frac{\Delta(t_{i+1}) - \Delta(t_i)}{t_{i+1} - t_i}\right].$$

Then $f_i^{n,\varepsilon} \to f_i$ as $n \to \infty$, for each $\varepsilon > 0$. By Lemma 3.1, (3.2) and the lower semicontinuity of $L(\cdot,\cdot)$ and its continuity on $\{x, \beta : \beta \in \overline{U}^\delta(x)\}$,

$$\begin{aligned} B_1 &\ge -T(\delta_1 + \delta_0) + \sum_i (t_{i+1} - t_i) L(\phi(t_i), f_i) \\ (3.3) \qquad &\xrightarrow[\varepsilon_0 \to 0]{} -T\delta_1 + \int_0^T L(\phi(s), \dot\phi(s) + \dot\Delta(s))ds. \end{aligned}$$

Finally, letting $\varepsilon \to 0$, $\delta \to 0$ and again using the lower semicontinuity of $L(\cdot,\cdot)$, yields the theorem. Q.E.D.

We next treat an unbounded $U(x)$ case.

(A3.5) Let (a) $\inf_{x \in G_1} \dfrac{L(x,\beta)}{|\beta|} \to \infty$ as $|\beta| \to \infty$

(b) (nondegenerate) $\sup_{|\beta| \le B} \sup_{x \in G_1} L(x,\beta) < \infty$, all $B < \infty$.

(degenerate), let $\beta_1 = \bar{b}_1(x)$, and take sup only over $|\beta_2| \le B$.

The conditions hold for (1.1),(1.2) if (non-degenerate case) $\sigma(x)\sigma'(x)$ is uniformly positive definite, and (degenerate case) if $\sigma_2(x)\sigma_2'(x)$ is uniformly positive definite.

Theorem 3.3: (Proof omitted). Under (A1.3),(A3.1),(A3.5), the conclusions of Theorem 3.2 hold.

Limits of $\{S_n\}$. The functional H_n corresponds to a system of one of the types (1.1),(1.2), with dynamical terms $b,\bar{b},\sigma$ subscripted by n and ξ_k^n replacing ξ_k, where the 'mean' dynamical term is $b_n(\cdot)$. As $n \to \infty$, $\bar{b}_n(x) \to \bar{b}(x)$ and many types of assumptions on the behavior of $\dot{x} = \bar{b}(x)$ can be dealt with. Here we simply assume (A3.6).

(A3.6) The system corresponding to H-functional H_n satisfies (A1.1), but where $\bar{x}_n$ replaces $\bar{x}_0$ and $\bar{x}_n \to \bar{x}_0$ as $n \to \infty$.

For the degenerate case, we need the 'controllability' condition (A3.7). In the non-degenerate case, with the unbounded $U(x)$, (A3.7) always holds if the conditions $|\dot{\phi}_2| \le M$ and $\dot{\phi}_1 = \bar{b}_1(\phi)$ are replaced by $|\dot{\phi}| \le M$. In the non-degenerate case with bounded $U(x)$, (A3.7) always holds if the condition $\dot{\phi}_2(t) \in \overline{U}_2^{\delta}(\phi(t))$ is replaced by $\dot{\phi}(t) \in \overline{U}^{\delta}(\phi(t))$.

(A3.7) (Unbounded $U(x)$ case.) There is an $M < \infty$ such that for each small $\varepsilon_2 > 0$ and each $y \in N_{\varepsilon_2}(\bar{x}_0)$, there is a function $\phi(\cdot) = (\phi_1(\cdot),\phi_2(\cdot))$ such that $\phi(0) = \bar{x}_0$, $\phi(t_y) = y$ for some $t_y \le T$, where $T \to 0$ as $\varepsilon_2 \to 0$, and $\dot{\phi}_1 = \bar{b}_1(\phi), |\dot{\phi}_2| \le M$.

(Bounded $U(x)$ case.) Simply replace M and $|\dot{\phi}_2| \le M$ by $\dot{\phi}_2(t) \in \overline{U}_2^{\delta}(\phi(t))$, for some $\delta > 0$.

Theorem 3.4: (Unbounded $U(x)$ case). Assume (A3.1),(A3.5),(A3.6) and (A3.7) (for the degenerate case) and (A1.1),(A1.3). Then

$$S_n \to S_0.$$

<u>Proof</u>: Fix $\varepsilon > 0$, let $S_0 < \infty$ and let $\phi^\varepsilon(\cdot)$ be an ε-optimal path for $S(\cdot)$ with $\phi^\varepsilon(0) = \bar{x}_0$. Write $T_\varepsilon = T(\phi^\varepsilon)$. For small $\varepsilon_3 > 0$, $\phi^\varepsilon(\cdot)$ can be selected such that it is defined until T'_ε, the exit time from $N_{\varepsilon_3}(G)$, and $S(T'_\varepsilon, \phi^\varepsilon) \le S_0 + 3\varepsilon$ and for some $K < \infty$, $|\phi^\varepsilon(t)| \le K$, and $\phi^\varepsilon(\cdot)$ is not tangent to any of the boundary curves at the exit point from G (see [17], proof of Theorem 4.4). In this part of the proof we do only the (more difficult) degenerate problem.

Define $\phi_n^\varepsilon = (\phi_{1n}^\varepsilon, \phi_{2n}^\varepsilon)$ by (in the non-degenerate case we would set $\phi_n^\varepsilon(0) = \bar{x}_n$, $\dot{\phi}_n^\varepsilon(t) = \dot{\phi}^\varepsilon(t)$)

$$\phi_{1n}^\varepsilon(t) = \bar{x}_{1n} + \int_0^t \bar{b}_{1n}(\phi_n^\varepsilon(s))ds$$

$$\phi_{2n}^\varepsilon(t) = \bar{x}_{2n} + \int_0^t \dot{\phi}_2^\varepsilon(s)ds,$$

where $\bar{x}_n$ is defined in (A3.6), and $\bar{b}_n = (\bar{b}_{1n}, \bar{b}_{2n})$. Recall that $\bar{b}(\cdot)$ is Lipschitz continuous. Then, by the properties of ϕ^ε assumed in the last paragraph, $T_n^\varepsilon = T(\phi_n^\varepsilon) < \infty$ for large n and $T_n^\varepsilon \to T$ as $n \to \infty$. By the boundedness of $\phi^\varepsilon(\cdot)$ and the uniform convergence of $L_n(x;\bar{b}_{1n}(x),\beta_2)$ to $L(x;\bar{b}_1(x),\beta_2)$ on bounded (x,β_2) sets,

$$\text{(3.7a)} \qquad S_n \equiv S_n(\bar{x}_n) \le S_n(T_n^\varepsilon, \phi_n^\varepsilon) \le \int_0^{T_n^\varepsilon} L_n(\phi_n^\varepsilon(s);\bar{b}_{1n}(\phi_n^\varepsilon(s)),\dot{\phi}_2^\varepsilon(s))ds$$

$$\longrightarrow \int_0^{T_\varepsilon} L(\phi^\varepsilon(s);\bar{b}_1(\phi^\varepsilon(s)),\dot{\phi}_2^\varepsilon(s))ds \le S_0 + 3\varepsilon.$$

Thus

$$\text{(3.7b)} \qquad \overline{\lim_n}\, S_n \le S_0.$$

Now, to get the reverse inequality to (3.7b) for either the degenerate or the non-degenerate case, let $\sup_n S_n < \infty$ and let $\phi_n^\varepsilon(\cdot)$ be the ε-optimal path for $S_n(\phi)$. We can select $\phi_n^\varepsilon(\cdot)$ such that $T_n^\varepsilon = T(\phi_n^\varepsilon) \to T < \infty$. Let $I_\varepsilon^{K,n}(\cdot)$ denote the indicator of the set where $|\dot{\phi}_n^\varepsilon(t)| \ge K$. By (A3.5), the convexity of $L_n(x,\cdot)$ and $L(x,\cdot)$ and the uniform convergence on bounded sets, for each large $N < \infty$ there is a $K_N < \infty$ such that

$$S_n(\phi_n^\varepsilon) \ge N \int_0^{T_n^\varepsilon} |\dot{\phi}_n^\varepsilon(t)| I_\varepsilon^{K_N,n}(t)dt$$

for large n. Thus, the set $\{\phi_n^\varepsilon(\cdot),\ n \text{ large},\ \varepsilon > 0\}$ is uniformly absolutely continuous. Extract a convergent subsequence, indexed by n, and with limit $\bar{\phi}^\varepsilon(\cdot)$, where $\bar{\phi}^\varepsilon(0) = \bar{x}_0$. By Theorem 3.3,

$$\text{(3.10)} \qquad \varepsilon + \underline{\lim_n}\ S_n(\bar{x}_n) \ge \underline{\lim_n}\ S_n(\phi_n^\varepsilon) \ge S(\bar{\phi}^\varepsilon) \ge S_0,\ \underline{\lim_n}\ S_n \ge S_0.$$

Thus, $S_n \to S_0$. Q.E.D.

A useful special case is given by Theorem 3.5. See also Theorem 3.6.

Theorem 3.5: Let the H-functionals satisfy $H_n(x,\alpha) \downarrow H(x,\alpha)$, each x,α. Then $S_n \leq S_0$ and under the conditions of Theorems 3.2 or 3.3, $S_n \to S_0$ as $n \to \infty$.

The theorem is obvious, since $L_n(x,\beta) \leq L(x,\beta)$. A case of particular interest is where $b(x,\xi_n) = \tilde{b}(x,\xi_n) + \hat{b}_n(x,\xi_n)$, and $\{\tilde{\xi}_n\}$ and $\{\hat{\xi}_n\}$ are independent of one another and $\hat{H}_n(x,\alpha) \to 0$, uniformly on bounded (x,α) sets (where $\hat{H}$ and $\tilde{H}$ are the H-functionals corresponding to $\hat{b}$ and $\tilde{b}$, respectively). Then if the system corresponding to $\tilde{b}$ satisfies the conditions of Theorems 3.2 or 3.3, $S_n \to \tilde{S}_0$ as $n \to \infty$.

The H-function for (1.1) or (1.2) takes the form (where H^0 is the H-functional for $\sigma = 0$)

$$H^{\sigma}(x,\alpha) = H^0(x,\alpha) + \alpha'\sigma(x)\sigma'(x)\alpha/2.$$

Theorem 3.6: Let $H_n^{\sigma}(x,\alpha) = H_n^0(x,\alpha) + \alpha'\sigma(x)\sigma'(x)\alpha/2$, where we assume the conditions of Theorem 3.4 with H^{σ} and H_n^{σ} replacing H and H_n resp. Then $S_n^{\sigma} \to S_0^{\sigma}$ as $n \to \infty$. Furthermore, if H^0 satisfies (A3.1 to 4), (A1.3) in the bounded $U(x)$ case or (A1.3),(A3.1),(A3.3) in the unbounded $U(x)$ case, then $S_0^{\sigma} \to S_0^0$ as $\sigma \to 0$.

The theorem follows from Theorems 3.2 to 3.5. Thus, when the system contains (independent) Gaussian noise, the exit times are robust with respect to changes in the other system noises. Also, the addition of small Gaussian noise changes the exit times only slightly under broad conditions. The remaining theorems are stated without proof.

Theorem 3.7: (Bounded $U(x)$). Assume (A1.1,3) and (A3.1,2,3,4,6,7). Suppose that for each $\varepsilon > 0$, there is a $\delta > 0$ such that there is an ε-optimal path $\phi^{\varepsilon}(\cdot)$ (with $\phi^{\varepsilon}(0) = \bar{x}_0$) for $S(\cdot)$, with $\phi^{\varepsilon}(t) \in \overline{U^{\delta}}(\phi^{\varepsilon}(t))$. Then $S_n \to S_0$ as $n \to \infty$. Under (A3.1 to 3) and (A1.1 to 4), there is such an ε-optimal path.

4. Examples of convergence of H_n to Gaussian H-functional.

4.1 Let $\bar{b}_n(\cdot)$ and $b_n(\cdot,\xi)$ be Lipschitz continuous, uniformly in ξ. Let $N_n \to \infty$ as $n \to \infty$, and let $\{\xi_{ki}, i \geq 0, k \geq 0\}$ be i.i.d. with $E\,\tilde{b}_n(x,\xi_{ki}) = 0$, and define $\tilde{b}_n(x,\xi_k^n) = \sum_{i=1}^{N_n} \tilde{b}_n(x,\xi_{ki})/\sqrt{N_n}$. Define $\{x_k^{\gamma}\}$ by (suppress the n index on x_k^{γ}) by

$$x_{k+1}^{\gamma} = x_k^{\gamma} + \gamma\bar{b}_n(x_k^{\gamma}) + \gamma\tilde{b}_n(x_k^{\gamma},\xi_k^n). \tag{4.1}$$

Let $\tilde{H}_n$ denote the H-functional when $\bar{b}_n(x) = 0$. For convergence to the Gaussian H-functional $H(x,\alpha) = \alpha'\bar{b}(x) + \alpha'\Sigma(x)\alpha/2$ we need $\bar{b}_n(x) \to \bar{b}(x)$ and

$$\tilde{H}_n(x,\alpha) = N_n \log E \exp \alpha'\tilde{b}_n(x,\xi_{ki})/\sqrt{N_n} \to \alpha'\Sigma(x)\alpha/2,$$

uniformly in $x \in G_1$ for some smooth $\Sigma(x)$. If the ξ_{ki} are bounded, then clearly $\Sigma(x) = \lim_{n\to\infty} E\,\tilde{b}_n(x,\xi_{ki})\tilde{b}_n'(x,\xi_{ki})$. But in general, the convergence or lack of it depends on the higher moments of $\tilde{b}_n(x,\xi_{ki})$.

4.2. Now let $\tilde{b}_n(x,\xi_k^n) = \sum_{i=1}^{N_n} \tilde{b}_n(x,\xi_{ki}^n)$, where N_n is Poisson with parameter λ_n, and for each n, $E\,\tilde{b}_n(x,\xi_{ki}^n) = 0$ and $\{\xi_{ki}^n,\ k \geq 0,\ i \geq 0\}$ are i.i.d. for each n. Then

$$\tilde{H}_n(x,\alpha) = \lambda_n[E \exp \alpha'\tilde{b}_n(x,\xi_{ki}^n) - 1].$$

Let $\lambda_n \to \infty$ and $\lambda_n E\,\tilde{b}(x,\xi_{ki}^n)\tilde{b}'(x,\xi_{ki}^n) \to \Sigma(x)$ uniformly in $x \in G_1$, as $n \to \infty$. Then, for H_n to converge to the Gaussian H-functional, it is sufficient that $\bar{b}_n(x) \to \bar{b}(x)$ and that $\lambda_n \sum_{\ell=3}^{\infty} |\alpha|^\ell E|\tilde{b}_n(x,\xi_{ki}^n)|^\ell/\ell! \to 0$ uniformly in bounded α-sets, as $n \to \infty$. This depends on the higher moments of $\tilde{b}_n(x,\xi_{ki}^n)$. If $|\tilde{b}_n(x,\xi_{ki}^n)| \leq \delta_n \to 0$ as $n \to \infty$ and $\overline{\lim_n}\, \lambda_n\delta_n^2 < \infty$, then the sum converges to zero as desired.

4.3. Consider the continuous parameter case

$$dx^\gamma = \bar{b}(x^\gamma)\,dt + \sigma(x^\gamma)dJ^n(t/\gamma), \tag{4.2}$$

where $J^n(\cdot)$ is a centered Poisson jump process with rate λ_n and jump random variables $\{\xi_i^n\}$. Then

$$\begin{aligned} H_n(x,\alpha) &= \alpha'\bar{b}(x) + \lim_{\gamma\to 0} \gamma \log E \exp \alpha'\int_0^{1/\gamma} \sigma(x)dJ^n(t) \\ &= \alpha'\bar{b}(x) + \lambda_n[E \exp \alpha'\sigma(x)\xi_i^n - 1], \end{aligned} \tag{4.3}$$

and the comments made in the discrete parameter case also apply here.

4.4. Let $J(\cdot)$ be a jump process with jump intervals $c > 0$ and bounded i.i.d. jump random variables $\{\psi_i\}$ with $E\psi_i = 0$ and consider the system

$$\dot{x}^\gamma = \bar{b}(x^\gamma) + v(x^\gamma)\xi(t/\gamma), \tag{4.4}$$

where $\xi(\cdot)$ is the filtered noise

$$\xi(t) = \int_0^t h(t-s)dJ(s). \tag{4.5}$$

For computational simplicity, let $h(s) = \exp - as$, $a > 0$ and set $K_\gamma = 1/c\gamma$. Then

$$\int_0^{1/\gamma} \xi(t)dt = \int_0^{1/\gamma} dt \int_0^t h(t-s)dJ(s)$$

$$= \int_0^{1/\delta} dJ(s) \int_s^{1/\gamma} h(t-s)dt = \sum_{i\leq 1/c\gamma} \frac{\psi_i}{a}\,[1 - \exp - ac(\tfrac{1}{c\gamma} - i)]\,.$$

Thus

$$\lim_{\gamma\to 0} \gamma \log E \exp \alpha v(x) \int_0^{1/\gamma} \xi(s)ds$$

$$(4.6) \qquad = \lim_{\gamma\to 0} \gamma \log \left(E \exp \frac{\alpha v(x)}{a} \psi\right)^{K_\gamma}$$

$$= \frac{1}{c} \log E \exp \frac{\alpha v(x)}{a} \psi.$$

Now, replace (c,a,ψ_i) by (c_n,a_n,ψ_i^n), let $|\psi_i^n| \leq \delta_n$, where $\delta_n/a_n \to 0$ as $n \to \infty$. Let $\lim E(\psi^n)^2/a_n^2c_n = u^2 > 0$. Then as $n \to \infty$, (4.6) converges to the Gaussian form $\alpha^2v^2(x)u^2/2$. If the deterministic intervals c were replaced by i.i.d. and exponentially distributed intervals, the (4.4),(4.5) would be close to actual physical noise models. We expect that the same conclusions would hold in this case, suggesting that the Gaussian approximation is indeed useful for a large class of physical noise models.

REFERENCES

[1] R. Azencott, "Grandes deviations et applications", Lecture Notes in Math., No. 774, 1980, Springer-Verlag, New York.

[2] R. Azencott, G. Ruget, "Melanges d'equations differentialles et grands écartes à la loi des grand nombres", A. Wahr. 38, 1977, pp. 1-54.

[3] A. D. Ventsel, M. I. Freidlin, "Some problems concerning stability under small random perturbations", Theory of Prob. and Applic., 17, 1972, pp. 269-283.

[4] A. D. Ventsel, "Rough limit theorems on large deviations for Markov processes", I, II, Theory of Prob. andApplic., 21, 1976, pp. 227-242.

[5] M. I. Freidlin, "The averaging principle and theorems on large deviations", Russian Math. Surveys, 33, July - Dec., 1978, pp. 117-176.

[6] M. Donsker, S. R. S. Varadhan, "Asynptotic evaluation of certain Markov process expectations for large time", I, II, III, Comm. Pure Appl. Math. 28 , 1975, pp. 1-47, 229-301, 389-461.

[7] S. Karlin, "Positive operators", J. Math. and Mech., 8, 1959, pp. 907-937.

[8] T. Kato, Perturbation Theory for Linear Operators, 2nd ed., Springer, Berlin, 1976.

[9] N. Dunford, J. T. Schwartz, Linear Operations, Part I, Wiley, New York, 1966.

[10] A. J. Viterbi, Principles of Coherent Communication, McGraw-Hill, New York, 1966.

[11] F. L. Gardner, Phaselock Techniques, 2nd ed., 1979, Wiley, New York.

[12] Z. Schuss, Theory and Applications of Stochastic Differential Equations, 1980, Wiley, New York.

[13] D. Ludwig, "Persistence of dynamical systems under random perturbations", SIAM Rev. 17, 1975, pp. 605-640.

[14] G. Ruget, "Some tools for the study of channel-sharing algorithms", CISM publication, ed. Longo, Springer, 1980.

[15] H. J. Kushner, "A cautionary note on the use of singular perturbation methods for 'small noise' models", Stochastics, 6, 1982, pp. 117-120.

[16] B. Z. Bobrovsky, A. Schuss, "A singular perturbation method for the computation of the mean first passage time in a non-linear filter", SIAM J. on Appl. Math., 42, 1982, p. 174-187.

[17] H. J. Kushner, "Robustness and approximations of escape times and large deviations estimates for systems with small noise effects", LCDS Report 1982, Lefschetz Center for Dynmaical Systems, Brown Unviersity.

ON STRONG SOLUTIONS OF STOHASTIC EQUATIONS WITH RESPECT TO SEMIMARTINGALES

A.V.Melnikov
Steklov Mathematical Institute, Moscow

In the theory of stochastic integral equations with respect to semimartingales the existence of strong solutions is known under Liptshitz-type conditions (see [1] - [5]).The purpose of the paper is to investigate the existence and the weak convergence of strong solutions of stochastic equations with respect to semimartingales with nonliptshitz-type coefficients.

Let (Ω , $\mathcal{F}$, $\mathbb{F}=(\mathcal{F})_{t\geqslant 0}$, P) be a filtered probability space with usial assumptions.

I.EXISTENCE OF STRONG SOLUTIONS

Consider the stohastic equation

$$\xi_t = \xi_0 + \int_0^t f(\xi_s)dA_s + \int_0^t g(\xi_s)dM_s , \tag{I}$$

where ξ_0 is a finite $\mathcal{F}_0$ - measurable random variable, $A=(A_t)_{t\geqslant 0}$ is a continuous increasing process, $M=(M_t)_{t\geqslant 0}$ is acontinuous local martingale, $f(x)$, $g(x)$ are continuous real function, $x\in R^1$.

<u>Theorem I.</u> Let $f(x)$ be a continuous bounded function, $g(x)$ satisfies Liptshitz condition : for each $x, y \in R^1$

$$|g(x)-g(y)| \leqslant G|x-y|.$$

Then the equation (I) have a strong solution.

Proof. Using V.A.Steklov smoothing method by means of functions

$$f_h(x) = \frac{1}{h}\int_0^h f(x+y)dy , \; h>0,$$

we construct two sequences of local Liptshitz functions $(f_n^{(i)}(x))_{n\geqslant 1}$ such that

$$f_n^{(1)} \downarrow\downarrow f , \; f_n^{(2)} \uparrow\uparrow f , \; n\uparrow\infty .$$

Now consider two sequences of stochastic equations ($i=1,2$)

$$\xi_n^{(i)}(t) = \xi_0 + \int_0^t f_n^{(i)}(\xi_n^{(i)}(s))dA_s + \int_0^t g(\xi_n^{(i)}(s))dM_s . \tag{2i}$$

It is well-known (see [1] - [5]) that these equations have continuous strong solutions $\xi_n^{(i)}(t,\omega)$ and by virtue of the comparison theorem (see [5]) for each $t \geqslant 0$ (P-a.s.)

$$(3) \quad -\infty < \ldots \leqslant \xi_n^{(2)}(t,\omega) \leqslant \ldots \leqslant \xi_n^{(1)}(t,\omega) < \infty .$$

Denote by $\xi^{(i)}(t,\omega), i=1,2$ the limits of the sequences.
Now we show that the processes satisfy the stochastic equation (1). Note here that (P-a.s.) for each $t \geqslant 0$ $\xi^{(1)}(t,\omega) \geqslant \xi^{(2)}(t,\omega)$. By virtue of the following relation

$$\xi^{(i)}(t,\omega) = mes \times P\text{-}\lim_n \xi_n^{(i)}(t,\omega) , \quad i=1,2 ,$$

processes $\xi^{(i)}(t,\omega)$ are predictable.Then it follows from conditions of the theorem 1 and (3) the following integrales are well defined

$$\int_0^t f(\xi^{(i)}(s))dA_s , \quad \int_0^t g(\xi^{(i)}(s))dM_s , \quad i=1,2.$$

Denote by $(\gamma_m)\ m \geqslant 1$ a sequence of stopping times from definition of the local martingale M .Let $\tau_m = \sigma_m \wedge \gamma_m \uparrow \infty,\ m\uparrow\infty$, where

$$\sigma_m = \inf\{t : \xi_1^{(1)+}(t) \vee \xi_1^{(2)-}(t) > m\} .$$

Then by properties of stochastic integrales

$$(4) \quad E\Big(\int_0^{t\wedge\tau_m} g(\xi_n^{(i)})dM_s - \int_0^{t\wedge\tau_m} g(\xi^{(i)})dM_s\Big)^2 = E\int_0^{t\wedge\tau_m}\big(g(\xi_n^{(i)}) - g(\xi^{(i)})\big)^2 d\langle M\rangle_s .$$

We have (because g is a continuous function) that (P-a.s.)

$$g(\xi_n^{(i)}(t,\omega)) \xrightarrow[n]{} g(\xi^{(i)}(t,\omega)) .$$

By the dominated convergence theorem the last integral in (4) tends to zero for each $m \geqslant 1$ and we get from (4) that

$$(5) \quad \int_0^t g(\xi_n^{(i)}(s,\omega))dM_s \xrightarrow[n]{P} \int_0^t g(\xi^{(i)}(s,\omega))dM_s .$$

Now we prove that for each $t \geqslant 0$, $\omega \in \Omega'$, $P(\Omega')=1$,

$$(6) \quad \int_0^t f_n(\xi_n^{(i)}(s,\omega))dA_s(\omega) \xrightarrow[n]{} \int_0^t f(\xi^{(i)}(s,\omega))dA_s(\omega) .$$

It is suffitient to prove that

$$(7) \quad f_n(\xi_n^{(i)}(s,\omega)) \xrightarrow[n]{} f(\xi^{(i)}(s,\omega)) .$$

For each $\omega \in \Omega'$ we have ($i=1,2$)

$$(8) \quad |f_n(\xi_n^{(i)}(s,\omega)) - f(\xi^{(i)}(s,\omega))| \le |f_n(\xi_n^{(i)}(s,\omega)) - f(\xi_n^{(i)}(s,\omega))| + |f(\xi_n^{(i)}(s,\omega)) - f(\xi^{(i)}(s,\omega))| = I_1^i(n) + I_2^i(n).$$

It is clear that $I_2^i(n) \longrightarrow 0$ .Now by virtue of the uniform convergence $f_n^{(i)}(x) \xrightarrow[n]{} f^n(x)$ on each compact we get that

$$|I_1^1(n)| \le \max_{x \in [\xi_1^{(1)}(s,\omega),\, \xi^{(1)}(s,\omega)]} |f_n^{(1)}(x) - f(x)| \xrightarrow[n]{} 0,$$

$$|I_1^2(n)| \le \max_{x \in [\xi^{(2)}(s,\omega),\, \xi_1^{(2)}(s,\omega)]} |f_n^{(2)}(x) - f(x)| \xrightarrow[n]{} 0.$$

Then (6) follows from (7) and (8).It means together with (5) that are strong solutions of the stochastic equation (1).

The theorem 1 is established.

<u>Remark 1</u>.The theorem 1 is an analog of Zvonkins theorem (see [6]) for stochastic equations with respect to semimartingales.Our approach to the proof of the existence of strong solutions (on the base of the comparison theorem) was presented in the paper [11] for diffusion equations.Here we consider equations with continuous coefficients.But by the presented method we can prove the existence of strong solutions of the equation (1) in the case when f is a bounded discontinuous function ander an additional condition (see [12]).

II.WEAK CONVERGENCE OF STRONG SOLUTIONS

Let us give stochastic equations

$$(9) \quad \xi_t^n = \xi_0 + \int_0^t f_n(s,\xi^n)\,dA_s + \int_0^t g(s,\xi^n)\,dM_s,$$

$$(10) \quad \xi_t = \xi_0 + \int_0^t f(s,\xi)\,dA_s + \int_0^t g(s,\xi)\,dM_s,$$

where A is a nonrandom continuous finite variation function; M is a continuous Gaussian martingale with the characteristic $\langle M \rangle$; $f(s,x)$, $g(s,x)$, $f_n(s,x)$ – $\mathcal{B}_s$ - measurable funtionales on the space $C[0,\infty)$ and $\mathcal{B}_s = \sigma\{x \in C[0,\infty): x_u,\ u \le s\}$. Denote $\xrightarrow{P}$ ($\xrightarrow{\mathcal{D}}$) a convergence in probability (a weak convergence of finite dimentional distributions).

The following theorem gives sufficient conditions (in terms of coefficients) for the weak convergence of strong solutions of stochastic equations (9) and (10).

Theorem 2. Let ξ^n, ξ be strong and unique solutions of stochastic equations (9) and (10) respectively. Let there are $\mathcal{B}_t$- measurable functionales $\gamma_n(t,x)$ such that for each $t>0$

1) $\int_0^t g(s,\xi^n)\gamma_n(s,\xi^n)d\langle M\rangle_s = \int_0^t [f(s,\xi^n)-f_n(s,\xi^n)]dA_s$;

2) $\int_0^\infty \gamma_n^2(s,\xi^n)d\langle M\rangle_s < \infty$;

3) $\sup_n E\int_0^t \gamma_n^2(s,\xi^n)d\langle M\rangle_s < \infty$;

4) $\int_0^t \gamma_n^2(s,\xi^n)d\langle M\rangle_s \xrightarrow[n]{P} 0$.

Then $\xi^n \xrightarrow[n]{\mathcal{D}} \xi$.

Proof. Let μ^n and μ be distributions of the processes ξ^n and ξ and μ^n_T, μ_T be their restrictions to the σ-algebra $\mathcal{B}_T$. Using conditions 1) - 3) and the uniqueness of the solutions ξ^n and ξ we can prove (see [7] - [8]) that

$$Z^n_\infty(\xi^n) = \frac{d\mu}{d\mu^n}(\xi^n) = \exp\left\{\int_0^\infty \gamma_n(s,\xi^n)dM_s - \frac{1}{2}\int_0^\infty \gamma_n^2(s,\xi^n)d\langle M\rangle_s\right\},$$

(11)

$$Z^n_T(\xi^n) = \frac{d\mu_T}{d\mu^n_T}(\xi^n) = \exp\left\{\int_0^T \gamma_n(s,\xi^n)dM_s - \frac{1}{2}\int_0^T \gamma_n^2(s,\xi^n)d\langle M\rangle_s\right\}.$$

Now it is sufficient to prove using Cramer-Worlds method (see [9]) that for each $T>0$ and each bounded function of the type

$$F(t) = \sum_{K=1}^N I_{(s_K,t_K]}(t), \quad t_N \le T, \quad N \ge 1,$$

(12) $$E\exp\left\{-\int_0^T F_s d\xi^n_s\right\} \xrightarrow[n]{} E\exp\left\{-\int_0^T F_s d\xi_s\right\}.$$

The required relation (12) follws from

(13) $$E\left|Z^n_T(\xi^n)-1\right| \xrightarrow[n]{} 0$$

because of

$$E\exp\left\{-\int_0^T F_s d\xi_s\right\} = E\exp\left\{-\int_0^T F_s d\xi^n_s\right\} Z^n_T(\xi^n).$$

The assertion (13) follows by the lemma 6.7 [8] from

$$(14) \qquad Z_T^n(\xi^n) \xrightarrow[n]{P} 1 .$$

By Tchebyshovs inequality we have

$$P\left\{ \left| \int_0^T \gamma_n(s,\xi^n)\,dM_s \right| > \varepsilon \right\} \le \frac{1}{\varepsilon^2} E \int_0^T \gamma_n^2(s,\xi^n)\,d\langle M\rangle_s .$$

Using conditions 3),4) we get the relation (14).

The proof of the theorem 2 is finished.

Remark 2. For diffusion type equations we have $A_t = \langle M\rangle_t = t$ and $\gamma_n(t,x) = g^{-1}(t,x)\left[f(t,x) - f_n(t,x)\right]$. Therefore in the case of uniform bounded coefficients of equations (9) and (10) the condition 4) of the theorem 2 is reduced to the following condition : for each $t>0$

$$\int_0^t |f_n(s,\xi^n) - f(s,\xi^n)|\,ds \xrightarrow[n]{P} 0 .$$

REFERENCES

1. Doleans-Dade C. Existence and unicity of solutions of stochastic integral equations, Z.W-theory, 36(1976), 93-102.
2. Protter P.D. On the existence, uniqueness, convergence and explosions of solutions of stochastic integral equations, Ann.Probab.5,2(1977), 243-261.
3. Galtchouk L.I. On existence and uniqueness of solutions of stochastic equations with respect to semimartingales, Theory Probab.Appl. 23,4(1978), 782-795.
4. Krylov N.V., Rozovsky B.L. On evolutional stochastic equations, Modern problems of mathematics 14, Moscow, 1979, 71-146.
5. Melnikov A.V. On the theory of stochastic equations with respect to semimartingales, Math.sbornik 110,3(1979), 414-427.
6. Zvonkin A.K. Transformations of a diffusion process state space annihilating a drift, Math.sbornik 93,1(1974), 129-149.
7. Kabanov Yu.M., Liptser R.Sh., Shiryaev A.N. On absolute continuity of probability measures for Markov-Ito processes, Lectures Notes in Control and Inform.Sciences 25(1980), 114-126.
8. Liptser R.Sh., Shiryaev A.N. Statistics of random processes, Moscow, Nauka, 1974.
9. Billingsley P. Convergence of probability measures, Moscow, Nauka, 1977.

10.Kabanov Yu.M.,Liptser R.Sh.,Shiryev A.N. Some limit theorems for simple point processes (martingale approach),Stochastics 3,3(1980), 203-216.

11.Melnikov A.V.On strong solutions of stochastic differential equations with nonsmooth coefficients,Theory Probab.Appl.24,1(1979),146-149.

12.Melnikov A.V.On the existence of strong solutions of stochastic equations with respect to semimartingales,Third Vilnius Conference on Probability Theory and Mathematical Statistics,Vilnius,1981, 45-46.

INVERSE PROBLEMS IN STOCHASTIC RIEMANNIAN GEOMETRY

Mark A. Pinsky
Northwestern University
Evanston, IL 60201/USA

1. Introduction We first give a non-stochastic prototype of the problem under consideration here. Suppose that M is a surface imbedded in three-dimensional space R^3 such that every geodesic disc of radius ε has area $\pi\varepsilon^2$; then M is flat, in the (intrinsic) sense that every small disc can be mapped isometrically to a flat disc in the plane R^2. Indeed, the area of every small disc determines the Gaussian curvature which in turn gives the required isometry.

In trying to generalize this result to higher dimensional manifolds, geometers have encountered difficulties--beginning with the hypothesis that the volume of every geodesic ball of radius ε is the same as the volume of the ball of radius ε in Euclidean space R^n. Therefore it is natural to replace the volume by other measurements. In this report we give results on the first and second moments of the exit time of Brownian motion, obtaining more effective characterizations of Euclidean and rank one symmetric spaces. Much of the work is in collaboration with Alfred Gray [4], to whom go thanks for many useful conversations.

2. Statement of Results

Let M be an n-dimensional Riemannian manifold with Laplace-Beltrami operator Δ. The Brownian motion (X_t,P_x) is a diffusion process with infinitesimal generator Δ. This can be constructed, for example, by solving a stochastic differential equation on the orthonormal frame bundle of M [5]. The generator and the process are connected by Dynkin's identity [3], expressed as follows.

$$f(x) - E_x f(X_T) = - E_x \int_0^T \Delta f(X_s)ds$$

where f is a smooth function on M and T is a stopping time with $E_x(T)$ finite.

Let T_ε be the exit time from the geodesic ball of radius ε centered at $m \in M$:

$$T_\varepsilon = \inf \{t > 0: d(X_t,m) = \varepsilon\}$$

and let the normalized moments be defined by

Supported by Air Force Office of Scientific Research AFOSR 80-0252A.

$$u_0 = 1$$

$$u_k(x) = E_x\,(T_\varepsilon^k)/k! \qquad (k = 1,2,\ldots)$$

u_k coincides with the classical solution of $\Delta u_k = -u_{k-1}$ with the boundary condition that $u_k = 0$ on the boundary, $k = 1,2,\ldots$ Therefore, computation of these moments is reduced to solving these partial differential equations. In case M is a Euclidean or rank one symmetric space these equations have radial solutions $u = u(r)$, $r = d(x,m)$ which can be used to give closed formulas for the moments of the exit time [4].

On a general Riemannian manifold a closed form solution is not available. In the following section we develop a perturbation theory for the Laplacian on small geodesic balls. This is used to prove the following asymptotic expansions.

Theorem 2.1 We have for small $\varepsilon > 0$

$$E_m(T_\varepsilon) = c_0\varepsilon^2 + c_1\tau\,\varepsilon^4 + \varepsilon^6[c_2|R|^2 + c_3|\rho|^2 + c_4\tau^2 + c_5(\Delta\tau)] + O(\varepsilon^8)$$

$$E_m(T_\varepsilon^2) = d_0\varepsilon^4 + d_1\tau\,\varepsilon^6 + \varepsilon^8[d_2|R|^2 + d_3|\rho|^2 + d_4\tau^2 + d_5(\Delta\tau)] + O(\varepsilon^{10})$$

Here $c_0,\ldots,d_5$ are constants which depend only on $n = \dim M$; R, ρ, τ are respectively the curvature tensor, the Ricci tensor and the scalar curvature, computed at $m\varepsilon M$. (cf. Section 4)

Using the exact values of the constants, we have

Corollary 2.2 Suppose that M is a Riemannian manifold such that for every $m\varepsilon M$ and every $\varepsilon > 0$, $E_m(T_\varepsilon) = \varepsilon^2/2n$. Then M is flat provided that any of the following hold

(i) $n = \dim M \leq 5$
(ii) M is Einstein, or more generally
(iii) M has non-negative (or non-positive) Ricci curvature

One may ask if the restrictions imposed by Corollary 2.2 might be removed if one could compute the coefficient of ε^8 in the expansion of $E_m(T_\varepsilon)$. The following example shows that this program is fruitless.

Example 2.3 Let $M = G\times G^c$ where G is a compact semi-simple Lie group and G^c denotes the non-compact dual of G. Then $E_m(T_\varepsilon) = \varepsilon^2/2n + O(\varepsilon^{10})$

Finally, we have a positive result which is valid in all dimensions. Let $V_m(T_\varepsilon) = E_m(T_\varepsilon^2) - E_m(T_\varepsilon)^2$ be the variance of the exit time from the geodesic ball of radius ε.

Corollary 2.4 Suppose that M is a Riemannian manifold such that for every $m\varepsilon M$, and every $\varepsilon > 0$,

$$V_m(T_\varepsilon) = \varepsilon^4/2n^2(n+2)\ .$$

Then M is flat.

3. Perturbation theory of the Laplacian

Let M_m be the tangent space at $m \varepsilon M$, $\exp_m$ the exponential mapping $\exp_m\colon M_m \to M$. We define a mapping on functions by

$$(S_\varepsilon f)(\exp_m \varepsilon x) = f(x) \qquad\qquad f \varepsilon C^\infty(M_m)$$

The map $f \to S_\varepsilon f$ is a linear isometry in the sup-norm from $C^\infty(\overline{B}_1)$ to $C^\infty(B_\varepsilon)$ where $\overline{B}_1$ is the unit ball in M_m and B_ε is the geodesic ball of radius ε in M. The following proposition defines the successive "curvature corrections" to the Laplacian.

Proposition 3.1 *For every* $f \varepsilon C^\infty(M_m)$ *and* $N = 1,2,\ldots$ *we have*

$$S_\varepsilon^{-1} \Delta S_\varepsilon f = \varepsilon^{-2}\Delta_{-2} f + \Delta_0 f + \ldots + \varepsilon^N \Delta_N f + O(\varepsilon^{N+1})$$

The operators $\Delta_{-2}, \Delta_0, \ldots$ which are invariantly defined, may be computed in normal coordinates $(x_1,\ldots,x_n)$ as follows:

$$\Delta_{-2} f = \sum_{i=1}^{n} \frac{\partial^2 f}{\partial x_i^2} \qquad\qquad \text{(the Euclidean Laplacian in } M_m)$$

$$\Delta_0 f = \frac{1}{3} \sum_{i,j,a,b} R_{iajb}\, x_a x_b \frac{\partial^2 f}{\partial x_i \partial x_j} - \frac{2}{3} \sum_{i,a} \rho_{ia}\, x_a \frac{\partial f}{\partial x_i}$$

In general Δ_k is a second order differential operator which maps homogeneous polynomials of degree j to homogeneous polynomials of degree $k + j$.

To find an approximate solution of $\Delta f = -1$. we try $f = S_\varepsilon(\varepsilon^2 f_0 + \varepsilon^4 f_2 + \varepsilon^5 f_3 + \varepsilon^6 f_4)$ where

$$\Delta_{-2} f_0 = -1$$

$$\Delta_{-2} f_2 + \Delta_0 f_0 = 0$$

$$\Delta_{-2} f_3 + \Delta_1 f_0 = 0$$

$$\Delta_{-2} f_4 + \Delta_0 f_2 + \Delta_2 f_o = 0$$

f_0, f_2, f_3, f_4 are required to be zero on the boundary of $\overline{B}_1$.
Explicit computation yields

$$f_0 = (1 - r^2)/2n$$

$$f_2 = \frac{1}{6n(n+4)} \rho \, (1-r^2) + \frac{\tau}{6n^2(n+4)} (1-r^2) - \frac{\tau}{12n(n+2)(n+4)} (1-r^4)$$

where $r^2 = \sum_{i=1}^{n} x_i^2, \rho = \sum_{i,j=1}^{n} \rho_{ij} x_i x_j$ and $\tau = \sum_{i=1}^{n} \rho_{ii}$. We have $f_3(0) = 0$ from parity. The explicit form of f_4 is not needed, since we only need $f_4(0)$, which can be computed from $\Delta_0 f_2 + \Delta_2 f_0$ and the solution of the Poisson equation in R^n. With some work, we have

$$f_4(0) = \frac{1}{180n^2(n+2)} [|R|^2 - |\rho|^2 + (5/n)\tau^2 + 6(\Delta\tau)]$$

Comparing this with the statement of theorem 2.1, we have

$$c_0 = f_0(0) = 1/2n \qquad c_1 = f_2(0) = 1/12n^2(n+2)$$

$$c_2 = 1/180n^2(n+2) \qquad c_3 = -1/180n^2(n+2)$$

$$c_4 = 1/36n^3(n+2) \qquad c_5 = 1/30n^2(n+2)$$

Dynkin's formula shows that $f(x) = E_x(T_\varepsilon) + O(\varepsilon^8)$, which gives the first part of Theorem 2.1.

To prove the second part, we look for an approximate solution g of the equation $\Delta g = -f$, in the form

$$g = S_\varepsilon(\varepsilon^4 g_0 + \varepsilon^6 g_2 + \varepsilon^7 g_3 + \varepsilon^8 g_4)$$

where

$$\Delta_{-2} g_0 = -f_0$$

$$\Delta_{-2} g_2 + \Delta_0 g_0 = -f_2$$

$$\Delta_{-2} g_3 + \Delta_1 g_0 = -f_3$$

$$\Delta_{-2} g_4 + \Delta_0 g_2 + \Delta_2 g_0 = -f_4$$

and g_0, g_2, g_3, g_4 are zero on the boundary of $\overline{B}_1$. Similar computations give the values of $g_0, g_2, g_3(0), g_4(0)$. To complete the identification with $E_x(T_\varepsilon^2)$ we use the "stochastic Taylor formula" [1,2]:

$$g(x) - E_x g(X_{T_\varepsilon}) = - E_x\{T\Delta g(X_{T_\varepsilon})\} + E_x \int_0^{T_\varepsilon} s\Delta^2 g(X_s)ds$$

From this it follows that $g(x) = \frac{1}{2} E_x(T_\varepsilon{}^2) + 0(\varepsilon^{10})$, from which the second part of theorem 2.1 follows.

4. Geometric Consequences

Using the expansions obtained, we now illustrate the proof of corollary 2.2. Let $(e_1,\ldots,e_n)$ be an orthonormal basis of M_m and let $R_{ijkl} = R(e_i,e_j,e_k,e_l)$ be the components of the curvature tensor in this basis. Then $\rho_{ij} = \sum_{k=1}^{n} R_{ikjk}$ is the Ricci tensor and $\tau = \sum_{i=1}^{n} \rho_{ii}$ is the scalar curvature. Define tensors A and B by

(4.11)

$$(A) = A_{ijkl} = \tau(\delta_{ik}\,\delta_{jl} - \delta_{il}\,\delta_{jk})$$

$$(B) = B_{ijkl} = \rho_{ik}\,\delta_{jl} - \rho_{jk}\,\delta_{il} + \rho_{jl}\,\delta_{ik} - \rho_{il}\,\delta_{jk}$$

Defining the inner product of two tensors by $\langle A,B\rangle = \sum_{ijkl} A_{ijkl}\,B_{ijkl}$ we have by direct computation

$$\langle A,B\rangle = 0$$
$$\langle A,A\rangle = 2n(n-1)\tau^2$$
$$\langle B,B\rangle = 4\tau^2 + 4(n-2)|\rho|^2$$
$$\langle R,A\rangle = 2\tau^2$$
$$\langle R,B\rangle = 4|\rho|^2$$

Assuming that $|\rho| \neq 0$, we can define a new tensor C by

$$C = R - \frac{\langle R,A\rangle}{\langle A,A\rangle} A - \frac{\langle R,B\rangle}{\langle B,B\rangle} B \tag{4.2}$$

$$C_{ijkl} = R_{ijkl} - \frac{1}{n(n-1)} A_{ijkl} - \frac{|\rho|^2}{\tau^2+(n-2)|\rho|^2} B_{ijkl}$$

(C_{ijkl} is closely related, but not identical to the Weyl conformal curvature tensor.) Computing the norm, we have

$$0 \leq \langle C,C\rangle = \langle R,R\rangle - \frac{2\tau^2}{n(n-1)} - \frac{4|\rho|^4}{\tau^2+(n-2)|\rho|^2} \tag{4.3}$$

Thus to prove corollary 2.2 suppose that $n > 2$ and that $E_m(T_\varepsilon) = \varepsilon^2/2n$ for all $m \in M$ and all $\varepsilon > 0$. The coefficient of ε^4 must be zero, hence $\tau = 0$. Equating to zero the coefficient of ε^6, we have $R^2 - |\rho|^2 = 0$. If $|\rho| = 0$ we have also $|R| = 0$ and we are done. Otherwise, we use the above inequality (4.3) in the form

$$\langle R,R\rangle = |R|^2 \geq \frac{2\tau^2}{n(n-1)} + \frac{4|\rho|^4}{\tau^2+(n-2)|\rho|^2}$$

which implies

$$|\rho|^2 \geq \frac{4}{(n-2)} |\rho|^2$$

This can only happen if $n \geq 6$, which proves the first part of the corollary. To prove the other parts, note that if either M is Einstein or has non negative or non posi- Ricci curvature, then $\tau = 0$ implies $|\rho| = 0$, but we have in addition $|R|^2 - |\rho|^2 = 0$, hence $|R| = 0$ and M must be flat in these cases also.

To prove corollary 2.4 we need the values of the constants $d_0, d_1, d_2, d_3, d_4, d_5$ which are too cumbersome to list here. A lengthy computation shows that the variance $V_m(T_\varepsilon)$ has a similar expansion which in the coefficient of ε^8 all of the numerical constants have the same sign. This has the immediate consequence that whenever this entire coefficient is zero, each of the terms $|R|^2$, $|\rho|^2$, τ^2 must separately be zero. Hence M must be flat in this case.

References

1. H. Airault and H. Follmer, Relative densities of semimartingales, Inventiones Mathematicae 27(1974), 299-327.

2. K. B. Athreya and T. G. Kurtz, A generalization of Dynkin's identity, Annals of Probability 1(1973), 570-579.

3. E. B. Dynkin, Markov Processes, 2 volumes, Springer Verlag. 1965.

4. A. Gray and M. Pinsky, Mean exit time from a small geodesic ball in a Riemannian manifold, submitted for publication.

5. N. Ikeda and S. Watanabe, Stochastic Differential Equations and Diffusion Processes, Tokyo: Kodansha. Amsterdam, New York, Oxford: North Holland.

SOME RESULTS ON LIKELIHOOD RATIOS FOR TWO-PARAMETER PROCESSES

Eugene Wong and Moshe Zakai

I. Introduction

Let $\mathbf{R}_+^2$ denote the positive quadrant in the plane. $z = (s,t)$, $\zeta = (\sigma,\tau)$ will denote points in $\mathbf{R}_+^2$; introduce the partial order $z_1 < z_2$ $(z_1 \leq z_2)$ if $s_1 < s_2$ and $t_1 < t_2$ $(s_1 \leq s_2$ and $t_1 \leq t_2)$. R_z will denote the rectangle $\{\zeta\colon (0,0) \leq \zeta \leq z\}$. Let (Ω,F,P) be a probability space and let F_z, $z \leq z_o$ be a filtration on F. W_z, $z \leq z_o$ will denote the Brownian sheet on R_{z_o}. Cf. [2] and [8] for additional notation and assumptions for results concerning integration with respect to W_z and for the representation of two parameter martingales with respect to the Brownian sheet. $(\Omega,F_z,z \leq z_o,P)$ will be said to satisfy the conditional independence or the F-4 property if F_{s,t_o} and $F_{s_o,t}$ are conditionally independent given $F_{s,t}$. The F-4 property plays an important role in the theory of integration with respect to two-parameter martingales, however, it is not invariant under an equivalent transformation of measures.

A sketch is given here of some recent results on likelihood ratios with respect to the measure induced by two-parameter processes and particularly by the Brownian sheet. In the next section it is shown that under reasonable assumptions and similarly to the one parameter case, non negative martingales of the Brownian sheet satisfy the equation $\Lambda_z = 1 + \int_{R_z} \Lambda_\zeta dM_\zeta$; however, unlike the one-parameter case, not every solution of this equation is necessarily non-negative. Section III presents characterizations of transformations of measures which leave the F4 property invariant. It is pointed out that every likelihood ratio with respect to the Brownian sheet can be decomposed into two successive transformations of measure, the first leaving the F-4 property invariant followed by a transformation which can be considered as "strictly non F-4", this transformation transforms strong martingales into weak martingales. The last section presents sufficient conditions for the probability law of a two-parameter process to be transformable to that of the Brownian sheet under an equivalent transformation of measures, correcting a result given in the final section of [9].

Some references to previous work on likelihood ratios with respect to the measure induced by W_z are: [5], [9], and chapter 8 of [4]. Cf [1] for the Gaussian case, and [7] for the two-parameter Poisson case.

II. Likelihood Ratios and Stochastic Differential Equations in the Plane

Let $(\Omega;F_z, z \leq z_o; P_o)$ be the probability measure and filtration induced by the Brownian sheet and let P_1 be a probability measure which is equivalent to P_o. Set

$$E_o\left(\frac{dP_1}{dP_o}(\omega)\,|\,F_z\right) = \Lambda_z \quad .$$

Assume now that Λ_{z_o} is a square integrable random variable, therefore ([8]) we can represent Λ_z for $z \leq z_o$ by

$$\Lambda_z = 1 + \int_{R_z} \alpha_\zeta dW_\zeta + \iint_{R_z \times R_z} \beta_{\zeta,\zeta'} dW_\zeta dW_{\zeta'} \quad .$$

And since the measures P_1 and P_o are equivalent we may divide by Λ_z:

$$\Lambda_z = 1 + \int_{R_z} \Lambda_\zeta \left(\frac{\alpha_\zeta}{\Lambda_\zeta}\right) dW + \iint_{R_z \times R_z} \Lambda_{\zeta \vee \zeta'} \frac{\beta_{\zeta,\zeta'}}{\Lambda_{\zeta \vee \zeta'}} dW_\zeta dW_{\zeta'}$$

$$= 1 + \int_{R_z} \Lambda_\zeta \phi_\zeta dW_\zeta + \iint_{R_z \times R_z} \Lambda_{\zeta \vee \zeta'} \psi_{\zeta,\zeta'} dW_\zeta dW_{\zeta'} \quad . \tag{1}$$

The resulting random functions ϕ_ζ, $\psi_{\zeta,\zeta'}$ need not be square integrable but they are locally square integrable [3]. For given ϕ and ψ, equation (1) can be considered as an integral equation in Λ. Explicit solutions to (1) are not available except for the particular case where $\psi_{\zeta,\zeta'} = \phi_\zeta \cdot \phi_{\zeta'}$ and then [9]

$$\Lambda_z = \mathrm{Exp}\left(\int_{R_z} \phi_\zeta d\zeta - \frac{1}{2}\int_{R_z} \phi_\zeta^2 d\zeta\right) \quad .$$

Let M denote the martingale

$$M_z = \int_{R_z} \phi_\zeta dW_\zeta + \iint_{R_z \times R_z} \psi_{\zeta,\zeta'} dW_\zeta dW_{\zeta'} \quad .$$

Then (1) becomes

$$\Lambda_z = 1 + \int_{R_z} \Lambda_\zeta dM_\zeta \quad . \tag{2}$$

Equation (2) is similar in form to the corresponding stochastic differential equation of the exponential formula for the one-parameter continuous sample case. Sufficient conditions for the existence and uniqueness of solutions to (1) can be derived following the standard one-parameter arguments. In particular, assume that ϕ and ψ satisfy a.s.

$$\phi_\zeta^2 + \psi_{\zeta,\zeta'}^2 \leq K \quad ,$$

for all ζ,ζ' in R_z. Set

$$\Lambda_z^{(n)} = 1 + \int_{R_z} \Lambda_\zeta^{(n-1)} dM_\zeta \ , \quad \Lambda_z^{(o)} = 1 \quad ,$$

then

$$E\left(\Lambda_z^{(n)}-\Lambda_z^{(n-1)}\right)^2 \le E\int_{R_{z_o}} K\left(\Lambda_\zeta^{(n-1)} - \Lambda_\zeta^{(n-2)}\right)^2 d\zeta$$

$$\le \frac{K^n}{(n!)^2}(s_o \cdot t_o)^n \cdot E\int_{R_z}(1+M_\zeta-1)^2 d\zeta \quad .$$

Since a maximal inequality is also available for the two-parameter case ([2]), it follows by a standard argument that $\Lambda_z^{(n)} \to \Lambda_z$.

A major difference between the one and the two-parameter cases appears when we consider the question whether the solution to (2) is necessarily nonnegative. Unlike the one parameter case, the solution to (2) may be negative as the following example shows:

Let $\phi = 0$, then

$$\Lambda_z = 1 + \iint_{R_z \times R_z} \Lambda_{\zeta,\zeta'} \psi_{\zeta,\zeta'} dW_\zeta dW_{\zeta'}$$

assume that $\psi_{\zeta,\zeta'}^2 \le K$. A necessary condition for Λ_z to be nonnegative is $\theta \ge 0$, where

$$\theta = E_o\left(\left(\int_{R_z} h_\zeta dW_\zeta\right) \cdot \Lambda_z\right) \quad .$$

Now,

$$\theta = E_o\left\{\left(\int_{R_z} h_\zeta^2 d\zeta + 2\int_{R_z} I_\zeta dW_\zeta + \iint_{R_z \times R_z} h_\zeta h_{\zeta'} dW_\zeta dW_{\zeta'}\right) \cdot \right.$$

$$\left. \cdot \left(1 + \iint_{R_z \times R_z} \psi\Lambda_{\zeta \vee \zeta'} dW_\zeta dW_{\zeta'}\right)\right\} \quad .$$

Assume that h_ζ is deterministic, then

$$\theta = E_o\left\{\int_{R_z} h_\zeta^2 d\zeta + \iint_{R_z \times R_z} \psi_{\zeta,\zeta'} \Lambda_\zeta h_\zeta h_{\zeta'} d\zeta d\zeta'\right\} \quad .$$

Set $h = 1$, $\psi_{\zeta,\zeta'} = c\, 1_{\zeta \wedge \zeta'}$ and note that $E_o \Lambda_z = 1$ (even if Λ changes signs), therefore

$$\theta = \int_{R_z} d\zeta + c \iint_{R_z \times R_z} 1_{\zeta \wedge \zeta'} d\zeta d\zeta' \quad .$$

Setting c negative enough, yields a negative result for θ and therefore the solution to (2) may turn out to be negative.

III. Some Results on Transformations of Measures

The first two propositions characterize transformation of measures which leave the F-4 property invariant. The last proposition presents a factorization of the likelihood ratio of a general transformation of measure into two successive transformations of measure, the first transformation leaving the F-4 property invariand followed by a transformation of measure which can be considered as being "strictly non F-4". The proofs of the results of this section will be published elsewhere.

Proposition 1: Let $(\Omega, F_z, z \le z_o, P_o)$ be a probability space and filtration, assume that F-4 is satisfied under P_o and let P_1 be equivalent to P_o. Let

$$E_o\left(\frac{dP_1}{dP_o}(\omega)\middle| F_z\right) = \Lambda_z$$

$$= E_o(\Lambda_{z_o} | F_z) \quad ,$$

and

$$\lambda_z = E_o\left(\Lambda_{z_o} \middle| F_z^1 \vee F_z^2\right) \quad ,$$

where E_o (E_1) denotes expectation with respect to P_o (P_1). A necessary and sufficient condition that F-4 satisfies under P_1 is that for all z in R_{z_o}

$$\lambda_z = \frac{\Lambda_{z \otimes z_o} \cdot \Lambda_{z_o \otimes z}}{\Lambda_z}$$

where, for $z_i = (s_i, t_i)$, $z_1 \otimes z_2 = (s_1, t_2)$.

Proposition 2: Let F_z, $z \le z_o$ be generated by the process W_z which is a Brownian sheet under P_o. Let P_1 be equivalent to P_o. Let

$$\Lambda_z = E_o\left(\frac{dP_1}{dP_o}(\omega)\middle| F_z\right) \quad ,$$

and assume that $E\Lambda_{z_o}^2 < \infty$. If F-4 is satisfied under P_1 then there exists a predictable process ϕ_z such that

$$\Lambda_z = \operatorname{Exp}\left(\int_{R_z} \phi_\zeta dW_\zeta - \frac{1}{2}\int_{R_z} \phi_\zeta^2 d\zeta\right) \quad , \tag{3}$$

with

$$\int_{R_{z_o}} \phi_\zeta^2 d\zeta < \infty \quad \text{a.s.}$$

Remark: This is the converse to the well known result that if Λ_z is as given by (3) then F-4 is satisfied under P_1.

Proposition 3: Let F_z, $z \le z_o$ be generated by the process W_z which is a Brownian sheet under P_o, let P_1 be equivalent to P_1 with

$$\Lambda_z = E_o\left[\frac{dP_1}{dP_o}(\omega) \,\middle|\, F_z\right]$$

$$= 1 + \int_{R_z} \Lambda_\zeta \phi_\zeta dW_\zeta + \iint_{R_z \times R_z} \Lambda_{\zeta \vee \zeta'} \gamma_{\zeta,\zeta'} dW_\zeta dW_{\zeta'}$$

Then

$$\Lambda_z = \ell_z \cdot L_z$$

where ℓ_z satisfies

$$\ell_z = 1 + \int_{R_z} \ell_\zeta \phi_\zeta dW_\zeta + \iint_{R_z \times R_z} \ell_{\zeta \vee \zeta'} \phi_\zeta \phi_{\zeta'} dW_\zeta dW_{\zeta'}$$

$$= \operatorname{Exp} \int_{R_z} \phi_\zeta dW_\zeta - \frac{1}{2}\int_{R_z} \phi_\zeta^2 d\zeta \quad ,$$

and L_z solves, for some $\psi_{\zeta,\zeta'}$, the equation

$$L_z = 1 + \iint_{R_z \times R_z} \psi_{\zeta,\zeta'} L_{\zeta \vee \zeta'} (dW_\zeta - \phi_\zeta d_\zeta)(dW_{\zeta'} - \phi_{\zeta'} d\zeta')$$

Remark: Let P'_o be defined by $(dP'_o/dP_o) = \ell_{z_o}$, then $W_z - \int \phi_\zeta d\zeta$ is Wiener under P'_o and the result of the proposition yields the decomposition

$$\frac{dP_1}{dP_o}(\omega) = \frac{dP'_o}{dP_o}(\omega) \cdot \frac{dP_1}{dP'_o}(\omega)$$

$$= \ell_{z_o} \cdot L_{z_o} \quad ,$$

where L_{z_o} represents a transformation of measure from P'_o (which satisfies F-4) to P_1 which is "strictly non F-4" in the following sense. It is easy to show that every strong P'_o martingale transforms into a weak martingale under P'_o while a transformation of measure which leaves F-4 invariant and transforms strong martingales into weak martingales is the identity transformation.

IV. Equivalence with Respect to a Measure Satisfying the F-4 Assumption

Let $\{\Omega, F_z, z \le z_o, P\}$ denote a probability space not satisfying F-4. The question arises whether there exists a probability measure P_o on $\{\Omega, F_{z_o}\}$ which is equivalent to P and such that $\{\Omega, F_z, z \le z_o, P_o\}$ does satisfy F-4. A particular result in this direction is

Proposition 4: Let $M_{s,t}$, $(s,t) \le (s_o,t_o)$ be a sample continuous, $F_{s.t}$ adapted process. Assume that

$$M_{s,t} = m_{s,t} + \int_0^t \int_0^s \alpha_{\sigma,\theta} d\sigma d\theta \tag{4}$$

where $m_{s,t}$ is $F_{\cdot,t_o}$ adapted and F_{s,t_o} one parameter local martingale in s and $a_{s,\theta}$ is measurable F_{s,t_o} adapted. Further assume that for every s, $M_{s,t}$ is a semimartingale in the t and s directions, $M_{s,t}$ is of orthogonal increments in the s direction, that is, $0 = \langle (M_{\cdot,t_4} - M_{\cdot,t_3}),(M_{\cdot,t_2} - M_{\cdot,t_1})\rangle_s$ whenever $t_4 > t_3 \geq t_2 > t_1$, that

$$E \int_0^{t_o} m^2_{s_o,\theta} d\theta < \infty \tag{5}$$

and

$$E\left[\operatorname{Exp} \frac{1}{2} \int_0^{s_o} \int_0^{t_o} \alpha^2_{\sigma,\theta} d\sigma d\theta\right] < \infty \quad , \tag{6}$$

then there exists a probability measure P_o on (Ω, F_z) such that P_o is equivalent to P and $M_{s,t}$ is Wiener on its natural σ-fields under P_o; in particular, if F_z are the σ-fields generated by M_z then $(\Omega, F_z, z \leq z_o, P_o)$ satisfies F-4.

<u>Remarks:</u> (1) This result replaces the incorrect Section V of [10]. (2) It follows from [9] that the decomposition (3) is a necessary condition for the existence of a measure under which M is Wiener. (3) From the arguments of final part of the proof it follows that under the measure P, $m_{s,t}$ is Wiener on its natural σ-fields; however, in general $m_{s,t}$ need not be adapted to $F_{s,t}$.

<u>Proof:</u> Assume that $s_o = t_o = 1$ and let ϕ^i be a complete orthonormal set of functions on [0,1]. Let

$$\mu^i_s = \int_0^1 \phi^i_\theta dm_{s,\theta} \quad ,$$

$$\beta^i_s = \int_0^1 \alpha_{s,\theta} \phi^i_\theta d\theta \quad .$$

Then, by (4) $(\mu^i_s, F_{s,1})$ is a continuous one-parameter martingale and

$$\langle \mu^i, \mu^j \rangle_s = \begin{cases} s, & i = j \\ 0, & i \neq j \end{cases}$$

Let $Y^{(n)}_s$ denote the continuous 1-martingales

$$Y^{(n)}_s = \sum_1^n \int_0^s \beta^i_\sigma d\mu^i_\sigma \quad ,$$

then

$$E(Y_s^{(n)})^n = \sum_1^n E\int_0^1 (\beta_\sigma^i)^2 d\sigma \le \int_0^1\int_0^1 \alpha_{\sigma,\theta}^2 d\sigma d\theta \quad ,$$

and $Y_s^{(n)}$ converges in the mean to a continuous martingale Y_s. The quadratic variation of Y is, therefore,

$$\langle Y\rangle_s = \int_0^s\int_0^1 \alpha_{\sigma,\theta}^2 d\theta d\sigma \quad .$$

Set $Z_s = \mathrm{Exp}(-Y_s - \frac{1}{2}\langle Y\rangle_s)$, then by (6) and a result of Novikov (cf. [6]), $EZ_s = 1$, $s \le 1$.

Setting

$$\frac{dP_0}{dP} = Z_1 \quad .$$

Then P_0 is absolutely continuous with respect to P. In order to show that $P_0 \sim P$ let A be an event such that $P_0(A) = 0$ and $P_1(A) \ne 0$ then $Z_{s_0}(\omega) = 0$ for a.a. (P) $\omega \in A$. However, since $E\langle Y\rangle_{s_0} < \infty$, both $\max_s Y_s$ and $\langle Y\rangle_{s_0}$ are finite a.s. (P), therefore $P\{\omega: Z_{s_0}(\omega) = 0\} = 0$ therefore we must have $P_1(A) = 0$ and therefore $P_0 \sim P$.

Consider now

$$E\left(m_{s,t} - \sum_1^n\left(\int_0^t \phi_i(\theta)d\theta\right)\mu_s^i\right)^2 =$$

$$= E\left\{(m_{s,t})^2 - 2\sum_1^n\left(\int_0^t \phi_i(\theta)d\theta\right)\cdot E(m_{s,t}\mu_s^i)\right.$$

$$\left. + \sum_1^n\left(\int_0^t \phi_i(\theta)d\theta\right)^2 E(\mu_s^i)^2\right\}$$

$$= s\cdot t - \sum_1^n\left(\int_0^t \phi_i(\theta)d\theta\right)^2\cdot s \quad .$$

Let t be fixed, set $f(\tau) = 0$, $\tau > t$, $f(\tau) = 1$, $\tau < t$, then by Riesz-Fischer:

$$\underset{n\to\infty}{\mathrm{l.i.m.}} \sum_1^n\left(\int_0^t \phi_i(\theta)d\theta\right)\cdot\phi_i(\tau) = \begin{cases} 1, & \tau < t \\ 0, & \tau > t \end{cases} \quad ,$$

hence, by the Parceval theorem:

$$\sum_1^\infty\left(\int_0^t \phi_i(\theta)d\theta\right)^2 = \int_0^t d\tau = t \quad .$$

It follows that

$$\sum_{1}^{n}\left(\int_{0}^{t}\phi_i(\theta)d\theta\right)\mu_s^i \xrightarrow[n\to\infty]{q.m} m_{s,t} \quad ,$$

and

$$<m_{.,t},Y_.>_s = \sum_i \int_0^t \phi_i(\theta)d\theta \int_0^s \beta_\sigma^i d\sigma$$

$$= \sum_i \int_0^t \phi_i(\theta)d\theta \cdot \int_0^s \left(\int_0^1 \alpha_{\sigma,\eta}\phi_\eta^i d\eta\right) d\sigma$$

$$= \int_0^s \left(\sum_i \left(\int_0^t \phi_i(\theta)d\theta\right)\left(\int_0^1 \alpha_{\sigma,\eta}\phi_\eta^i d\eta\right)\right) d\sigma$$

$$= \int_0^s \int_0^t \alpha_{\sigma,\eta} d\sigma d\eta \quad .$$

Therefore, by Girsanov's theorem

$$m_{s,t} - <m_{.,t},Y_.>_s = m_{s,t} + \int_0^t \int_0^s \alpha_{\sigma,\theta} d\sigma d\theta = M_{s,t}$$

is an F_{s,t_0} 1-martingale under P_0. Since $M_{s,t}$ is $F_{s,t}$ adapted, $M_{s,t}$ is also an adapted 1-martingale under P_0 with orthogonal increments. It follows that for every bounded, deterministic, simple function G_ξ,

$$E_0 \mathrm{Exp}\left(\int_{R_{z_0}} GdM - \frac{1}{2}\int_{R_{z_0}} G^2 d\xi\right) = 1 \ ,$$

where E_0 denotes expectation with respect to P_0. Hence

$$E_0 \mathrm{Exp}\int GdM = \mathrm{Exp}\,\frac{1}{2}\int G_\xi^2 d\xi \ , \ ,$$

and therefore M is Wiener on its own σ-fields under P_0.

References

[1] C. Bromely and G. Kallianpur: Gaussian Random Fields, Appl. Math. Optim. 6, 361-376 (1980).

[2] R. Cairoli and J.B. Walsh: Stochastic Integrals in the Plane. Acta Math. 134, 111-183 (1975).

[3] R. Cairoli and J.B. Walsh: Régions d'arrêt, localizations et prolongements de martingales. Z. für Wahr. verw. G. 44, 279-306 (1978).

[4] X. Gruyon and B. Prum: Semi-martingale à indice dans $\mathbb{R}^2$. Thèse et Publication Université Orsay (1980).

[5] B. Hajek and E. Wong: Representation and Transformation of Two-Parameter Martingales under a Change of Measure. Z. für Wahr. verw. G. 54, 313-330 (1980).

[6] N. Kazamaki and T. Sekiguchi: On the Transformation of Some Classes of Martingales by a Change of Law. Tôhoku Math. J. 31, 261-279 (1979),

[7] G. Mazziotto and J. Szpirglas: Equations du filtrage pour un processus de Poisson mélangé à deux indixes. Stochastics, 4, 89-119 (1980).

[8] E. Wong and M. Zakai: Martingales and Stochastic Integrals for Processes with a Multidimensional Parameter. Z. für Wahr. verw. G. 29, 109-122 (1974).

[9] E. Wong and M. Zakai: Likelihood Ratios and Transformations of Probability Associated with Two-Parameter Wiener Processes. Z. für Wahr. verw. G., 40, 283-308 (1977).

[10] M. Zakai: Some Remarks on Integration with Respect to Two-Parameter Martingales, in Lecture Notes Math. No. 863, Springer Verlag, Berlin, pp. 149-161 (1981).

Eugene Wong
Electronic Research Laboratory
University of California
Berkeley, California 94720
U.S.A.

Moshe Zakai
Dept. of Electrical Engineering
Technion - Israel Inst. of Technology
Haifa, 32000
Israel.

CONTROLLABILITY OF STOCHASTIC SYSTEMS

J. Zabczyk
Institute of Mathematics
Polish Academy of Sciences
Sniadeckich 8, 00-950 Warszawa

0. Introduction

The object of the paper is to present recent results on controllability of stochastic processes. Paper is devided into 2 parts. In part I devoted to controllability of Markov processes, such concepts like recurrence and positive recurrence are studied in the context of controlled Markov processes. Beside of general, necessary and sufficient or sufficient conditions for various types of controllability, a special attention is paid to controlled processes with independent increments and to controlled linear systems. Problems discussed in this part of the paper seem to be of some importance because they reflect an attempt to treat in a unified way recurrence of stochastic processes and controllability of deterministic systems. They are also related to a probabilistic potential theory which aim is to extend, at least some, of the results of the monograph [2], to the more general situation of controlled Markov processes, see [10].

Part II of the paper is devoted exclusively to linear systems controlled through affine feedback strategies. The resulting process is then both Gaussian and Markov, although not homogenous in time. The main question one is asking here is how rich is the set of attainable distributions at deterministic moments.

Most of the results discussed in the paper are taken from the report [15] where proofs and more complete discussion can be found. Related material is contained also in papers [1] and [6].

This paper has been preapared while the author was a guest of Bremen University in 1982. He wishes to thank Prof. L. Arnold and members of his group for the hospitality and stimulating discussions on topics presented in the paper.

I. Controllability of Markov processes

I.1. Potential theoretic characterization of controllable systems

Let $X = (\Omega, F_t, x(t), \theta_t, P_x, E)$ be a standard Markov process on a state space E and let (P_t) be the corresponding Markov semigroup, see [3]. Process X is said to be *recurrent* if and only if for arbitrary $s \in E$ and open set $B \subset E$, $B \neq \emptyset$,

$$P_x(D_B<+\infty) = 1 ,$$

where D_B is the first entry time of B :

$$D_B = \inf\{t \geq o;\ x(t) \in B\}.$$

A universally measurable function $f \geq o$ defined on E is *excissive* if and only if

$$P_t f(x) \leq f(x), \quad \text{for all} \quad t \geq o \quad \text{and} \quad x \in E,$$

and $P_t f(x) \uparrow f(x)$, as $t \downarrow o$. If $h \geq o$ is a universally measurable function on E then *potential* Gh:

$$Gh(x) = E_x\left(\int_o^{+\infty} h(x(t))dt\right), \quad x \in E$$

is an excessive function.

Assuming that excessive functions are lower semicontinous (lsc) the following statements are equivalent, see [8],

(1) X is recurrent

(2) Excessive functions are constant

(3) If for a Borel set $B \subset E$ and $x \in E$, $GI_B(x) > o$ then $GI_B \equiv +\infty$.

More specific characterizations of recurrent processes exist for special classes of Markov processes. For instance necessary and sufficient conditions for recurrence of processes with independent increments in terms of the so called Levy exponential are known. In particular a truly n-dimensional process with independent increments $n \geq 3$ is not recurrent.

Let us assume now that a family of Markov semigroups (P_t^u), $u \in U$ is given and let Π^d and P_x^π, $\pi \in \Pi^d$, $x \in E$ denote respectively the set of all dyadic strategies [11] and corresponding distributions on the canonical Skorokhod space of trajektories Ω. Controlled system $\Sigma = (\Omega, F_t, x(t), O_t, P_x^\pi, \Pi^d, E, U)$ is called *weakly controllable* if for all open sets $B \neq \emptyset$ and $x \in E$:

$$\sup_{\pi \in \Pi^d} P_x^\pi(D_B < +\infty) > o,$$

and *controllable* if

$$\sup_{\pi \in \Pi^d} P_x^\pi(D_B < +\infty) = 1.$$

The characterization (3) can not be extended to the controlled case. To see this let $E = \{\ldots,-1,o,1,\ldots\}$, $U = \{o,1\}$ and let X^o be a compound Poisson process with the jump measure ν; $o < \nu\{-1\} = q < \nu\{1\} = p$, $p + q = 1$. If X^1 is the trivial Markov process staying uninterruptly in the initial state, then it is easy to see, that for arbitrary set $B \neq \emptyset$, $B \subset E$ and state $x \in E$, there exists a strategy $\pi \in \Pi^d$ such that $E_x(I_B(x(t))dt) = +\infty$ but the controlled system is certainly not controllable.

To check whether the characterization (2) has its controlled counterpart an appropriate concept of excessivity is needed. The most natural and the proper one is the following. A non-negative, universaly measurable function f defined on E is said to be *U-excessive* if it is excessive with respect to all semigroups (P_t^u), $u \in U$. The following result holds true, see [15].

Theorem 1 If for arbitrary $u \in U$ the semigroup (P_t^u) is Feller then the stochastic system Σ is controllable if and only if the only lsc, U-excessive function is the constant one.

U-excessive functions can be introduced in a different and equivalent way, see Proposition 1 below and [11]. Let S_t and R_λ be operators defined for Borel functions $f \geq o$ as follows:

$$S_t f(x) = \sup_{\pi \in \Pi^d} E_x^\pi (f(x(t))) ,$$

$$R_\lambda f(x) = \sup_{\pi \in \Pi^d} E_x^\pi \left(\int_o^{+\infty} e^{-\lambda t} f(x(t))dt\right), \quad t \geq o, \ a > o, \ x \in E.$$

Proposition 1 Under the same assumptions as in Thm 1, a lsc function $f \geq o$ is U-excessive if and only if

(4) $\quad S_t f \leq f$, for $t \leq o$

$\quad S_t f \uparrow f$, as $t \downarrow o$

and if and only if

(5) $\quad \lambda R_\lambda f \leq f$, for $\lambda > o$

$\quad \lambda R_\lambda f \uparrow f$ as $\lambda \uparrow +\infty$.

The next proposition is useful when proving controllability of specific systems. Let $\Pi \supset \Pi^d$ denotes a set of non-anticipating processes defined on Ω and with values in U and let M be a set of (F_t) stopping times. Set Π is called *M-closed* if for every finite family of strategies $\pi_1, \pi_2, \ldots, \pi_n \in \Pi$ and stopping times $T_o \equiv o, T_1, \ldots, T_n \in M$, there exists a strategy $\pi \in \Pi$ such that for every $k = o, 1, \ldots, n-1$ and random variable $\xi \geq o$, F_{T_k}-measurable

(6) $\quad E_x^\pi(\theta_{S_k}(\xi) \mid F_{S_k}) = E_{x(S_k)}^{\pi_k}(\xi)$, P_x^π a.e. on $S_k < +\infty$,

$\quad$ where $S_o = o$, $S_{k+1} = S_k + \theta_{S_k}(T_k)$.

Proposition 2 Assume that B and C are closed subsets of E and $\Sigma = (\Omega, F_t, x(t), \theta_t, P_x^\pi, \Pi, E, U)$ is an M-closed system. If for some strategies $\pi_1, \pi_2 \in \Pi$, finite stopping times $\tau_1, \tau_2 \in M$ and a positive number α :

$$P_x^{\pi_1}(x(\tau_1) \in B) \geq \alpha \quad \text{for all} \quad x \in E$$

$$P_x^{\pi_2}(x(\tau_2) \in C) \geq \alpha \quad \text{for all} \quad x \in B ,$$

then

$$\sup_{\pi\in\Pi} P_x^{\pi}(D_B<+\infty, D_C<+\infty) = 1, \text{ for every } x \in E.$$

A similar principle is helpful for a study of strong controllability. A system Σ is called *strongly controllable* if and only if

$$\inf_{\pi\in\Pi} E_x^{\pi}(D_B) < +\infty ,$$

for arbitrary open set $B \neq \Phi$ arbitrary $x \in E$. A family Π is *strongly M-closed* if the identity (6) holds for any countable family of strategies $\pi_k \in \Pi$ and stopping times $T_k \in M$, $k = 0,1,\ldots$

<u>Proposition 3</u> If in addition to assumptions of Proposition 2 the set Π is strongly M-closed and

(7) $\qquad E_x^{\pi_1}(D_B) = h(x) < +\infty \quad$ for every $\quad x \in E,$

(8) $\qquad E_x^{\pi_2}(\tau_2+h(x(\tau_2))) < K \quad$ for a

constant K and arbitrary $x \in B$,

then there exists a strategy $\pi \in \Pi$ such that

$$E_x^{\pi}(D_B) < +\infty , \quad \text{for} \quad x \in E.$$

I.2 Controlled processes with independent increments

Let us assume that for every $n \in U$, the semigroup (P_t^u) corresponds to a stochastic process with independent increments on the state space $E = R^n$. Let us define for functions f bounded and continous with their first and second derivatives, shortly $f \in C_b^2$, operators A^u :

$$A^u f(x) = \langle b^u, f_x'(x)\rangle + \frac{1}{2}\,\text{Trace}\; a^u f_{xx}''(x)$$

$$+ \int_{|y|\le 1} (f(x+y) - f(x) - \langle y, f_y'(x)\rangle)\nu^u(dy)$$

$$+ \int_{|y|>1} (f(x+y) - f(x))\nu^u(dy), \; x \in E,$$

where (b^u,a^u,ν^u) are infinitesimal characteristics of processes X^u.

The next proposition is an easy corollary of Theorem 1.

<u>Proposition 4</u> System Σ of controlled processes with independent increments is controllable if and only if the only function $f \in C_b^2$ such that

$$f \ge 0, \; A^u f \le 0 \quad \text{for all} \quad u \in U$$

is the constant one.

Example 1 The following system on $E = R^n$ was introduced in [7] in connection with Dirichlet problem for the Monge-Ampere equation, (see also [11]). Let the set U of control parameters be composed of all symmetric, positive nxn matrices σ such that $\det \sigma \geq 1$ and let the corresponding Markov process X^σ be equivalent to

$(x+\sigma w_t)_{t\geq 0}$, $x \in R^n$, where $(w_t)_{t\geq 0}$ is a standard Wiener process on R^n. Then this system is controllable. This follows either from direct probabilistic considerations or from Proposition 4 and the following Lemma.

Lemma 1 If a function $f \geq o$ belongs to $C_b^2(R^n)$ and for every $\sigma \in U$:

$$\text{Trace } \sigma f''_{xx} \leq o ,$$

then f is constant.

Remark 1 An equivalent formulation of Lemma 1 is the following. If

$$f \in C_b^2, \quad f \geq o \quad \text{and} \quad f''_{xx} \leq o ,$$

then f constant.

Example 2 Let a controlled system Σ be given in the form of a stochastic equation on $E = R^3$:

$$(9) \qquad dx(t) = u(t)e\,dt + dw_t, \quad t \geq o,$$

where $e = (1,o,o)$ and U is a closed subinterval of R^1. The interesting special cases are $U = [-1,1]$, $U = [o,+\infty)$ and $U = [o,1]$.

If $U = [-1,1]$, then system Σ is controllable. An appropriate strategy is:

$$u(t,x(\cdot)) = -\text{sgn}x_1(t), \; t \geq o, \quad \text{see [15]}.$$

Taking into account Proposition 4 one can thus state:

If $\quad f \in C_b^2(R^n), \quad f \geq o$

and $\quad \frac{1}{2} \Delta f + \left|\frac{\partial f}{\partial x_1}\right| \leq o \quad$ on $\quad R^3,$

then $\quad f \quad$ constant.

If $U = [o,+\infty)$, then system Σ is also controllable. To see this let us fix a ball $C = B(\hat{x},r)$, $r > o$, $\hat{x} \in R^3$ and let B be the half cylinder

$$B = \{x\ R^3;\ x_1 \leq \hat{x}_1, (x_2-\hat{x}_2)^2+(x_3-\hat{x}_3)^2 \leq \frac{\tau^2}{2}\}.$$

An elementary argument implies that Wiener process on R^3, starting from an arbitrary initial condition, will hit B with probability 1. Moreover one can construct a "high speed" control such that starting from an arbitrary state in B the trajectory of (9) will hit C , at a fixed time $\hat{t}$, with probability $\geq 1/2$. It is enough now to apply Propostion 2.

Controllability of system Σ if $U = [o,+\infty)$ implies that if $f \in C_b^2$, $f \geq o$, $\frac{\partial f}{\partial x_1} \leq o$ and $\Delta f \leq o$, then f constant.

If $U = [o,1]$ the answer is unknown to the author. The analytic version of the controllability question reads in this case as follows:

Does there exists $f \geq o$, $f \in C_b^2(R^3)$ and such that

$$\frac{1}{2}\Delta f + \left(\frac{\partial f}{\partial x_1}\right) + \leq o \quad ?$$

Remark 2 Neither of systems Σ considered in Example 2 is strongly controllable.

I.3 Controllability of linear, finite dimensional systems

Let F,G and H be nxn, nx 1 and nxm matrices and $(w_t)_{t\geq o}$ an m-dimensional Wiener process. A complete characterization of controllable linear systems of the form:

(10) $$dx(t) = Fx(t)dt + Gu(t)dt + Hdw_t$$
$$x(o) = x \in R^n$$

is given in Theorem 2 below, see [13]. In its formulation $\mathcal{F}$ denotes the transformation F reduced to the quotient space R^n/K, where K is the controllability subspace of the deterministic system $\dot{x} = Fx + Gu$.

Theorem 2 For system (10) the following characterizations hold true:

(i) System (10) is weakly controllable if and only if

(11) $$\text{Rank } [G,FG,\ldots,F^{n-1}G,H,FH,\ldots,F^{n-1}H] = n.$$

(ii) System (10) is controllable if and only if the rank condition (11) holds and the Jordan representation of the transformation $\mathcal{F}$ is of the form:

$$\mathcal{F} = \begin{pmatrix} \mathcal{F}_o, & o \\ o, & \mathcal{F}_1 \end{pmatrix} ,$$

where eigenvalues of $\mathcal{F}_1$ have negative real parts and $\mathcal{F}_o$ is at most of dimension 2 and of the form:

$$\mathcal{F}_o = [o] \quad \text{or} \quad \mathcal{F}_o = \begin{pmatrix} \alpha, & o \\ o, & -\alpha \end{pmatrix}, \quad \alpha \text{ a real number } \neq o ,$$

(in the case $\dim \mathcal{F}_o \geq 1$).

We will sketch a proof of Thm 2 using results of section I.1. It will be clear from the presentation that the proof can be modified to cover the *discrete-time case* not discussed in [13].

The proof of part (i) is straightforward as one can restrict all considerations to deterministic strategies only. To prove (ii) let us assume that $\dim K = k \geq 1$. On can rewrite (10) in the form:

(12) $$dy = (F_1y+F_2z)dt + G_1udt + H_1dw_t$$
$$dz = F_3zdt + H_2dw_t,$$

where processes $(y(t))_{t\geq o}$, $(z(t))_{t\geq o}$ are of dimensions k and $n-k$ respectively and the pair (F_1,G_1) is controllable. If system (10) is controllable then necessarily the rank condition (11) holds and the process $(z_t)_{t\geq o}$ is recurrent. To show the reverse implication let us fix closed balls $B = B(o,r_1)$ and $C = B(x^2,r_2)$. The strategies π_1 and π_2 required in the formulation of Proposition 2 can be described as follows. *Strategy* π_1: wait until the z-component enters the set $\{z \in R^{n-k};\ |z| \le r_1/2\}$ and then use a "high speed" deterministic control which steers the y-component into the neighbourhood $\{y \in R^k;\ |y| \le r_1/2\}$ in time t_1. *Strategy* π_2: use a determinstic control function which steers the state O into the set C in time t_2. Moreover moments τ_1 and τ_2 can be defined as:

$$\tau_1 = t_1 + \inf\{t \ge o;\ |z(t)| \le r_1/2\}$$

$$\tau_2 = t_2 .$$

One can show, by taking $r_1 > o$ sufficiently small, that all assumptions of Proposition 2 are satisfied and therefore:

$$\sup_{\pi \in \Pi} P_x^{\pi}(D_C < +\infty) = 1 .$$

Since the recurrence of the z-component is equivalent to the property of transformation F expressed in (ii), see [4], and [5] and [13] therefore the characterization (ii) is true. A similar construction and the use of Proposition 3 instead of Proposition 2 imply part (iii).

I.4 Controllability of linear infinite-dimensional systems

Let us consider the following stochastic delay system on R^n.

(13) $$dz(t) = \left(\int_{-h}^{o} z(t+s)dF(s)\right)dt + Gudt + Hdw_t,$$

where $F(\cdot)$ is a matrix valued function of bounded variation. Such systems are natural generalizations of those considered in section I.3. It is well knows, see [3] and [12], that (13) can be treated as an infinite dimensional system of the form:

(14) $$dx(t) = Ax(t)dt + Gudt + Hdw_t$$

where A is the infinitesimal generator of a C_o-semigroup $(S(t))_{t \ge o}$ acting on a Hilbert space E; G and H are bounded linear operators from Hilbert spaces U and V into E and $(w_t)_{t \ge o}$ is a Wiener process on V. Thus, it is of some interest, to study controllability of general systems (14), to cover delay systems (13) and also some parabolic and hyperbolic stochastic equations. Such a study however requires explicit characterizations of recurrent or positive recurrent processes of the form:

(15) $$dx(t) = Ax(t)dt + Hdw_t .$$

In this direction only partial answers are known, see [14], therefore the results presented below are less satisfactory then those in section I.3.

We consider only the so called mild version of (14):

$$(16)\quad x(t) = S(t)x + \int_0^t S(t-s)Gu(s)ds + \int_0^t S(t-s)Hdw_s .$$

Admissible strategy $\pi = (u(t,\cdot))_{t\geq o}$ is any U-valued nonanticipating process such that the integral equation (16) has exactly one solution with continous trajectories.

The following proposition characterizes weakly controllable systems of the form (14).

Proposition 5 System (14) is weakly controllable if and only if the deterministic system $\dot{x} = Ax + Gu + Hv$, where $(u(\cdot),v(\cdot))$ is a new control function, is approximately controllable starting from arbitrary initial state in E .

Let E_1 be the controllability space of the deterministic system:

$$(17)\quad \dot{x} = Ax + Gu,\ E_1 = \overline{\text{lin}}\ \{\int_o^t S(r)Gu(r)dr;\ u(\cdot)\in L^2[o,t;U],\ t > o\}$$

and let E_2 be its (orthogonal) complement. Thus if $x \in E$ then $x = y + z$, $y \in E_1$ and $z \in E_2$ and the decomposition is unique.

Proposition 6 The mild solution (16) has decomposition $x(t) = (y(t),z(t))$, $t \geq o$ satisfying the following relation:

$$(18)\quad y(t) = S_1(t)y_o + \int_0^t S_1(t-s)Gu(s)ds$$

$$+ \int_0^t S_2(t-s)H_1dw_s + \int_0^t S_3(t-s)H_2dw_s$$

$$(19)\quad z(t) = S_2(t)z_o + \int_0^t S_2(t-s)H_2dw_s,\ t \geq o$$

where G,H_1 and H_2 are bounded operators, S_1 and S_2 are C_o-semigroups on E_1 and E_2 respectively and S_3 is a C_o-continous family of linear operators from E_2 into E_1 with $S_3(o) = I$. Moreover the deterministic part of (18) is on E_1 approximately controllable from the state o .

Theorem 3 Let us assume that system (14) is weakly controllable.

(i) If (14) is controllable then process $(z(t))_{t\geq o}$ is recurrent.

(ii) If the process $(z(t))_{t\geq o}$ is recurrent and the deterministic system (17) reduced to E_1 is controllable to arbitrary neighbourhood of the origin in a time dependent only on the neighbourhood and not on the starting point, then (14) is controllable.

The first part of the Theorem is obvious. Assumptions made in the part (ii) are exactly those which make all considevations of the proof of the finite dimensional result valid.

An analogous theorem holds for the strong controllability, see [15].

Theorem 4 Let us assume that system (14) is controllable.

(i) If (14) is strongly controllable then process $(z(t))_{t\geq o}$ is positive recurrent.

(ii) If in additions to assumptions of part (ii) Theorem 3, the semigroup S_2 is exponentially stable, then (14) is strongly controllable.

II. Controllability of Gaussian processes

In the second part of the paper we are concerned with linear stochastic systems:

(20) $$dx(t) = Fx(t)dt + Gudt + Hdw_t$$
$$x(o) = \xi$$

introduced in section I.3 but with a more narrow notion of a strategy. *An affine strategy* π is a pair $(K(\cdot),r(\cdot))$ of a locally integrable, lxn matrix valued function K and a locally integrable n-vector valued function r . The pair $\pi = (K,r)$ defines a feedback strategy:

(21) $$u(t) = K(t)x(t) + r(t), \quad t \geq o .$$

The solution of (20) corresponding to π will be denoted as $(x^{\pi}(t))_{t\geq o}$. Thus $(x^{\pi}(t))$ is the unique solution of the equation

(22) $$dx^{\pi}(t) = (F+GK(t))x^{\pi}(t)dt + Gr(t)dt + Hdw(t)$$
$$x^{\pi}(o) = \xi .$$

By $\gamma(m,Q)$ we denote the Gaussian distribution on R^n with mean value m and the covariance matrix Q . It is clear that if the distribution of ξ is $\gamma(m_o,Q_o)$, then process x^{π} is both Gaussian and Markov.

In a private conversation Professor S. Mitter and Professor A. Moro posed a problem of characterizing those systems (20) which are controllable in the following sense:

For artitrary distributions $\gamma(m_o,Q_o)$, $\gamma(m_1,Q_1)$ and $\varepsilon > o$ there exist an affine strategy π and $t > o$ such that the distribution of $x^{\pi}(o)$ is $\gamma(m_o,Q_o)$ and $|E(x^{\pi}(t)) - m_1| < \varepsilon$ $|\text{Covariance } x^{\pi}(t) - Q_1| < \varepsilon$.

A satisfactory answer to this question is unknown to us. Here we give a complete characterization of systems (20) having a weaker property. Namely system (20) is called *affine controllable*, shortly a-controllable, if for arbitrary initial distribution $\gamma(m_o,Q_o)$, vector $m_1 \in R^n$ and $\varepsilon > o$ there exist a strategy $\pi = (K,r)$ and $t > o$ such that:

$$P(|x^{\pi}(t)-m_1|>\varepsilon) < \varepsilon .$$

An equivalent definition is the subject of the following proposition.

Proposition 7 System (20) is a-controllable if and only if for arbitrary initial distribution $\gamma(m_o,Q_o)$, final vector m_1 and $\varepsilon > o$, there exist a strategy π and $t > o$ such that

$$|E(x^{\pi}(t))-m_1| < \varepsilon$$

$$|\text{Covariance } x^{\pi}(t)| < \varepsilon .$$

A complete characterization of a-controllable systems is given by the following theorem, see [15].

<u>Theorem 5</u> System (20) is a-controllable if and only if the pair (F,G) is controllable.

Let us assume now that the pair (F,G) is controllable. An interesting question whether, in this case, one could restrict the classe of admissible strategies to those with constant feedback matrices K, still preserving a-controllability, was studied in [6]. In general the answer is negative as Example 3.2 in [6] shows. Paper [6] formulates also sufficient conditions for this property to hold. This question is closely related to the problem of characterizing those triplets of matrices (F,G,H) for which the closure of the set of all solutions Γ of the equation:

$$(F+GK)\Gamma + \Gamma(F+GK)^* + HH^* = 0,$$

$$\Gamma \geq 0,\ F + GK \text{ - stable matrix}$$

is identical with the set of all nonegative $n \times n$ matrices or at least contains the matrix 0.

References

[1] Arnold, L. and Kliemann, W., Qualitative theory of stochastic systems in: Probabilistic analysis and related topics, Vol. 3., Bharucha-Reid, A. T. (ed), New York, Academic Press 1982.

[2] Blumenthal, R. M. and Getoor, R. K., Markov processes and potential theory, Academic Press, New York 1968.

[3] Chojnowska-Michalik, A., Stochastic differential equations in Hilbert spaces, Banach Center Publications, vol. 5, PWN, Warsaw 1979

[4] Dym, H., Stationary measures for the flow of a linear differential equation driven by white noise, TAMS 123 (1966), 130-164

[5] Erickson, R. V., Constant coefficient linear equations driven by white noise, Ann. Math. Stat. 2 (1971), 820-823

[6] Ehrhardt, M. and Kliemann, W., Controllability of linear stochastic systems, Report 50, Universität Bremen, 1981

[7] Gaveau, B., Methods de control optimal en analyse complexe, Resolution d'equations de Monge-Ampere, Jour. Funct. Analy. 25 (1977), 391-411

[8] Getoor, R. K., Transience and recurrence of Markov processes. Lecture Notes in Mathematics 784, Seminaire de Probabilité XIV (1980), p. 397-409

[9] Helmes, K. and Zabczyk, J., Recurrence for controlled Markov chains, Preprint 333, Universität Bonn, SFB 72, 1980

[10] Hordijk, A., Dynamic programming and Markov potential theory, Mathematische Centrum, Amsterdam 1974

[11] Nisio, M., Lectures on stochastic control theory, ISI Lecture Notes No. 9, 1981

[12] Vinter, R. B., A representation of solutions to stochastic delay equations, Imperial College Report, 1979

[13] Zabczyk, J., Controllability of stochastic linear systems, Systems and Control Letters, vol. 1, No. 1, 1981, 25-31

[14] Zabczyk, J., Structural properties and limit behaviour of linear stochastic systems in Hilbert spaces, to appear in Banach Center Publications vol. 13.

[15] Zabczyk, J., Controllability of stochastic processes, Preprint 263, Institute of Mathematics, Polish Academy of Sciences, 1982

STOCHASTIC DIFFERENTIAL SYSTEMS

PART II : FILTER THEORY

Solving the Zakai Equation by Ito's Method

V. E. Beneš

Bell Laboratories
Murray Hill, New Jersey 07974

1. Introduction

Investigation of the Zakai equation for the evolution of the conditional density used in nonlinear filtering has depended heavily on methods from partial differential equations.[1] This approach, while completely successful, has nevertheless overlaid the essential simplicity of Zakai's equation with often ponderous PDE machinery. Even the "pathwise" construction[2] of solutions, which proceeds by using a device of Rozovsky[3] to factor the density into a natural martingale times a function satisfying a related nonstochastic PDE in which the observation is a parameter, is left with solving a parabolic equation with some random drifts and potentials.

Thus it is of interest to have an existence theory akin to that of Ito for stochastic DEs: a direct proof that Picard-type iterations converge to a solution, uniformly on intervals a.s. Also, for the purposes of stochastic control, and in order to replace the (Fokker-Planck, adjoint) operator A^* that usually occurs in the density formulation by the less skittish generator A, it is desirable to solve a "Zakai" equation for an unnormalized conditional *measure*. Such an equation is the natural translation

$$d(f,\sigma_t\mu) = (Af,\sigma_t\mu)dt + (hf,\sigma_t\mu)dy_t \qquad \sigma_0\mu = \mu$$

of the usual density form

$$dp = A^*pdt + hpdy_t \qquad p_0 = \frac{d\mu}{dx}$$

Here $\sigma_t\mu$ is the conditional measure of the diffusion x_t described by A based on observing $dy_t = h(x_t)dt + db_t$, with b an independent Brownian noise.

2. Preliminaries

A specially chosen metric for the weak topology of signed bounded measures on R^d will be our principal tool. A common way to pick such a metric is to choose a set $\{\phi_n\}$ of bounded uniformly continuous functions which are dense in the space U of bounded uniformly continuous functions, and to use as a metric the sum

$$d(\mu,\nu) = \sum_{n=1}^{\infty} 2^{-n} \frac{|(\phi_n,\mu) - (\phi_n,\nu)|}{\|\phi_n\|}$$

We shall show that $\{\phi_n\}$ can be chosen to be in C^∞, and so in the domain of A, and instead of 2^{-n} we shall use a different set of weights $\{w_n\}$ which have suitable properties relative to the ϕ_n and to A. We next give the background leading to a choice of $\{w_n\}$.

Let $\{\phi_n\}$ be such as would appear in the metric shown above. The convolutions

$$\psi_{mn}(x) = \frac{m^{1/2}}{(2\pi)^{d/2}} \int_{R^d} e^{-\frac{1}{2}|x-z|^2 m} \phi_n(z)dz$$

are bounded, uniformly continuous, and belong to C^∞. To show that they are dense in U, let $\{\phi_{n_j}\}$ be a subsequence of $\{\phi_n\}$ which converges to $\psi \in U$ uniformly. Then $\sup_j \|\phi_{n_j}\| < \infty$, and

$$\psi_{mn_j}(x) - \phi_{n_j}(x) = \frac{m^{1/2}}{(2\pi)^{d/2}} \int_{R^d} e^{-\frac{1}{2}|x-z|^2 m} [\phi_{n_j}(z) - \phi_{n_j}(x)] dz$$

Pick *first* $\delta_j \to 0$ and *then* $m_j \to \infty$ so that both

$$\frac{(m_j)^{1/2}}{(2\pi)^{d/2}} \int_{|x-z|>\delta_j} e^{-\frac{1}{2}|x-z|^2 m_j} dz \to 0 \quad \text{unif. in } z$$

$$\sup_{|x-z|\le\delta_j} |\phi_{n_j}(z) - \phi_{n(x)}| \to 0$$

This is possible because each ϕ_{n_j} is uniformly continuous and $\sup_j \|\phi_{n_j}\| < \infty$. Then $\psi_{m_j n_j} \to \psi$, so the ψ_{mn} are dense in U. We reorder them in a single sequence and rename them ϕ_n.

Assuming that the coefficients in A are bounded, we can now pick numbers w_n to use as weights in an equivalent metric

$$\sum_{n=1}^{\infty} w_n \frac{|(\phi_n,\mu) - (\phi_n,\nu)|}{\|\phi_n\|}$$

for the weak topology, to have the additional properties

(i) $\sum_{n=1}^{\infty} w_n^2 \|A\phi_n\|^2$

(ii) $\sum_{n=1}^{\infty} w_n^2 2^{4n} < \infty$.

3. Successive approximations

In order to obtain a "factorial" convergence similar to that in the classical Picard-Ito iteration scheme for stochastic DEs, it is useful to introduce the semigroup T_t, $t \ge 0$, whose generator is A, and to rewrite the Zakai equation suitably. Starting with

$$d(f,\sigma_t\mu) = (Af,\sigma_t\mu)dt + (hf,\sigma_t\mu)dy_t$$

we remark that y_t will be a translation of a Brownian motion, so it is enough to solve the equation with y_t replaced by a Wiener process w_t. We then view the last term as inhomogeneous, and bring in $T_t = \exp tA$ as a fundamental solution. Extending the Zakai equation to f depending C^1 on t by the derivation formula

$$d(f_t,\sigma_t\mu) = \left(\frac{\partial f}{\partial t}, \sigma_t\mu\right)dt + d(g,\sigma_t\mu)_{g=f_t}$$

we choose $f_t = T_{u-t}f$, $u > t$, and calculate that

$$d(T_{u-t}f,\sigma_t\mu) = -(AT_{u-t}f,\sigma_t\mu)dt + d(g,\sigma_t\mu)_{g=T_{u-t}f}$$

$$= (hT_{u-t}f,\sigma_t\mu)dw_t$$

or integrating and putting t for u, s for t,

$$(f,\sigma_t\mu) = (T_tf,\mu) + \int_0^t (hT_{t-s}f,\sigma_s\mu)dw_s \tag{1}$$

This formula exhibits the unnormalized conditional expectation of f as its unconditional expectation, the homogeneous term (T_tf,μ), plus a stochastic "variation of parameters" term given by the action of h and w on the fundamental solution.

To define a sequence of successive approximations, put $\sigma_t^0\mu \equiv \mu$ and then let

$$(f,\sigma_t^{n+1}\mu) = (f,\mu) + \int_0^t (Af,\sigma_s^{n+1}\mu)ds + \int_0^t (hf,\sigma_s^n\mu)dw_s \tag{2}$$

or equivalently

$$(f,\sigma_t^{n+1}\mu) = (T_tf,\mu) + \int_0^t (hT_{t-s}f,\sigma_s^n\mu)dw_s \tag{3}$$

Note the raised index $n+1$ in the Lebesgue term on the right of (2); if it were n as in the usual case (3) would not be equivalent. We define the differences $\Delta\sigma_t^n\mu$ as $\sigma_t^{n+1}\mu - \sigma_t^n\mu$, and find by linearity that

$$(f,\Delta\sigma_t^{n+1}\mu) = \int_0^t (hT_{t-s}f,\Delta\sigma_s^n\mu)dw_s$$

whence, assuming bounded h, so the integrals exist,

$$E(f,\Delta\sigma_t^{n+1}\mu)^2 = \int_0^t E(hT_{t-s}f,\Delta\sigma_s^n\mu)^2 ds$$

4. Convergence

Now $\sigma_t^1\mu$ is given by

$$(f,\sigma_t^1\mu) = (T_tf,\mu) + \int_0^t (hT_{t-s}f,\mu)dw_s$$

so that

$$(f,\Delta\sigma_t^0\mu) = (T_tf,\mu) - (f,\mu) + \int_0^t (hT_{t-s}f,\mu)dw_s$$

$$E(f,\Delta\sigma_t^0\mu)^2 \leq \|f\|^2\|\mu\|^2(4+\|h\|^2t)$$

The norm of μ here is the variational, the others are L_∞. We can restrict attention to t in the interval $[0,1]$. Thus there is a constant c such that $E(f,\Delta\sigma_t^0\mu)^2 \leq c\|f\|^2$. Assume as a hypothesis of induction that for bounded continuous g

$$E(g,\Delta\sigma_t^n\mu)^2 \leq c\|g\|^2\frac{\|h\|^{2n}}{n!}$$

Then by (3), putting $g = hT_{t-s}f$,

$$E(f, \Delta\sigma_t^{n+1}\mu)^2 M = c \int_0^t \|hT_{t-s}f\|^2 \frac{\|h\|^{2n}}{n!} ds \quad , \quad 0 \le t \le 1$$

$$\le c\|f\|^2 \frac{\|h\|^{2n+2}}{(n+1)!} .$$

By Doob's inequality, the martingales $M_t^n = \int_0^t (hf, \Delta\sigma_s^n\mu) dw_s$ satisfy

$$Pr\{\max_{t\le1} |M_t^n| \ge \ell\} \le \ell^{-2} E|M_1^n|^2$$

$$\le c\|fh\|^2 \frac{\|h\|^{2n}}{n!}$$

Also

$$Pr\left\{\max_{t\le1} \left| \int_0^t (Af, \Delta\sigma_s^{n+1}\mu) ds \right| \ge \ell\right\}$$

$$\le Pr\left\{\int_0^1 (Af, \sigma_s^{n+1}\mu)^2 ds \ge \ell\right\}$$

$$\le \ell^{-2} \int_0^1 E(Af, \Delta\sigma_s^{n+1}\mu)^2 ds$$

$$\le \ell^{-2} c\|Af\| \frac{\|h\|^{2n+2}}{(n+1)!}$$

Hence by (2), putting $f = \phi_m$

$$Pr\{\max_{t\le1} |(\phi_m, \Delta\sigma_t^{n+1})| \ge 2\ell\}$$

$$\le \ell^{-2} c \frac{\|h\|^{2n}}{n!} \left(\|\phi_m h\|^2 + \frac{\|A\phi_m\| \cdot \|h\|^2}{n+1}\right)$$

Since this inequality is valid for all $\ell > 0$ we can replace ℓ by $\frac{1}{2} w_m^{-2} 2^{-m} \|\phi_m\|^{-1} \ell$ to obtain first

$$Pr\{\max_{t \le 1} w_m |(\phi_m, \Delta\sigma_t^{n+1}\mu)| \ge \ell \cdot 2^{-m} \|\phi_m\|^{-1}$$

$$\le \ell^{-2} \frac{2^{4m} w_m^2}{\|\phi_m\|^2} 4c \frac{\|h\|^{2n}}{n!} \left(\|\phi_m h\|^2 + \frac{\|A\phi_m\|^2 \cdot \|h\|^2}{n+1} \right)$$

and then

$$Pr\{\max_{t \le 1} d(\sigma_t^{n+1}\mu, \sigma_t^n\mu) \ge \ell\}$$

$$\le 4c\ell^{-2} \left(\sum_{m=1}^{\infty} w_m^2 \frac{2^{4m}}{\|\phi_m\|^2} \left[\|\phi_m h\|^2 + \frac{\|A\phi_m\|^2 \cdot \|h\|^2}{n+1} \right] \right) \frac{\|h\|^{2n}}{n!}$$

The series is the large parentheses is a finite constant, so we can put $\ell = (n-2)!^{-\frac{1}{2}}$ and use the Borel-Cantelli lemma to conclude that

$$Pr\{\max_{t \le 1} d(\sigma_t^{n+1}\mu, \sigma_t^n\mu) \le (n-2)!^{-\frac{1}{2}} \text{ eventually}\} = 1$$

Thus the $\sigma_t^n\mu$ converge uniformly for $t \le 1$ to a measure-valued process $\sigma_t\mu = \mu + \sum_{j=0}^{\infty} \Delta\sigma_t^j\mu$.

To see that $\sigma_t\mu$ is a solution of (1) we write (3) as

$$(f, \sigma_t\mu) + (f, \sigma_t^{n+1}\mu - \sigma_t\mu) = \int_0^t (hT_{t-s}f, \sigma_s^n\mu - \sigma_s\mu)dw_s + (T_tf, \mu)$$

$$+ \int_0^t (hT_{t-s}f, \sigma_s\mu)dw_s$$

The terms in n approach zero uniformly on intervals, a.s. Uniqueness follows by a standard Gronwall argument.

References

1. E. Pardoux, Nonlinear filtering, prediction, and smoothing, Stochastic Systems, M. Hazewinkel, ed., NATO Advanced Study Institute Series, Reidel, Dordrecht.
2. M. H. A. Davis, Pathwise nonlinear filtering, ibid.
3. B. L. Rozovsky, Stochastic partial differential equations arising in nonlinear filtering problems, Uspekhi. Mat. Nauk., Vol. 27 (1972), pp. 213-214.

AUTHOR'S NOTE: During the conference J. Szpirglas raised the question whether the successive approximations $\sigma_t^n\mu$ really defined a sequence of measure-valued processes. It is not hard to see that if $\sigma_t^n\mu$ is such a process then

$$\int_0^t (hT_{t-s} 1_A, \sigma_t^n\mu) \, dw_s$$

defines a (random) countably additive set function of A which will also be bounded. Since

$$\sigma_t^{n+1}\mu (A) = (T_t 1_A, \mu) + \int_0^t (hT_{t-s} 1_A, \sigma_s^n\mu) \, dw_s$$

it follows that all the $\sigma^n\mu$ are well-defined.

SIMPLE AND EFFICIENT LINEAR AND NONLINEAR FILTERS

BY REGULAR PERTURBATION METHODS*

B. Z. Bobrovsky** and H. P. Geering
Laboratory for Measurement and Control,
ETH-Zentrum
CH-8092 Zurich, SWITZERLAND

1. Introduction

Consider the following problem: the signal process x_t and the observations y_t satisfy the Ito equations

$$dx_t = m(x_t,t)dt + \sigma(x_t,t)dw_t$$

$$dy_t = h(x_t,t)dt + \sqrt{N_0}\, d\nu_t \tag{1}$$

where w_t and ν_t are independent standard Brownian motions. The filtering problem is to construct the optimal estimator $\hat{x}_t = E\,(x_t|y_0{}^t)$. Since no exact construction of $\hat{x}_t$ is known we shall use a perturbation technique to construct an approximation to $\hat{x}_t$. We assume the measurements noise intensity N_0 is small (this is the case in many practical applications [1])and use this fact for constructing a regular perturbation scheme for the approximation of $\hat{x}_t$. Asymptotic error bounds in this case were presented in [1] and [2], and error expressions for the linear case were given in [3] and [4]. In this work we apply a regular perturbation method for calculating the filter gain and the filtering error for (1). Error expressions which are between the bounds of [1] and [2] are presented in Section 2, and the results of [4] are generalized for the linear, time varying case in Section 3. Furthermore, it turns out that the structure of the filter (both for the nonlinear and the linear time varying cases) is much simpler than what was believed to be the case since the on line integration of the corresponding Riccati-like differential equation can be avoided, resulting in fairly fast filters.

Another filter which avoids the integration of Riccati's equation is presented in [5], where the filter gain is approximated by a constant. This approximation may be inefficient in some cases. We illustrate this by some examples. We show that for $N_0 << 1$ a constant gain may not be close to the optimal solution.

Finally, we note that in the linear case, our filter can approximate the optimal one

* This work was supported by research grant No. 330342-10 of the ETH, Zurich.

** On leave from The Department of Electrical Engineering, Tel Aviv University, Ramat Aviv 69978, ISRAEL.

to higher degree of approximation if we take more terms in the asymptotic expansion of the filter gain in powers of N_0. In the nonlinear case, higher order approximations are presented in [6] and [7].

2. Asymptotic one-dimensional nonlinear filtering

Let x_t and y_t in (1) be scalars and, for simplicity, let $m(x_t,t) = m(x_t)$, $\sigma(x_t,t) = \sigma$, $h(x_t,t) = h(x_t)$. We assume $m(\cdot)$ and $h(\cdot)$ to be analytic functions. Setting $e_x = x_t - \hat{x}_t$ and $e_h = h(x_t) - \hat{h}(x_t)$ we can write the optimal filter equation in the form [8]

$$d\hat{x}_t = \hat{m}(x_t)\,dt + \widehat{e_x e_h}\,N_0^{-1}\,(dy_t - \hat{h}(x_t)\,dt) \tag{2}$$

We make the following assumptions: (i) $\widehat{e_x^2}$ and $\widehat{e_h e_x}$ have asymptotic expansions in powers of $\sqrt{N_0}$ as $N_0 \to 0$, and (ii) $\widehat{|e|^n} = 0\,(\widehat{e^2}^{n/2})$. Under these assumptions we obtain the following result $\hat{x}_t = x_t^* + 0\,(\sqrt{N_0})$ where

$$dx_t^* = m(x_t^*)\,dt + \frac{\sigma}{\sqrt{N_0}}\,(dy_t - h(x_t^*)\,dt) \tag{3}$$

The filter (3) is obviously a constant gain filter. Next we outline the derivation of (3). First we observe that

$$\widehat{e_x e_h} = \sqrt{N_0}\,\sigma + 0(N_0) \quad \text{as} \quad N_0 \to 0 \tag{4}$$

Indeed, let

$$K = \widehat{e_x e_h} \sim \sqrt{N_0}\,K_0 + N_0\,K_1 + \dots$$

then, by [8]

$$\begin{aligned} dK &= d\,\widehat{xh} - d(\hat{x}\,\hat{h}) = \\ &= \{\widehat{hm} + \widehat{xh'm} + \tfrac{1}{2}\sigma^2\,(2\widehat{h'} + \widehat{xh''}) - \hat{h}\hat{m} - \hat{x}\,\widehat{h'm} - \tfrac{1}{2}\sigma^2\,\hat{x}\,\widehat{h''} \\ &\quad - \frac{\widehat{e_h^2}\,K}{N_0}\}\,dt + \frac{\widehat{e_h^2 e_x}}{\sqrt{N_0}}\,d\mathfrak{J} \end{aligned} \tag{5}$$

where $d\mathfrak{J} = \dfrac{dy - \hat{h}dt}{\sqrt{N_0}}$ (thus $\mathfrak{J}$ is the normalized innovation process).

The first term in the asymptotic expansion of K satisfies the equation

$$dK_0 \simeq \{K_0\,[m'(\hat{x}) + h''(\hat{x})\,m(\hat{x}) + h'(\hat{x})\,m'(\hat{x})] + \sigma^2\,[h'(\hat{x}) + \widehat{e_x^{\,2}}\,h'''(\hat{x})]$$

$$- \frac{K_0^{\,2} h'(\hat{x})}{N_0}\,\}\,dt + \frac{\widehat{e_h^{\,2} e_x}}{\sqrt{N_0}}\,d\mathcal{J} \qquad (6)$$

Now, by (i), for $N_0 \to 0$ the expression in the first brackets vanishes, and so does $\widehat{e_x^{\,2}}\,h'''(\hat{x})$. By (ii) the last term can be neglected, hence (4) follows. For $h'(x) > 0$ we have

$$\widehat{e_x^{\,2}} \simeq \frac{K}{h'(\hat{x}_t)} \simeq \frac{\sqrt{N_0}\,\sigma}{h'(\hat{x}_t)} \qquad (7)$$

If $E[\frac{1}{h'(x_t)}] < \infty$ we also have by (i)

$$E[e_x^{\,2}] = \overline{(x_t - \hat{x}_t)^2} = \sqrt{N_0}\,\sigma\,E[\frac{1}{h'(x_t)}] + O(N_0). \qquad (8)$$

Note that the first term in (8) is between the upper and lower bounds of [1]. A more precise derivation of (3), (7) and (8), containing higher order terms is given in [6] and [7].

Example: Let $m(x) = -x$, $\sigma = 1$ and $h(x) = G\,x^3 + \frac{1-G}{1.65}\,x$, $0 \leq G \leq 1$.

For $G=1$ this is the cubic sensor problem [9]. We ran simulations for various values of G of the Extended Kalman Filter, the Gaussian Filter [8], the Constant Gain Filter (3), and for $G=1$ we used the results of [9] for the optimal filter. The Constant Gain Filter was found to be within the distance of 1dB to the optimal and the Gaussian filters, for values of N_0 for which

$\frac{\overline{e_x^{\,2}}}{\overline{x^2}} < 0.25$. Similar results were observed for $h(x) = G\,x^5 + \frac{1-G}{4.37}\,x$.

The filters (3) were seen to be much faster than the others.

3. Asymptotic time dependent linear filtering

We consider now the system of vector equations

$$dx_t = A(t)\, x_t\, dt + B(t)\, dw_t$$

$$dy_t = H(t)\, x_t\, dt + \sqrt{N_0}\, d\nu_t \tag{9}$$

where the matrices A, B, and H are assumed analytic functions of t. The solution of the corresponding Riccati equation is assumed to have an asymptotic expansion as $N_0 \to 0$. Neglecting the initial layer we also assume that the outer solution is symmetric and positive definite, which is unique. We note that unlike the case of "cheap control" here the initial layer has negligible influence on the results. We construct a solution to Riccati's equation by assuming the covariance matrix P has a series expansion in powers of N_0. We note, however, that each matrix element $[P]_{ij}$ may have an expansion in different powers of N_0. This procedure can be carried out off line for any matrices A, B, and H. General expressions are not easy to get in terms of A, B, H, see [3]. However, for the canonical case of [4], it can be shown that the leading term in the expansion of P is exactly the one in [4], although the derivation there was done for time independent matrices. The reason for this is that for $N_0 \ll 1$, the memory of the filter is "short" so A, B, and H can be considered constant. Since the details in [4] are tedious, we illustrate this claim with an example.

Example: Let

$$A = \begin{bmatrix} 0 & 1 \\ -a_0 & -a_1 \end{bmatrix} \qquad B = \begin{bmatrix} 0 \\ b \end{bmatrix} \qquad H = [\, h \,,\, 0 \,]$$

P is given by

$$P = \begin{bmatrix} \frac{\sqrt{2b}\, N_0^{3/4}}{h^{3/2}} & \frac{b\sqrt{N_0}}{h} \\ \\ \frac{b\sqrt{N_0}}{h} & \frac{\sqrt{2}b^{3/2}\, N^{1/4}}{\sqrt{h}} \end{bmatrix} +$$

$$+ \begin{bmatrix} \left(\frac{\dot{h}}{h^3} - \frac{a_1}{h^2}\right) N_0 & \left(\frac{\sqrt{2b}\,\dot{h}}{4\,h^{5/2}} - \frac{a_1\sqrt{2b}}{h^{3/2}}\right) N_0^{3/4} \\ \\ \left(\frac{\sqrt{2b}\,\dot{h}}{4\,h^{5/2}} - \frac{a_1\sqrt{2b}}{h^{3/2}}\right) N_0^{3/4} & \left(\frac{b\dot{h}}{2h^2} - \frac{2a_1 b}{h}\right) \sqrt{N_0} \end{bmatrix} + \ldots \tag{10}$$

Note that the leading matrix is the same as in [4]. The leading terms in the filter gain vector $\underline{k}$ are given by

$$\underline{k} = PH^T N_0^{-1} = \begin{bmatrix} \sqrt{\frac{2b}{h}} \cdot \frac{1}{N_0^{1/4}} \\ \\ \frac{b}{\sqrt{N_0}} \end{bmatrix} + \; . \; . \; . \; . \; . \tag{11}$$

It is clear in this case that we can assume neither P nor $\underline{k}$ to be constant. In this example the excess of poles over zeroes is two hence each term in the expansion differs from the next one by $N_0^{1/4}$ only. It follows that for realistic values of N_0, more than one term in the expansion (11) has to be taken to obtain a significant degree of accuracy.

References

[1] B. Z. Bobrovsky, M. Zakai, "Asymptotic bounds on the minimal mean square error of nonlinear filtering," in Stochastic Systems: The Mathematics of filtering and Identification and Applications (Hazewinkel and Willems, editors), 1981, D. Reidel Publishing Co., pp. 573-581.

[2] ___, "Asymptotic a Priori estimates for the error in the nonlinear filtering Problem," IEEE Trans. on Information Theory, vol. I.T. 28, No. 2, March: pp. 371-376, 1982.

[3] R. E. O'Malley, Jr., "A more direct solution of the nearly singular linear regulator problem," SIAM J. Control, vol. 14, No. 6, November: pp. 1063-1077, 1976.

[4] U. Shaked, B. Z. Bobrovsky, "The Asymptotic minimum variance estimate of stationary linear single output processes," IEEE Trans. on A.C., vol. 26, No. 2, April: pp. 498-504, 1981.

[5] M. G. Safonov, M. Athans, "Robustness and computational aspects of nonlinear stochastic estimators and regulators," IEEE Trans. on A.C., vol. 23, No. 4, August: pp. 717-725, 1978.

[6] R. Katzur, B. Z. Bobrovsky, Z. Schuss, "Asymptotic analysis of the optimal filtering problem for one dimensional diffusions measured in a low noise channel, Part I," submitted to SIAM J. Appl. Math.

[7] ___, "Asymptotic analysis of the optimal filtering problem for one dimensional diffusions measured in a low noise channel, Part II."

[8] A. H. Jazwinski, "Stochastic processes and filtering theory," Academic Press, New York, 1971.

[9] R. S. Bucy, J. Pages, "A priori error bounds for the cubic sensor problem," IEEE Trans. on A.C., vol. 23, No. 1, February: pp. 88-91, 1978.

THE NON LINEAR FILTERING EQUATIONS

Robert J. Elliott

University of Hull, England.

1. INTRODUCTION.

There have been many derivations of the equations of nonlinear filtering. See, for example, the books of Kallianpur [3] and Lipster and Shiryayev [4] . A central part of these proofs is the determination of an integrand process in the equation giving the nonlinear filter. Below we observe, adapting a method of Wong [6], that this integrand process can be obtained immediately using the unique decomposition of certain special semimartingales.

We shall assume that all processes are defined on a fixed probability space (Ω, F, P) for time $t \in [0,T]$. We suppose there is a right continuous filtration $\{F_t\}$ of sub σ-fields of F , and that the filtration is complete. The signal process is adapted to $\{F_t\}$, as is the observation process $\{y_t\}$, $t \in [0,T]$ .Writing $Y_t = \sigma\{y_s : s \leq t\}$ we have $Y_t \subset F_t$, the inclusion being in general strict.

NOTATION

If $\{\eta_t\}$ is any process write

$$\hat{\eta}_t = E[\eta_t | Y_t] .$$

OBSERVATION PROCESS 1.2

We shall suppose that the OBSERVATION PROCESS $\{y_t\}$, $0 \leq t \leq T$, is an m-dimensional semimartingale of the form:

$$y_t = \int_0^t z_u \, du + \int_0^t \alpha(y_u) dw_u, \; y_0 = 0 \in R^m .$$

Here: a) $\{w_t\}$ is a standard m-dimensional Brownian motion,

b) $\alpha : C([0,T];R^m) \to L(R^m,R^m)$ is a nonsingular matrix function such that

$\|\alpha(y)\| \geq \delta > 0$ for some δ, and which satisfies a Lipschitz condition of the form $\|\alpha(y_t) - \alpha(y'_t)\| \leq K|y - y'|_t^*$,

c) $\{z_t\}$ is some functional of the signal process, (for example, $z_t = h(x_t)$), and for simplicity we suppose that $E[\int_0^T z_u^2 du] < \infty$.

DEFINITION 1.3

The process $\{\nu_t\}$, $0 \leq t \leq T$, defined by $\nu_t = \int_0^t \frac{dy_u}{\alpha(y_u)} - \int_0^t \frac{\hat{z}_u}{\alpha(y_u)} du$, $\nu_0 = 0 \in R^m$, is the INNOVATIONS PROCESS.

LEMMA 1.4

$\{\nu_t\}$ is a standard Brownian motion with respect to the filtration $\{Y_t\}$.

PROOF.

We first prove that $\{\nu_t\}$ is a $\{Y_t\}$ martingale. For $s \leq t$:

$$E[\nu_t - \nu_s | Y_s] = E[\int_s^t \frac{dy_u - \hat{z}_u du}{\alpha(y_u)} \quad | Y_s]$$

$$= E[\int_s^t \frac{z_u - \hat{z}_u}{\alpha(y_u)} du + w_t - w_s | Y_s] = 0 \text{ a.s.}$$

by Fubini's Theorem. With respect to the filtration $\{F_t\}$, $\nu_t = (\nu'_t, \ldots, \nu^m_t)$ is an m-dimensional semimartingale. By the differentiation rule

$$\nu^i_t \nu^j_t = \int_0^t \nu^i_u d\nu^j_u + \int_0^t \nu^j_u d\nu^i_u + \langle w^i, w^j \rangle_t .$$

Therefore, with respect to the $\{F_t\}$ filtration $\langle \nu^i, \nu^j \rangle_t = \langle w^i, w^j \rangle_t = \delta_{ij} t$. This process is deterministic, so with respect to the $\{Y_t\}$ filtration $\langle \nu^i, \nu^j \rangle = \delta_{ij} t$.

$\{\nu_t\}$ is continuous, so by Lévy's characterization of Brownian motion $\{\nu_t\}$ is an m-dimensional Brownian motion with respect to the filtration $\{Y_t\}$.

The martingale representation result of Fujusaki, Kallianpur and Kunita, [2], in this context states the following:

THEOREM 1.5

If $\{N_t\}$ is a locally square integrable martingale with respect to the filtration $\{Y_t\}$ then there is a $\{Y_t\}$ predictable m-dimensional process $\{\gamma_s\}, 0 \le s \le T$, and an increasing sequence of stopping times $\{T_n\}$ such that:

$$\lim_n T_n = T \quad a.s.$$

$$E[\int_0^{T_n} |\gamma_u|^2 du] < \infty$$

$$N_t = E[N_o] + \int_0^t \gamma_u d\nu_u \quad . \quad a.s.$$

2. THE FILTERING EQUATION

To obtain the general filtering equation we shall consider a real $\{F_t\}$ semimartingale $\{\xi_t\}$, $0 \le t \le T$, and obtain a stochastic differential equation satisfied by $\hat{\xi}_t$. The kind of semimartingale we have in mind is some real valued function ϕ of the signal process $\{x_t\}$.

THEOREM 2.1

Suppose $\{\xi_t\}$, $0 \le t \le T$, is a real $\{F_t\}$ semimartingale of the form

$$\xi_t = \xi_o + \int_0^t \beta_u du + N_t ,$$

where $\{N_t\}$ is an $\{F_t\}$ martingale ,(either continuous or square integrable). The observation process is of the form:

$$dy_t = z_t dt + \alpha(y_t) dw_t = \hat{z}_t dt + \alpha(y_t) d\nu_t , \qquad y_o = 0 \in R^m ,$$

where $\{w_t\}$, $0 \le t \le T$, is a standard m-dimensional Brownian motion and $\langle N, w^i \rangle_t = \int_0^t \lambda_u^i du$, $1 \le i \le m$. Then $\{\hat{\xi}\}$ is given by the stochastic differential equation

$$\hat{\xi}_t = \hat{\xi}_o + \int_0^t \hat{\beta}_u du + \int_0^t (\hat{\lambda}_u + \alpha^{-1}(y_u)(\widehat{\xi_u z_u} - \hat{\xi}_u \hat{z}_u))' d\nu_u \, . \tag{2.1}$$

PROOF

The proof below is an extension of an idea of Wong [6] and uses the unique decomposition of special semimartingales.

Define

$$M_t = \hat{\xi}_t - \hat{\xi}_0 - \int_0^t \hat{\beta}_u du .$$

Then for $0 \geq s \geq T$:

$$E[M_t - M_s | Y_s] = E[\hat{\xi}_\tau - \hat{\xi}_s - \int_s^t \hat{\beta}_u du | Y_s] .$$

However

$$E[\hat{\xi}_t - \hat{\xi}_s | Y_s] = E[\xi_t - \xi_s | Y_s] = E[\int_0^t E[\beta_u | Y_u] du | Y_s] + E[E[N_t - N_s | F_s] Y_s]$$

$$= E[\int_s^t \hat{\beta}_u du | Y_s]$$

because $\{N_t\}$ is an $\{F_t\}$ martingale. Therefore, $\{M_t\}$ is locally a square integrable martingale, so by Theorem 1.5 there is a $\{Y_t\}$ predictable process $\{\gamma_u\}$ such that

$$M_t = \int_0^t \gamma_u' d\nu_u \quad a.s.$$

and we can write

$$\hat{\xi}_t = \hat{\xi}_0 + \int_0^t \hat{\beta}_u du + \int_0^t \gamma_u' d\nu_u \quad . \tag{2.2}$$

(The prime here denotes the transpose of column vector γ_u to row vector γ_u').

We now wish to determine γ_u . $\xi \in R$ and $y \in R^m$, but by the Ito differentiation rule

$$\xi_t y_t = \xi_0 y_0 + \int_0^t \xi_u (z_u du + \alpha(y_u) dw_u) + \int_0^t y_u (\beta_u du + dN_u) + \langle N, y \rangle_t . \tag{2.3}$$

However,

$$\langle N, y \rangle = \int_0^t \alpha(y^u) \lambda_u du .$$

The integrals

$$H_1(t) = \int_0^t \xi_u \, \alpha(y_u) dw_u \, , \; H_2(t) = \int_0^t y_u dN_u$$

are local martingales with respect to the filtration $\{F_t\}$. Therefore, the processes $\hat{H}_1$, $\hat{H}_2$ are local martingales with respect to the filtration $\{Y_t\}$.

Consider the processes

$$K_1(t) = \int_0^t \xi_u z_u du, \; K_2(t) = \int_0^t y_u \beta_u du, \; K_3(t) = \int_0^t \alpha(y_u) \lambda_u du .$$

Then by a calculation similar to that for M above the processes:

$$\tilde{K}_1(t) = \hat{K}_1(t) - \int_0^t \widehat{\xi_u z_u}\, du, \quad \tilde{K}_2(t) = \hat{K}_2(t) - \int_0^t y_u \hat{\beta}_u du,$$
$$\tilde{K}_3(t) = \hat{K}_3(t) - \int_0^t \alpha(y_u)\hat{\lambda}_u du$$

are local martingales with respect to the filtration $\{Y_t\}$.

Therefore, from (2.3)

$$\widehat{\xi_t y_t} = \hat{\xi}_t y_t = \hat{\xi}_o y_o + \hat{H}_1(t) + \hat{H}_2(t) + \tilde{K}_1(t) + \int_0^t \widehat{\xi_u z_u}\, du + \tilde{K}_2(t) + \int_0^t y_u \hat{\beta}_u du$$

$$+ \tilde{K}_3(t) + \int_0^t \alpha(y_u)\hat{\lambda}_u du . \tag{2.4}$$

Because this represents $\hat{\xi}_t y_t$ as the sum of local martingales plus continuous, (and so predictable), bounded variation processes, we see $\hat{\xi}_t y_t$ is a special semimartingale with respect to the filtration $\{Y_t\}$. However, using (2.2) and the differentiation rule

$$\hat{\xi}_t y_t = \hat{\xi}_o y_o + \int_0^t \hat{\xi}_u(\hat{z}_u du + \alpha(y_u)d\nu_u)$$
$$+ \int_0^t y_u(\hat{\beta}_u du + \gamma'_u d\nu_u) + \int_0^t \alpha(y_u)\gamma_u du . \tag{2.5}$$

The integrals with respect to $\{\nu_t\}$ are again local martingales, and the remaining integrals give continuous, and so predictable, processes. The two canonical decompositions of the special semimartingale $\hat{\xi}_t y_t$ must be the same, so equating the integrands in the bounded variation terms:

$$\widehat{\xi_u z_u} + \alpha(y_u)\hat{\lambda}_u = \hat{\xi}_u \hat{z}_u + \alpha(y_u)\gamma_u \quad a.s.$$

Therefore

$$\gamma_u = \alpha(y_u)^{-1} (\widehat{\xi_u z_u} - \hat{\xi}_u \hat{z}_u) + \hat{\lambda}_u .$$

Substituting in (2.2) the result follows.

THEOREM 2.2.

Suppose that the signal process $\{X_t\}$ is the unique strong solution of the stochastic differential equation

$$dX_t = g(X_t)dt + \sigma(X_t)dB_t , \quad 0 \le t \le T,$$

with initial condition x_o independent of $F^B_{o,T}$ and where the coefficients satisfy appropriate Lipschitz and growth conditions. Suppose that ϕ is a twice continuously differentiable function on R^d so that

$$\phi(X_t) = \phi(X_o) + \int_o^t L\phi(X_u)du + \int_o^t \nabla\phi.\sigma(X_u)dB_u ,$$

where

$$L = \tfrac{1}{2} \sum_{i,j=1}^{d} a^{ij}(x) \frac{\partial^2}{\partial x^i \partial x^j} + \sum_{i=1}^{d} g^i(x) \frac{\partial}{\partial x^i}$$

is the infinitesimal generator of $\{X_t\}$.

The observation process $\{y_t\}$ will be as above with $z_t = h(X_t)$, and suppose

$$\langle B^i, w^j \rangle_t = \int_o^t \rho_u^{ij}\, du, \quad 1 \le i \le d, \quad 1 \le j \le m \quad .$$

For $f \in L^2_{\text{loc}}(R^d)$ write

$$\Pi_t(f) = E[f(X_t)|Y_t] .$$

Then

$$\Pi_t(\phi) = \Pi_o(\phi) + \int_o^t \Pi_u(L\phi)du$$

$$+ \int_o^t (\Pi_u(\nabla\phi.\sigma.\rho) + \alpha^{-1}(y_u)(\Pi_u(\phi h) - \Pi_u(\phi)\Pi_u(h)))' d\nu_u \tag{2.6}$$

PROOF.

$\phi(X_t)$ plays the role of the semimartingale ξ_t in Theorem 2.1 above.

$N_t = \int_o^t \nabla\phi.\sigma(X_u)dB_u$ is a martingale, and for $1 \le j \le m$

$\langle N, w^j \rangle_t = \int_o^t \nabla\phi.\sigma(X_u).\rho_u^j.du$ where $\rho_u^j = (\rho_u^{1j}, \dots, \rho_u^{dj})$.

Therefore, in the notation of Theorem 2.1

$$\lambda_u^j = \nabla\phi.\sigma(X_u).\rho_u^j \quad .$$

Also

$$\hat{z}_u = \Pi_u(h) .$$

Substituting in the formula of Theorem 2.1 gives the stated result.

3. THE ZAKAI EQUATION.

This approach was first considered by Zakai [7] .

The signal process $\{X_t\}$ will be, as above, a d-dimensional homogeneous Markov process, which is the unique strong solution of the stochastic differential equation

$$dX_t = g(X_t)dt + \sigma(X_t)dB_t \, , \quad 0 \le t \le T \, , \tag{3.1}$$

with initial condition X_0 independent of $F^B_{0,T} = \sigma\{B_s : 0 \le s \le T\}$. The observation process will be as in Theorem 2.1.

In this section instead of $\Pi_t(\phi)$ we shall consider an 'unormalised' conditional expectation and show it satisfies a less complicated equation than (18.11). First we introduce a new probability measure P_0 on (Ω, F) defined by $\frac{dP_0}{dP} = \Lambda_T^{-1}$ where

$$\Lambda_t^{-1} = \exp\{- \int_0^t (\alpha^{-1}(y_u)h(X_u))' dw_u - \tfrac{1}{2} \int_0^t |\alpha^{-1}(y_u)h(X_u)|^2 du\}$$

$$= \exp\{- \int_0^t (\alpha^{-1}(y_u)h(X_u))' \alpha^{-1}(y_u)dy_u + \tfrac{1}{2} \int_0^t |\alpha^{-1}(y_u)h(X_u)|^2 du\} .$$

Consequently,

$$\Lambda_t = \exp\{\int_0^t (\alpha^{-1}(y_u)h(X_u))' \alpha^{-1}(y_u)dy_u - \tfrac{1}{2} \int_0^t |\alpha^{-1}(y_u)h(X_u)|^2 du\} .$$

By Girsanov's Theorem, under P_0 the processes $\nu_t = (\nu^1, \dots, \nu^d)$ and y_t are standard d and m dimensional Brownian motions, respectively where

$$d\nu_t^i = dB_t^i + \langle \rho^i, \alpha^{-1}h \rangle dt$$

and

$$d\tilde{y}_t = \alpha^{-1}dy = dw_t + \alpha^{-1}hdt = d\nu_t + \alpha^{-1}(y_t)\Pi_t(h)dt.$$

Furthermore, under P_0 $\langle \nu^i, y^j \rangle_t = \int_0^t \rho_u^{ij} \, du$.

Now Λ_t is an $\{F_t\}$ martingale under P_0 and $\Lambda_t = E_0[\frac{dP}{dP_0} | F_t]$ where E_0 denotes expectation with respect to P_0 . From Loève [5] §27.4 for any integrable function ϕ

$$\Pi_t(\phi) = E[\phi(X_t)|Y_t] = \frac{E_o[\Lambda_t\phi(X_t)|Y_t]}{E_o[\Lambda_t|Y_t]} = \frac{\sigma_t(\phi)}{\sigma_t(1)} \tag{3.2}$$

where

$$\sigma_t(\phi) = E_o[\Lambda_t\phi(X_t)|Y_t]$$

is the unnormalized conditional expectation we shall now investigate. By definition

$$\sigma_t(\phi) = \sigma_t(1) \cdot \Pi_t(\phi) ,$$

so, as we already have an expression (2.6) for $\Pi_t(\phi)$, we shall derive an equation for $\sigma_t(\phi)$ by obtaining an equation for $\sigma_t(1) = E_o[\Lambda_t|Y_t]$ and using the differentiation rule for the product.

NOTATION 3.1

Write $\hat{\Lambda}_t = E_o[\Lambda_t|Y_t]$.

(Note the expectation is with respect to measure P_o).

THEOREM 3.2

$$\hat{\Lambda}_t = \exp \{\int_0^t(\alpha^{-1}(y_u)\Pi_u(h))'\alpha^{-1}(y_u)dy_u - \tfrac{1}{2}\int_0^t|\alpha^{-1}(y_u)\Pi_u(h)|^2du\}$$

Here

$$\Pi_u(h) = E[h(X_u)|Y_u] ,$$

and is an expectation with respect to measure P .

PROOF

We have already noted that, under P_o , Λ_t is an $\{F_t\}$ martingale and

$$\Lambda_t = 1 + \int_0^t \Lambda_u(\alpha^{-1}(y_u)h(X_u))'\alpha^{-1}(y_u)dy_u \quad . \tag{3.3}$$

Now $\hat{\Lambda}_t$ is an $\{Y_t\}$ martingale under P_o . Therefore, from Theorem 1.5 there is a locally square integrable, $\{Y_t\}$-predictable, m-dimensional, process $\{\eta_s\}$, such that for all $t \in [0,T]$:

$$\hat{\Lambda}_t = 1 + \int_0^t \eta_u' . \alpha^{-1}(y_u) dy_u \quad . \tag{3.4}$$

As in Theorem 2.1 we shall identify $\{\eta_u\}$ by using the unique decomposition of special semimartingales. Using the differentiation rule for Λ_t and $y_t \in R^m$ we have from (3.1) and (3.3):

$$\Lambda_t y_t = \int_0^t \Lambda_u dy_u + \int_0^t y_u \Lambda_u (\alpha^{-1}(y_u) h(X_u))' \alpha^{-1}(y_u) dy_u$$

$$+ \int_0^t \Lambda_u (\alpha^{-1}(y_u) h(X_u))' \alpha(y_u) du. \tag{3.5}$$

The integrals

$$H_1(t) = \int_0^t \Lambda_u dy_u$$

$$H_2(t) = \int_0^t y_u \Lambda_u (\alpha^{-1}(y_u) h(X_u))' \alpha^{-1}(y_u) dy_u$$

are local martingales under measure P_0 with respect to the filtration $\{F_t\}$. Therefore, the processes $\hat{H}_1(t) = E_0[H_1(t) | Y_t]$, $\hat{H}_2(t) = E_0[H_2(t) | Y_t]$ are local martingales under measure P_0 with respect to the filtration $\{Y_t\}$. Write

$$K(t) = \int_0^t \Lambda_u (\alpha^{-1}(y_u) h(X_u))' \alpha(y_u) du$$

and

$$\hat{K}(t) = E_0[K(t) | Y_t].$$

Then

$$\tilde{K}(t) = \hat{K}(t) - \int_0^t (\alpha^{-1}(y_u) \widehat{\Lambda_u h}(X_u))' \alpha(y_u) du$$

is a local martingale under measure P_0 with respect to the filtration $\{Y_t\}$. From (3.5), therefore,

$$E_0[\Lambda_t y_t | Y_t] = \hat{\Lambda}_t y_t = \hat{H}_1(t) + \hat{H}_2(t) + \tilde{K}(t)$$

$$+ \int_0^t (\alpha^{-1}(y_u) \widehat{\Lambda_u h}(X_u))' \alpha(y_u) du \ .$$

This represents $\hat{\Lambda}_t y_t$ as the sum of a local martingale and a continuous (and so predictable) process of bounded variation. Consequently, $\hat{\Lambda}_t y_t$ is a special semimartingale and this representation is unique. However, from (3.4) and Definition (1.3):

$$\hat{\Lambda}_t y_t = \int_0^t \hat{\Lambda}_u dy_u + \int_0^t y_u (\eta_u' \alpha^{-1}(y_u) dy_u) + \int_0^t \eta_u' \alpha(y_u) du.$$

Again the first two integrals are local $\{Y_t\}$ martingales under measure P_0 . By the uniqueness of the decomposition of special semimartingales

$$\eta_u = \alpha^{-1}(y_u)\widehat{\Lambda_u h(X_u)} = \alpha^{-1}(y_u)E_0[\Lambda_u h(X_u)|Y_t] \quad .$$

However, from (3.2) this is

$$= \alpha^{-1}(y_u)\hat{\Lambda}_u\Pi_u(h) \quad .$$

Substituting in (3.4)

$$\hat{\Lambda}_t = 1 + \int_0^t \hat{\Lambda}_u(\alpha^{-1}(y_u)\Pi_u(h))'\alpha^{-1}(y_u)dy_u \quad . \tag{3.6}$$

From the exponential formula of Doléans-Dade [1] this equation has the unique solution

$$\hat{\Lambda}_t = \exp\{\int_0^t(\alpha^{-1}(y_u)\Pi_u(h))'\alpha^{-1}(y_u)dy_u - \tfrac{1}{2}\int_0^t|\alpha^{-1}(y_u)\Pi_u(h)|^2du\} ,$$

and the Theorem is proved.

THEOREM 3.3

For any $\phi \in C_b^2(R^d)$, $\sigma_t(\phi)$ satisfies the equation:

$$\sigma_t(\phi) = \sigma_0(\phi) + \int_0^t \sigma_u(L\phi)du + \int_0^t\{\sigma_u(\nabla\phi.\sigma.\rho) + \alpha^{-1}(y_u)\sigma_u(\phi h)\}'\alpha^{-1}(y_u)dy_u \quad . \tag{3.7}$$

PROOF

We have noted that

$$\sigma_t(\phi) = \hat{\Lambda}_t\Pi_t(\phi)$$

so from (2.6) and (3.6)

$$\hat{\Lambda}_t\Pi_t(\phi) = \sigma_0(\phi) + \int_0^t\hat{\Lambda}_u\Pi_u(L\phi)du$$

$$+ \int_0^t \hat{\Lambda}_u\{\Pi_u(\nabla\phi.\sigma.\rho) + \alpha^{-1}(y_u)(\Pi_u(\phi h) - \Pi_u(\phi)\Pi_u(h))\}'dv_u$$

$$+ \int_0^t\Pi_u(\phi)\hat{\Lambda}_u(\alpha^{-1}(y_u)\Pi_u(h))'\alpha^{-1}(y_u)dy_u$$

$$+ \int_0^t \hat{\Lambda}_u\{\Pi_u(\nabla\phi.\sigma.\rho) + \alpha^{-1}(y_u)(\Pi_u(\phi h) - \Pi_u(\phi)\Pi_u(h))\}'(\alpha^{-1}(y_u)\Pi_u(h))du.$$

$$= \sigma_0(\phi) + \int_0^t\sigma_u(L\phi)du + \int_0^t\{\sigma_u(\nabla\phi.\sigma.\rho) + \alpha^{-1}(y_u)\sigma_u(\phi h)\}'\alpha^{-1}(y_u)dy_u ,$$

by (3.2)

REMARKS 3.4

Note the much simpler form of the equation (3.7) for $\sigma_t(\phi)$ compared with (2.6) for $\Pi_t(\phi)$: (3.7) is linear σ_t , whereas (2.6) is quadratic in Π_t . In particular, when the signal noise $\{B_t\}$ is independent of the noise $\{w_t\}$ in the observation, so that the predictable quadratic covariation matrix $\rho = (\rho^{ij}) = (\langle B^i, w^j\rangle)$

is zero, the unormalized density $\sigma_t(\phi)$ satisfies the equation

$$\sigma_t(\phi) = \sigma_o(\phi) + \int_o^t \sigma_u(L\phi)du + \int_o^t \{\alpha^{-1}(y_u)\sigma_u(\phi h)\}' \alpha^{-1}(y_u)dy_u \quad .$$

REFERENCES

[1] Doléans-Dade, C. Quelques applications de la formula de changement des variables pour les semimartingales locales. Zeits für Wahrs. 16 (1970), 181-194.

[2] Fujisaki, Kallianpur and Kunita, H. Stochastic differential equations for the nonlinear filtering problem. Osaka J.Math. 9 (1972) 19-40.

[3] Kallianpur, G. Stochastic Filtering Theory.Applications of Mathematics. Vol. 13 Springer-Verlag, New York, Heidelberg, Berlin, 1980.

[4] Lipster, R.S. and Shiryayev A. N. Statistics of Random Processes, Vol. 1. Applications of Mathematics. Vol. 5 Springer-Verlag, New York, Heidelberg, Berlin, 1977.

[5] Loève, M. Probability Theory, Vol. II. 4th Edition. Springer-Verlag, New York, Heidelberg, Berlin, 1978.

[6] Wong, E. Recent progress in Stochastic Processes - a Survey I.E.E.E. Trans. in Information Theory 9 (1973), 262-275.

[7] Zakai, M. On the optimal filtering of diffusion process. Zeits für Wahrs 11 (1969), 230-243.

ON ROBUST APPROXIMATIONS IN NONLINEAR FILTERING

Giovanni B. Di Masi
LADSEB-CNR and
Istituto di Elettrotecnica
Università di Padova
I-35100 Padova (Italy)

Wolfgang J. Runggaldier
Seminario Matematico
Università di Padova
I-35100 Padova (Italy)

Abstract. We consider a nonlinear filtering problem with observations of the mixed type. With reference to such a problem we discuss the concept of robustness, describe an approximation approach and show its robustness properties.

1. INTRODUCTION

We consider the following nonlinear filtering problem, where for simplicity we restrict ourselves to the scalar case.

Let a partially observable process (x_t, y_t), $t \in [0,T]$, be given on a probability space $\{\Omega, \mathcal{F}, \mathbb{P}\}$. The unobservable component x_t, called the signal process, has known statistics and is partially observed through the observation process y_t modelled as

$$y_t = \int_0^t c(x_s)ds + v_t + N_t , \tag{1}$$

where $\{v_t\}$ is standard Wiener independent of $\{x_t\}$, and $\{N_t\}$ is doubly stochastic Poisson with rate $\lambda(x_t)$. Given a Borel function f, the problem is to evaluate

$$\hat{f}(x_t) = E\left\{ f(x_t) \mid y_s,\ 0 \leqslant s \leqslant t \right\} , \quad t \in [0,T] , \tag{2}$$

namely a functional of the observations that minimizes the mean square error $E\left\{(\psi_t(y)-f(x_t))^2\right\}$ among all adapted functionals $\psi_t(y)$. In the sequel we shall call optimal filter any version of $\hat{f}(x_t)$ in (2).

As the observation equation (1) is an idealized, mathematically convenient model for an actual physical observation process $\{y^p_t\}$, one would like to have an optimal filter that is "robust" in some suitable sense. In this paper we consider a filter to be robust if:

a) It can be computed as an explicit functional $F_t(y)$ of the observation process $\{y_t\}$.

b) If P and P^p are the distributions corresponding to the model and to the "true" physical situation respectively, then, if P and P^p are close in the sense of weak convergence, $E^p\left\{(F_t(y)-f(x_t))^2\right\}$ is close to $E\left\{(F_t(y)-f(x_t))^2\right\}$.

Robustness for nonlinear filtering problems has already been studied in the recent literature, see f.e. /1,2,5/. In /1,2/ the model for the observations corresponds to (1) with $N_t \equiv 0$ and the robustness consists essentially in requiring $F_t(y)$ to be locally Lipschitz in y. For our observation model (1), y_t has discontinuous trajectories and it is not possible to obtain a filter that, with respect to the usual

Skorokhod metric, is locally Lipschitz or even continuous everywhere (continuity could be preserved by choosing a more appropriate metric /5 /). On the other hand, it is reasonable to consider as robust a filter with only the weaker property b) above. In fact, since a filter is optimal if it minimizes the mean square error, it is natural to require that this criterion be approximately satisfied also when the filter is applied to an actual physical situation that might differ from that described by the model. Furthermore, to completely exploit the continuity property of a filter one should be able to construct a model that provides a "pathwise approximation" of an actual physical situation. In applications however, it appears more likely that one can obtain such an approximation in a weaker sense, namely in the sense of distributions. A sufficient condition for property b) above is that f in (2) be continuous and bounded and $F_t(y)$ continuous except on a set of P-measure zero; this is the case for our model as shown in Section 3.

Often a filter can be computed only in some approximate way. This then raises the problem of having a "robust approximation". In line with the above robustness properties a) and b), we shall consider a sequence of approximating filters $\{F_t^h(y)\}$ $(h \downarrow 0)$ to be a robust approximation if:

a') For all $h>0$, $F_t^h(y)$ can be computed as an explicit functional of the observation process $\{y_t\}$.

b') If P and P^p are close in the sense of weak convergence and h is close to zero, then $E^p \left\{(F_t^h(y)-f(x_t))^2\right\}$ is close to $E\left\{(F_t(y)-f(x_t))^2\right\}$.

In Section 2 we derive, for x_t a diffusion, a sequence $\{F_t^h(y)\}$ of approximating filters that are themselves optimal filters for a sequence of approximating filtering problems. In Section 3 we then show, under some assumptions on our model, that these approximations are equicontinuous except for a set of P-measure zero that does not depend on h and use this result to show that $\{F_t^h(y)\}$ form a robust approximation in the above sense. It then follows immediately that if $\{G_t^h(y)\}$ is any approximating sequence of filters (h varies over a countable set) that are versions of $F_t^h(y)$, then $\{G_t^h(y)\}$ too forms a robust approximation. The consequences of this fact are briefly discussed in the conclusive Section 4.

2. MEASURE TRANSFORMATION AND APPROXIMATING FILTERS

We start by defining some symbols that are used in the sequel. As usual we denote by $C[0,T]$ the space of continuous function on $[0,T]$ with the uniform metric and by $D[0,T]$ the space of real functions that are right-continuous and have left-hand limits, with the Skorokhod metric. Furthermore $D_1[0,T] := \{y \in D[0,T] \mid y$ has unit jumps$\}$ so that if y is given by (1), then $y \in D_1$; also $D_1^{dt} := \{y \in D_1 | y$ is discontinuous at $t\}$. Finally, let $\mathscr{F}_t^x := \sigma\{x_s, 0 \leqslant s \leqslant t\}$, $\mathscr{F}_t^y := \sigma\{y_s, 0 \leqslant s \leqslant t\}$, each completed with all $\mathbb{P}$-null sets in $\mathscr{F}$, and $\mathscr{F}_t := \mathscr{F}_T^x \vee \mathscr{F}_t^y$.

For the sequel we shall make the following assumptions:

A.1. The signal x_t is a diffusion process given by

$$dx_t = a(x_t)dt + b(x_t)dw_t \,, \qquad x_o \quad \text{a r.v.} \tag{3}$$

where w_t is Wiener.

A.2. For some m, $M>0$

$|c(x)|<M$, $m<\lambda(x)<M$, $|f(x)|<M$, for all $x\in\mathbb{R}$

and λ and f are continuous and c is of class C^2.

A.3. $\{v_t\}$ in (1) is independent of $\{x_t\}$ as well as of $\{N_t\}$.

For our approximation purposes the so-called measure transformation approach to filtering appears particularly convenient and in what follows we synthesize its main features for our model. Under the given assumptions, the measure $\mathbb{P}_o$ defined on $\{\Omega, \mathcal{F}\}$ by

$$\frac{d\mathbb{P}_o}{d\mathbb{P}} = \exp\left[-\int_0^T c(x_s)dv_s - \frac{1}{2}\int_0^T c^2(x_s)ds - \right.$$

$$\left. - \int_0^T \log\lambda(x_s)dN_s - \int_0^T [1-\lambda(x_s)]\,ds\right] \tag{4}$$

is (see f.e. /4/) a probability measure, mutually absolutely continuous with respect to $\mathbb{P}$ and such that:

i) $y_t^c := y_t - N_t$ is an $(\mathcal{F}_t, \mathbb{P}_o)$-standard Wiener process, N_t is $(\mathcal{F}_t, \mathbb{P}_o)$-standard Poisson and under $\mathbb{P}_o$ the processes $\{y_t\}$ and $\{x_t\}$ are independent.

ii) The restrictions of $\mathbb{P}$ and $\mathbb{P}_o$ to $\mathcal{F}_t^x$ $(t\in[0,T])$ are the same.

Defining

$$L_t := E_o\left\{\frac{d\mathbb{P}}{d\mathbb{P}_o} \,\middle|\, \mathcal{F}_t^x \vee \mathcal{F}_t^y\right\}, \tag{5}$$

which under our assumptions is given by

$$L_t = \exp\left[\int_0^t c(x_s)dy_s^c - \frac{1}{2}\int_0^t c^2(x_s)ds + \right.$$

$$\left. + \int_0^t \log\lambda(x_s)dN_s + \int_0^t [1-\lambda(x_s)]\,ds\right] \tag{6}$$

the following filter representation holds /9/

$$E\{f(x_t)\mid\mathcal{F}_t^y\} = \frac{E_o\{f(x_t)L_t\mid\mathcal{F}_t^y\}}{E_o\{L_t\mid\mathcal{F}_t^y\}} \qquad \mathbb{P}\text{-a.s.} \tag{7}$$

Notice that, by properties i) and ii) of the measure $\mathbb{P}_o$, we can think of the space $\{\Omega, \mathcal{F}, \mathbb{P}_o\}$ as a product space $\{\Omega' \times \Omega'', \mathcal{F}' \times \mathcal{F}'', \mathbb{P}' \times \mathbb{P}''\}$ with x_t and y_t depending only on $\omega' \in \Omega'$ and $\omega'' \in \Omega''$ respectively. Then (7) can also be written as

$$E\left\{f(x_t) \middle| \mathcal{F}_t^Y\right\} = \frac{\int f(x_t(\omega'))L_t(\omega', \omega'')d\mathbb{P}'(\omega')}{\int L_t(\omega', \omega'')d\mathbb{P}'(\omega')} \qquad \mathbb{P}\text{-a.s.} \tag{8}$$

so that the evaluation of the conditional expectation on the left reduces to an integration over ω'. Notice that the process L_t as given in (6) involves a stochastic integral in y, so that it cannot be computed for a given trajectory of y. However, by our assumptions, we can "integrate by parts" in (6) obtaining

$$L_t = \exp\left[y_t^c c(x_t) - \int_0^t y_s^c dc(x_s) - \frac{1}{2}\int_0^t c^2(x_s)ds + \right.$$

$$\left. + \int_0^t \log \lambda(x_s)dN_s + \int_0^t [1-\lambda(x_s)]\, ds \right] . \tag{9}$$

Denoting by $V_t(y;f)$ the numerator in (8) with L_t given by (9), we have

$$E\left\{f(x_t) \mid \mathcal{F}_t^Y\right\} = \frac{V_t(y;f)}{V_t(y;1)} := F_t(y) , \qquad P \text{ a.s.} , \tag{10}$$

where the r.h.s. is now an explicit functional of y. Even if L_t as given in (9) is defined for each given trajectory of y, there is still the difficulty to actually evaluate $F_t(y)$. This has led to approximation schemes, where the optimal filter for a diffusion is approximated by a sequence of filters for finite-state Markov chains. In what follows we shall sketch such an approximation based on work by H. Kushner /6/ and the authors /3/. We shall also require the following strenghtening of assumption A1.

A1'. Equation (3) has a unique solution in the weak sense and a and b in (3) are continuous and bounded.

Starting from the filter representation (10), the idea is to approximate the optimal filter by approximating the functional $V_t(y;f)$. This can be done as follows: First approximate the diffusion x_t by a sequence of weakly converging, continuous-time and finite-state Markov chains x_t^h. Then, analogously to $V_t(y;f)$ construct the sequence of approximating functionals

$$V_t^h(y;f) = \int f(x_t^h(\omega'))L_t^h(\omega', \omega'')d\mathbb{P}'(\omega') , \tag{11}$$

where L_t^h is as in (9) with x_t^h replacing x_t there, so that $V_t^h(y;f)$ is a well defined functional of y. Analogously to /3, Thm.3/ one can now prove that under the given assumptions

$$\lim_{h\downarrow 0} v_t^h(y;f) = v_t(y;f) \qquad t \in [0,T], \quad P\text{-a.s.} \tag{12}$$

so that the functionals $F_t^h(y) := v_t^h(y;f)/v_t^h(y;1)$ indeed approximate the optimal filter $F_t(y)$

Remark: There exists a measure $\mathbb{P}^h$ on $\{\Omega, \mathcal{F}\}$ such that

$$L_t^h = E_o \left\{ \frac{d\,\mathbb{P}^h}{d\,\mathbb{P}_o} \mid \mathcal{F}_t^{x^h} \vee \mathcal{F}_t^y \right\} \tag{13}$$

and, under $\mathbb{P}^h$ the distribution of the process x_t^h is the same as under either $\mathbb{P}_o$ or $\mathbb{P}$, while y_t admits the representation

$$y_t = \int_0^t c(x_s^h)\,ds + \tilde{v}_t + \tilde{N}_t \quad , \tag{14}$$

where $\tilde{v}_t$ is $(\mathbb{P}^h, \mathcal{F}_T^{x^h} \vee \mathcal{F}_t^y)$ - standard Wiener and $\tilde{N}_t$ is doubly stochastic Poisson with rate $\lambda(x_t^h)$. Furthermore, $F_t^h(y)$ is a version of the optimal filter $E^h \{f(x_t^h) | \mathcal{F}_t^y\}$, so that the approximating functionals are indeed optimal filters themselves.

3. ROBUST APPROXIMATION

The main result in this section is Theorem 1, which is an extension of an analogous theorem in /7/. An immediate consequence is Corollary 1 implying that the filter $F_t(y)$ in (10) is robust. Corollary 2 then shows that the approximating filters $F_t^h(y)$ derived in the previous section form a robust approximation in the sense specified under a') and b') in the Introduction.

Assuming in addition to A1.', A2., A3., that c has its first two derivatives bounded and uniformly continuous, we have

Theorem 1: For each $t \in [0,T]$, the sequence $\{F_t^h(y)\}$ is equicontinuous in $D_1 - D_1^{dt}$.

Proof. Let $\bar{y} \in D_1 - D_1^{dt}$ and denote by d_s the Skorokhod metric in D. We have to show that $|F_t^h(y) - F_t^h(\bar{y})| \to 0$ uniformly in h as $d_s(y,\bar{y}) \to 0$. We have

$$|F_t^h(y) - F_t^h(\bar{y})| = \left| \frac{v_t^h(y;f)}{v_t^h(y;1)} - \frac{v_t^h(\bar{y};f)}{v_t^h(\bar{y};1)} \right| \leqslant$$

$$\leqslant \left| \frac{v_t^h(\bar{y};1)\,[v_t^h(y;f) - v_t^h(\bar{y};f)]}{v_t^h(y;1)\,v_t^h(\bar{y};1)} \right| + \left| \frac{v_t^h(\bar{y};f)\,[v_t^h(\bar{y};1) - v_t^h(y;1)]}{v_t^h(y;1)\,v_t^h(\bar{y};1)} \right| .$$

Since $|f(x)| < M$ for all x and therefore $|V_t^h(y;f)| < M\,|V_t^h(y;1)|$ for all y, it will be enough to show that $|V_t^h(y;1) - V_t^h(\bar{y};1)| \to 0$ uniformly in h as $d_s(y,\bar{y}) \to 0$ and that $V_t^h(y;1)$ is bounded away from zero uniformly in h.

Define

$$M_t^h = y_t^c\, c(x_t^h) - \int_0^t y_s^c\, dc(x_s^h) - \frac{1}{2}\int_0^t c^2(x_s^h)\,ds + $$

$$+ \int_0^t \log \lambda(x_s^h)\,dN_s + \int_0^t [1-\lambda(x_s^h)]\,ds$$

and let $\bar{M}_t^h$ be the corresponding expression with $\bar{y}^c$ and $\bar{N}$ replacing y^c and N. Using the inequality $|\exp\alpha - \exp\beta| \leqslant |\alpha - \beta|\,(\exp\alpha + \exp\beta)$, we have, denoting by E' integration with respect to $\mathbb{P}'$,

$$|V_t^h(y;1) - V_t^h(\bar{y};1)| \leqslant E'\left\{ |M_t^h - \bar{M}_t^h|\,(\exp M_t^h + \exp \bar{M}_t^h)\right\} \leqslant$$

$$\leqslant \left[E'\,|M_t^h - \bar{M}_t^h|^2\right]^{1/2}\left[E'(\exp M_t^h + \exp \bar{M}_t^h)^2\right]^{1/2}.$$

Since the terms $-\frac{1}{2}\int_0^t c^2(x_s^h)\,ds$ and $\int_0^t [1-\lambda(x_s^h)]\,ds$ in M_t^h and $\bar{M}_t^h$ are bounded, there is a constant K such that

$$E'\{|M_t^h - \bar{M}_t^h|^2\} \leqslant K\,E'\Big\{ c^2(x_t^h)\,|y_t^c - \bar{y}_t^c|^2 + $$

$$+ \left|\int_0^t (y_s^c - \bar{y}_s^c)\,dc(x_s^h)\right|^2 + \left|\int_0^t \log\lambda(x_s^h)\,d(N_s - \bar{N}_s)\right|^2\Big\} \tag{15}$$

and

$$E'\{(\exp M_t^h + \exp \bar{M}_t^h)^2\} \leqslant K\,E'\Big\{\exp 2\left[y_t^c c(x_t^h) - \int_0^t y_s^c\,dc(x_s^h) + \int_0^t \log\lambda(x_s^h)\,dN_s\right] +$$

$$+ \exp 2\left[\bar{y}_t^c c(x_t^h) - \int_0^t \bar{y}_s^c\,dc(x_s^h) + \int_0^t \log\lambda(x_s^h)\,d\bar{N}_s\right]\Big\} \tag{16}$$

Furthermore, by /7, Thm.1/, there is a constant K_1 such that

$$E'\left\{\left|\int_0^t q_s\,dc(x_s^h)\right|^2\right\} \leqslant K_1\,\|q\|^2 \text{ for all } q \text{ in a bounded set of } C[0,T] \tag{17}$$

and moreover

$$E'\left\{\exp\left[\int_0^t q_s\, dc(x_s^h)\right]\right\}$$ is bounded uniformly in h in bounded sets of $q \in C[0,T]$.

Taking into account the fact that, if $d_S(y,\bar{y}) \to 0$ then $\| y^c - \bar{y}^c \| \to 0$ and (since $\bar{y} \notin D_1^{dt}$) $|N_t - \bar{N}_t| \to 0$, and using the boundedness of λ one has that the l.h.s. of (15) converges to zero uniformly in h and the l.h.s. of (16) is bounded uniformly in h, thus proving that $|V_t^h(y;1) - V_t^h(\bar{y};1)| \to 0$ uniformly in h as $d_S(y,\bar{y}) \to 0$. Finally we show that $V_t^h(y;1)$ is bounded away from zero uniformly in h. It is enough to consider only the term $-\int_0^t y_s dc(x_s^h)$ in M_t^h since all the other terms are bounded. By Jensen's inequality

$$E'\left\{\exp\left[-\int_0^t y_s dc(x_s^h)\right]\right\} \geq \exp\left[E'\left\{-\int_0^t y_s dc(x_s^h)\right\}\right]$$

and

$$\left|E'\left\{-\int_0^t y_s dc(x_s^h)\right\}\right|^2 \leq E'\left\{\left|\int_0^t y_s dc(x_s^h)\right|^2\right\}$$

from which the assertion follows using (17).

<u>Corollary 1</u>: For each $t \in [0,T]$, the filter $F_t(y)$ in (10) is P-a.s. continuous.

<u>Proof</u>. Immediate from Theorem 1, (12), and the fact that $P(D_1^{dt})=0$.

Letting P_n denote a sequence of distributions corresponding to possible physical situations, we also have

<u>Corollary 2</u>: If P_n converges weakly to P as $n \to \infty$, then, for each $t \in [0,T]$,

$$E_n\left\{(F_t^h(y) - f(x_t))^2\right\} \to E\left\{(F_t(y) - f(x_t))^2\right\}$$

as $n \to \infty$ and $h \to 0$.

<u>Proof</u>. Using Theorem 1 and the boundedness and continuity of f, it is possible to show by arguments similar to those in /8, Ch II, Thm.6.8/ that

$$E_n\left\{(F_t^h(y) - f(x_t))^2\right\} \to E\left\{(F_t^h(y) - f(x_t))^2\right\}$$

uniformly in h as $n \to \infty$. Furthermore, by (12)

$$E\left\{(F_t^h(y) - f(x_t))^2\right\} \to E\left\{(F_t(y) - f(x_t))^2\right\}$$

as $h \to 0$. The assertion then follows.

4. CONCLUSIONS

The robust approximating filters $F_t^h(y)$ considered in Sections 2 and 3, can be explicitely computed for all observations y, but not in a recursive way. On the other hand it is well known that recursive filters can be obtained from the so-called Zakai equation. In the case of our observation model (1), a Zakai equation for the approximations considered in Section 2 can be found in /4/. Following /1/ it is not difficult to obtain a solution to this Zakai equation leading to a sequence of filters $\{G_t^h(y)\}$ that are explicitely computable as a functional of y, so that, by the considerations made at the end of the Introduction, the approximating sequence $\{G_t^h(y)\}$ has the same robustness properties as $\{F_t^h(y)\}$. If, in particular, $N_t \equiv 0$, the recursive and nonrecursive approximating filters can be shown /4/ to coincide for all observations y, so that the approximate recursive filters satisfy the same local Lipschitz condition (uniformly in h) as the nonrecursive ones do in this case /7/.

REFERENCES

1. J.M.C. Clark, The design of robust approximations to the stochastic differential equations of nonlinear filtering" in J.K. Skwirzynski (ed.), "Communication systems and random process theory" Sijthoff and Noordhoff, 1978.

2. M.H.A. Davis, "Pathwise nonlinear filtering", in M. Hazewinkel and J.C. Willems (eds.), "Stochastic systems: the mathematics of filtering and identification and applications", Reidel 1981.

3. G.B. Di Masi and W.J. Runggaldier, "Continuous-time approximations for the nonlinear filtering problem", Appl. Math. and Opt. 7, 233-245, 1981.

4. G.B. Di Masi and W.J. Runggaldier, "On approximation methods for nonlinear filtering", Proc. CIME Summer School on "Nonlinear filtering and stochastic control" (A. Moro ed.), Springer-Verlag (to appear).

5. B. Grigelionis, "Stochastic nonlinear filtering equations and semimartingales", Proc. CIME Summer School on "Nonlinear filtering and stochastic control" (A.Moro ed.), Springer-Verlag (to appear).

6. H.J. Kushner, "Probability methods for approximations in stochastic control and for elliptic equations", Academic Press, 1977.

7. H.J. Kushner, "A robust discrete-state approximation to the optimal nonlinear filter for a diffusion", Stochastics 3, 75-83, 1979.

8. K.R. Parthasarathy, "Probability measures on metric spaces", Academic Press, 1967.

9. E. Wong, "Stochastic processes in information and dynamical systems", McGraw-Hill, 1971.

SMOOTHING OF A DIFFUSION PROCESS CONDITIONNED AT FINAL TIME

E. Pardoux
U.E.R de Mathématiques
Université de Provence
3,Pl.V.Hugo
13331 Marseille Cedex 3

Abstract Let $(X_t, t \in [0,1])$ be a diffusion process. Suppose we observe X_0, X_1, and $(Y_t, t \in [0,1])$, where Y_t is a noisy observation of $(X_s, s \in [o,t])$. We caracterise the conditional law of X_t (for $t \in]0,1[$), given these data, by means of a pair of stochastic PDEs.

1. Introduction :Let $(X_t, Y_t) t \in [0,1]$ satisfy :

(1.1) $dX_t = b(X_t)\, dt + \sigma(X_t) dW_t$

(1.2) $dY_t = h(X_t)\, dt + dv_t$

where $(W_t, t \geqslant o)$ and $(v_t, t \geqslant o)$ are possibly correlated standard Wiener processes, with values in $\mathbb{R}^d$ and $\mathbb{R}^k$ respectively. X_t and Y_t take values in $\mathbf{R}^d$ and $\mathbf{R}^k$ respectively. For the sake of simplifying the notations, we suppose that $k = 1$. The general case can be treated along the same lines.

We suppose that we observe X_0, X_1 and $(Y_t, t \in [0,1])$, and we want to caracterise the conditional law of X_t (for $t \in]0,1[$), given these data. In fact, as will be indicated below, the times 0 and 1 won't play symetric roles. We will consider the process $(X_t, t \in [0,1])$, starting from an initial condition (or an initial law), and look for the conditional law of X_t, given X_1 and $(Y_s, s \in [0,1])$.

In §2, we caracterise the conditional law of X_t, given only X_1, by means of both the backward and forward Kolmogorov equations. In particular, we verify that in general, the conditioned process obtained by a similar procedure, but after inverting the roles of times 0 and 1, is not the same as the first one : 0 and 1 don't play symetric roles. In §3, we show that the conditioned process $(X_t, t \in [0,1])$, given X_1, satisfies a backward Ito equation, thus recovering recent results by B.D.O. ANDERSON [1]. Finally, we solve the general smoothing problem indicated above, by means of both the Zakaï equation, and a backward stochastic PDE, which was already introduced by the author in [5],[7] and [8], and has been considered by others - see [4]

2. Conditional law of X_t, given X_1

We assume now that $b : \mathbb{R}^d \to \mathbb{R}^d$ and $\sigma : \mathbb{R}^d \to \mathbb{R}^{d^2}$ are C^∞ functions, bounded as well as all their derivatives (these codfficients could in fact as well depend on t). We make moreover the following ellipticity assumption :

(2.1) $\exists\, \alpha > o \quad \text{s.t.} \quad a(x) = \sigma\sigma^*(x) \geqslant \alpha\, I$

(2.1) could in fact be replaced by the weaker hypoellipticity assumption of Hörmander, but we don't want to go into such refinements here. We denote by L the infinitesimal generator of the Diffusion Process X_t given by (1.1), i.e.:

$$L = \frac{1}{2} \sum_{i,j} a_{ij}(x) \frac{\partial^2}{\partial x_i \partial x_j} + \sum_i b_i(x) \frac{\partial}{\partial x_i}$$

We denote by p_o the law of X_o. We suppose that either $p_o = \delta_{x_o}$, or p_o has a density w.r. to Lebesgue measure, denoted abusively by $p_o(x)$. In the latter case, we suppose that $p_o(.) \in L^2(\mathbb{R}^d)$. L^* will denote the formal adjoint of L.

We now introduce Kolmogorov forwark equation (know also as "Fokker-Planck equation"):

$$(2.2) \qquad \begin{cases} \dfrac{dp}{dt}(t) = L^* p(t) \ , \ t \geqslant o \\ \\ p(o) = p_o \end{cases}$$

and Kolmogorov backward equation :

$$(2.3) \qquad \frac{dv}{dt}(t) + L\,v\,(t) = 0, \quad t \leqslant 1$$

v^y will denote the solution of equation (2.3) satisfying the final condition :

$v^y(1) = \delta_y$

If $\psi \in C_K(\mathbb{R}^d)$, v^ψ will denote the solution of (2.3) satisfying the final condition :

$v^\psi(1,x) = \psi(x)$

It follows from the above assumptions, using standard arguments from PDE theory, that $\forall\, t \in]0,1]$, $x \to p(t,x)$ is C^∞, as well as $x \to v^y(t,x)$ and $x \to v^\psi(t,x)$, $\forall\, t \in [0,1[$. Moreover, $\forall (t,x) \in [0,1[\times \mathbb{R}^d$, $v^{X_1}(t,x)$ is a well defined random variable.

We now have :

Theorem 2.1 $\quad \forall\, t \in]0,1[$, $\dfrac{p(t,x)\, v^{X_1}(t,x)}{p(1,X_1)}$ is the uensity of the conditional law of X_t , given X_1 .

<u>Corollary 2.2</u> : $\forall t \in]0,1[$, $p(t,x)v^{X_1}(t,x)$ is an "unnormalised conditional density" of X_t, given X_1 .

<u>Remark 2.3</u> : Clearly, the result extends to t = o, in the case where p_0 has a density.

<u>Proof of 2.1</u> : The result is closely related to Bayes formula. Let $\varphi, \psi \in C_K(\mathbb{R}^d;\mathbb{R}_+)$. The following identity is well known :

$$E_{tX_t}[\psi(X_1)] = v^{\psi}(t,X_t)$$

We need only to compute the following :

$$\begin{aligned} E[\varphi(X_t)\ \psi(X_1)] &= E\{\varphi(X_t)E_{tX_t}[\psi(X_1)]\} \\ &= E[\varphi(X_t)v^{\psi}(t,X_t)] \\ &= \int \varphi(x)\ p(t,x)v^{\psi}(t,x)dx \\ &= \int\int \varphi(x)p(t,x)v^{y}(t,x)\ \psi(y)dx\ dy \\ &= \int\int \varphi(x)\ \frac{p(t,x)v^{y}(t,x)}{p(1,y)}\ \psi(y)p(1,y)dx\,dy \\ &= E[\ \psi(X_1)\int\varphi(x)\ \frac{p(t,x)v^{X_1}(t,x)}{p(1,X_1)}\ dx\ ] \end{aligned}$$

We have used the fact that $p(t,x)$ is the density of the law of X_t, and the obvious identity :

$$v^{\psi}(t,x) = \int v^{y}(t,x)\ \psi(y)dy$$

The result now follows from the freedom for the choices of φ and ψ □

<u>Remark 2.4</u> : It is apparently not possible to define the conditioned diffusion process $(X_t, t \in [0,1])$, given that $X_0 = x$ and $X_1 = y$, in a way where t = o and t = 1 would play symmetric roles. Suppose for simplicity that d = 1.

Let X_t be defined by :

(2.4) $$X_t = x + \int_0^t b(X_s)ds + \int_0^t \sigma(X_s)dW_s$$

Denoting by $X_t(x,\omega)$ the solution of (2.4), it is well known - see BISMUT's contribution in this volume - that we can choose it such that ω.a.s.,

$$x \rightarrow X_1(x,\omega)$$

is a diffeormorphism. The inverse of this map, that we denote.

$$y \rightarrow Y_0(y,\omega)$$

is-see e.g. KRYLOV-ROSOVSKII [3] - the value at time t = o of the solution of the following backward Ito equation (for the notation $\oplus$, see below):

$$Y_t = y + \int_t^1 [\sigma\sigma'(Y_s) - b(Y_s)]ds - \int_t^1 \sigma(Y_s) \oplus dW_s$$

Note that in "Stratanovitch language", the expressions for dX_t and dY_t would be "identical".

But now the conditional law of Y_t, given that $Y_0 = x$, differs in general from the conditional law of X_t, given that $X_1 = y$. There are exceptions, like the case of Ornstein Uhlenbeck processes.

□

3. Time-reversal of diffusion processes.

It is well-known that the process $\{X_t, t \in [0,1]\}$, starting at $t = 1$, and t running backward to 0, is a Markov process. Under certain conditions, it is again a diffusion process, and we want to write the corresponding stochastic differential equation. The same results appear in B.D.O. ANDERSON [1] (see also the bibliography there in). We follow here a line of reasoning from [6].

Choose $t \in [0,1[$, and $f \in C_K^\infty(\mathbb{R}^d)$. p denoting again the solution of Kolmogorov forward equation (2.2), we define $q(s,x)$, $s \geq t$, as the solution of the forward equation :

$$(3.1) \quad \begin{cases} \frac{dq}{ds}(s) = L^* q(s), \ s \geq t \\ \\ q(t,x) = p(t,x) f(x) \end{cases}$$

Comparing with equation (2.3), we get immediately [(.,.) denotes the scalar product in $L^2(\mathbb{R}^d)$]:

$$\frac{d}{ds} (q(s), v^\psi(s)) = 0, \quad s \in [t,1]$$

$$(q(1), \psi) = (q(t), v^\psi(t))$$

It then follows :

$$E\left[\frac{q(1,X_1)}{p(1,X_1)} \psi(X_1)\right] = E[f(X_t) v^\psi(t,X_t)]$$

$$= E[f(X_t)\psi(X_1)]$$

So that :

$$(3.2) \quad \frac{q(1,X_1)}{p(1,X_1)} = E[f(X_t)/X_1]$$

Define $u(s,x) = \dfrac{q(s,x)}{p(s,x)}$, $s \geq t$. From (3.2), u is bounded, and is the solution of :

$$(3.3) \quad \begin{cases} \frac{du}{ds}(s) = \tilde{L}\, u(s), \ s \geq t \\ u(t,x) = f(x) \end{cases}$$

where $\tilde{L} = \frac{1}{2} \sum_{i,j} a_{ij}(x) \frac{\partial^2}{\partial x_i \partial x_j} - \sum_i \tilde{b}_i \frac{\partial}{\partial x_i}$

with $\tilde{b}$ given by :

$$\tilde{b}_i(s,x) = b_i(x) - \frac{1}{p(s,x)} \sum_j \frac{\partial}{\partial x_j}[a_{ij}(x)p(s,x)]$$

Since $p(s,x) > 0$ for $s > 0$, $x \in \mathbf{R}^d$, $\tilde{b}$ is a well-defined measurable function of (s,x), at least for $s \in]0,1]$. We now suppose :

$$(3.4) \quad \forall n \in \mathbf{N}, \exists C_n \in \mathbf{R}_+ \quad \text{s.t.} \quad |\tilde{b}(s,x)| \leq C_n(1+|x|),$$
$$\forall (s,x) \in [\tfrac{1}{n},1] \times \mathbf{R}^d$$

Clearly, one can write conditions on $p(s,x)$, that are sufficient for (3.4) to hold. But we don't know sufficient conditions on the coefficients a and b.

Now (3.2) can be rewritten as :

$$u(1,X_1) = E[f(X_t)/X_1]$$

It follows that (3.3) is Kolmogorov backward equation associated to the time reversed process $(X_t, t \in [0,1])$ starting at final time $t=1$.

Suppose now that $\Omega = C(]0,1]; \mathbf{R}^d)$, and $X_s(\omega) = \omega(s)$. Define $\mathcal{G} = \sigma(X_s, s \in]0,1])$, and Q a probability measure on $(\Omega, \mathcal{G})$, which solves the backward martingale problem, associated with the infinitesimal generator $\tilde{L}$, and the "initial" condition X_1 (i.e. $Q(X_1 \in A) = \int_A p(1,x)dx$). Let $\mathring{\mathcal{G}}_t^1 = \sigma(X_s, t \leq s \leq 1)$, $\mathcal{G}_t^1$ be equal to $\mathring{\mathcal{G}}_t^1$, completed with the Q null set of $\mathcal{G}$.

There exists a $\mathcal{G}_t^1$ - Q "backward standard Wiener process"$(\tilde{W}_t, t \in [0,1])$ with values in $\mathbf{R}^d$, i.e. such that $\forall 0 \leq s < t \leq 1$, $\tilde{W}_s - \tilde{W}_t$ is a zero - mean gaussian vector with covariance $(t-s)I$, independent of $\mathcal{G}_t^1$; and such that :

$$(3.5) \quad dX_t = \tilde{b}(t,X_t)dt + \sigma(X_t) \oplus d\tilde{W}_t, \quad t \in]0,1]$$

where $\oplus$ stands for backward Ito integral, i.e. $\int_s^t \sigma(\theta,X_\theta) \oplus d\tilde{W}_\theta =$

$$\lim_{n \to \infty} \sum_{i=1}^n \sigma(t_i^n, X_{t_i^n})(\tilde{W}_{t_i^n} - \tilde{W}_{t_{i-1}^n})$$

with $t_i^n = s + (t-s)i/n$.

If we suppose moreover that $\sigma(x)$ is invertible, $\forall x \in \mathbb{R}^d$, it is then possible to express $\tilde{W}$ in terms of W, b, σ and p. Let us write equations (1.1) and (3.5) in the sense of stratanovitch :

$$dX_t = (b(X_t) - \frac{1}{2} \sum_j \frac{\partial \sigma}{\partial x_j} \sigma_{j\cdot}(X_t))dt + \sigma(X_t) \circ dW_t$$

$$dX_t = (\tilde{b}(X_t) + \frac{1}{2} \sum_j \frac{\partial\sigma}{\partial x_j} \sigma_{j\cdot}(X_t))dt + \sigma(X_t) \circ d\tilde{W}_t$$

We can now compare the two expressions for $\sigma^{-1}(X_t) \circ dX_t$ thus yielding :

$$(3.6) \quad d\tilde{W}_t = dW_t + \frac{1}{p(t,X_t)} \sum_j \frac{\partial}{\partial x_j} [p\, \sigma_{j\cdot}](t,X_t)dt$$

It must be emphasized that $\tilde{b}$ in (3.5), as well as the right hand side of (3.6), depends on the given law of X_o, through $p(t,x)$.

Now the conditional density of X_t, given that $X_1 = y$, that we have caracterised in §2, is also the value at time t of the following time-reversed Fokker-Planck equation :

$$\begin{cases} \frac{dr}{ds}(s) + \tilde{L}^* r(s) = 0, & s \leqslant 1 \\ r(1) = \delta_y & \end{cases}$$

Indeed, u being again the solution of (3.3),

$$\begin{aligned} E[f(X_t) \mid X_1 = y] &= u(1,y) \\ &= (f, r(t)) \end{aligned}$$

Since, again $\frac{d}{ds}(u(s),r(s))=0$, $s \in]0,1[$, and :

$$(u(s),r(s)) \to u(1,y), \text{ as } s \to 1$$

§4. The smoothing problem

Consider the stochastic differential system :

$$(4.1) \quad \begin{aligned} dX_t &= b(X_t)dt + \sigma(X_t)dW_t \\ dY_t &= h(X_t)dt + g(t)dW_t + \tilde{g}(t)d\tilde{W}_t \end{aligned}$$

where, in addition to the hypotheses of §2, we assume that $\tilde{W}$. is a scalar-valued standard Wiener process independent of $(X_o; W_t, t \in [0,1])$; g^* and $\tilde{g}$ are measurable, functions of $t \in [0,1]$, with values in $\mathbf{R}^d$ and $\mathbf{R}$ respectively. Finally, $h \in C^\infty(\mathbf{R}^d)$, and is supposed to be bounded, together with all its derivatives. We suppose that the "noise" in the second line of (4.1) has been normalised, so that :

$$(4.2) \quad g(t)g^*(t) + \tilde{g}(t)^2 = 1, \ \forall t \in [0,1]$$

and that :

$$(4.3) \quad \exists \beta > o \quad \text{s.t.} \quad \tilde{g}(t)^2 \geqslant \beta, \text{ t-a.e.}$$

(4.3), together with (2.1), implies a sort of ellipticity property for the SPDEs that we will consider. These assumptions could be replaced by weaker hypoellipticity assumptions, see BISMUT (this volume), and CHALEYAT-MAUREL-MICHEL [2].

Let now $\Omega = C([0,1]; \mathbb{R}^{d+1})$, $\binom{X_t(\omega)}{Y_t(\omega)} = \omega(t)$, P (resp. P_{sx}) is the solution of the martingale problem associated with (4.1), with initial condition.

$$\text{Law}\ (X_0) = p_0,\quad Y_0 = 0$$

$$(\text{resp. } X_\theta = x,\quad Y_\theta = 0\ ;\quad \theta \in [o,s])$$

where p_0 is as in §2.

Define $\mathcal{G}_t = \sigma\{X_s, Y_s\ ; 0 \leq s \leq t\}$, completed with the P-null sets of $\mathcal{G}_1$.

$$Z_t^s = \exp[\int_s^t h(X_\theta)dY_\theta - \frac{1}{2}\int_s^t h^2(X_\theta)d\theta],\ Z_t = Z_t^0$$

We define new probabilities $\overset{o}{P}$ and $\overset{o}{P}_{sx}$ by :

$$\frac{d\overset{o}{P}}{dP} = Z_1^{-1}\ ,\quad \frac{d\overset{o}{P}_{sx}}{dP_{sx}} = (Z_1^s)^{-1}\ ,\quad 0 \leq s < 1\ .$$

It follows from Girsanov theorem that $(Y_t, t \in [0,1])$ is a $\overset{o}{P}$ Wiener process, and $\exists$ a $\overset{o}{P}$ wiener process $(\tilde{Y}_t, t \in [0,1])$ with values in $\mathbb{R}^d$, which is independent of Y. - see e.g. [7]- such that :

$$dX_t = [b(X_t) - C\,h\,(X_t)\,]\,dt + C(X_t)dY_t + \tilde{C}(X_t)d\tilde{Y}_t$$

where $C = \sigma g^*$, $C = \sigma[I - g^* g]^{1/2}$.

We consider the following PDE operator :

$$B = h + \sum_i C_i \frac{\partial}{\partial x_i}$$

and the following stochastic PDEs :

$$(4.4)\quad \begin{aligned} dp(t) &= L^* p(t)dt + B^* p(t)dY_t \\ p(o) &= p_o \end{aligned}$$

$$(4.5)\quad dv(t) + L\,v\,(t)dt + B\,v\,(t) \oplus dY_t = 0$$

v^y will denote the solution of (4.5) with final condition $v^y(1) = \delta_y$; and for $\psi \in C_K(\mathbb{R}^d)$, v^ψ will denote the solution of (4.5) with final condition $v^\psi(1,x) = \psi(x)$.

For existence and uniqueness to (4.4) and (4.5), see [8]. Let $\mathcal{F}_t^s = \sigma\{Y_\theta - Y_s,\ s \leq \theta \leq t\}$, $\mathcal{F}_t = \mathcal{F}_t^0$.

<u>Theorem 4.1</u> [5] , [7] $v^\psi(t,x) = \overset{o}{E}_{tx}[\psi(X_1)Z_1^t / \mathcal{F}_1^t]$

Let us indicate a formal argument for Th 4.1 : let us rewrite everywhere the dY integrals in the sense of Stratonovitch :

$$dX_t = [\, b - Ch - \frac{1}{2}(\nabla C)C\,](X_t)dt + C(X_t)\circ dY_t + \tilde{C}(X_t)d\tilde{Y}_t$$

$$dv(t) + (L - \frac{1}{2}B^2)v(t)dt + Bv(t)\circ dY_t = 0$$

$$Z_1^t = \exp\{\int_t^1 h(X_s)\circ dY_s - \frac{1}{2}\int_t^1 [\nabla h\, C(X_s) + h^2(X_s)]ds\}$$

Now replacing everywhere dY_s by y(s)ds, y being considered as a fixed given function (we condition upon $\mathcal{F}_1^t$), and applying the classical Feynman-Kac formula, we get the Theorem.

The key to our result is the following consequence of Theorem 4.1 :

Lemma 4.2 $\quad \overset{o}{E}[\varphi(X_t)\psi(X_1)Z_1/\mathcal{F}_1]\int p(t,\ )\varphi(x)v^{\psi}(t,x)dx$

Ptoof : From Lemma 3.9 and 3.10 in [7], we can write :

$$\overset{o}{E}{}^{\mathcal{F}_1}[\varphi(X_t)Z_t\psi(X_1)Z_1^t] = \overset{o}{E}{}^{\mathcal{F}_1}[\varphi(X_t)Z_t\overset{o}{E}{}^{\mathcal{F}_1^t}_{tX_t}(\psi(X_1)Z_1^t)]$$

$$= \overset{o}{E}{}^{\mathcal{F}_1}[\varphi(X_t)Z_t\, v^{\psi}(t,X_t)\,]$$

$$= (p(t),\ \varphi\, v^{\psi}(t))$$

□

Theorem 4.3 $\quad \dfrac{p(t,x)v^{X_1}(t,x)}{p(1,X_1)}$ is the density of the conditional of X_t, given $\mathcal{F}_1 \vee \sigma(X_1)$.

Corollary 4.4 $\quad p(t,x)v^{X_1}(t,x)$ is an "unnormalised conditional density" of X_t, given $\mathcal{F}_1 \vee \sigma(X_1)$.

Proof of Theorem : The argument is similar to that of Theorem 2.1, using first lemma 4.2 :

$$\overset{o}{E}{}^{\mathcal{F}_1}[\varphi(X_t)\psi(X_1)Z_1] = (p(t),\varphi\, v^{\psi}(t))$$

$$= \iint p(t,x)\varphi(x)v^{y}(t,x)\psi(y)dxdy$$

$$= \iint \frac{p(t,x)\varphi(x)v^{y}(t,x)}{p(1,y)}\, p(1,y)\psi(y)d\,xdy$$

$$= \overset{o}{E}{}^{\mathcal{F}_1}[Z_1\ \psi(X_1)\int \frac{p(t,x)\varphi(x)v^{X_1}(t,x)}{p(1,X_1)}d\,x\,]$$

It then follows :

$$\overset{o}{E}[\varphi(X_t)Z_1/\mathcal{F}_1 \vee \sigma(X_1)] = \frac{\overset{o}{E}[Z_1/\mathcal{F}_1 \vee \sigma(X_1)]}{p(1,X_1)} \int p(t,x)\varphi(x)v^{X_1}(t,x)dx$$

and finally :

$$E[\varphi(X_t)/\mathcal{F}_1 \vee \sigma(X_1)] = \frac{\overset{o}{E}[\varphi(X_t)Z_1/\mathcal{F}_1 \vee \sigma(X_1)]}{\overset{o}{E}[Z_1/\mathcal{F}_1 \vee \sigma(X_1)]}$$

$$= (p(1,X_1))^{-1} \int p(t,x)v^{X_1}(t,x)\varphi(x)dx$$

□

As a consequence of Theorem 4.3, we recover the smoothing result of [7]: $p(t,x)v^1(t,x)$ is the "unnormalised conditional density" of X_t, given $\mathcal{F}_1$.

It is possible to generalise Theorem 4.3 - as well as Theorem 2.1 - to cases where X_1 is only partially observed. Let us just consider the case where we observe X_1^1, the first component of the random vector X_1 .

If $\psi \in C_b(\mathbb{R})$, we define $\overline{\psi} \in C_b(\mathbb{R}^d)$

by :

$$\overline{\psi}(y) = \psi(y^1)$$

From lemma 4.2 ,

$$\overset{o}{E}[\varphi(X_t)\psi(X_1^1)Z_1/\mathcal{F}_1] = \int p(t,x)\varphi(x)v^{\overline{\psi}}(t,x)dx$$

Now $\forall\, \alpha \in \mathbb{R}$, we define the measure μ^α on $(\mathbb{R}^d, B_d)$ by :

$$\mu^\alpha(A) = \int_{A \cap \{x^1=\alpha\}} dx^2 \ldots dx^n$$

It is possible to define v^{μ^α} as the solution of (4.5) with final condition :

$$v^{\mu^\alpha}(1) = \mu^\alpha$$

and we have the following identity :

$$v^{\overline{\psi}} = \int v^{\mu^\alpha}\, \psi(\alpha)\, d\alpha$$

so that, by the argument of Theorem 4.3, we have :

<u>Theorem 4.4</u> $p(t,x)v^{\mu^{X_1^1}}(t,x)$ is the "unnormalised conditional density" of X_t, given $\mathcal{F}_1 \vee \sigma(X_1^1)$

□

Bibliography

[1] B.D.O. ANDERSON .- Reverse-time diffusion equation models. *Stochastic Proc. and their Applic.* 12 (1982) pp. 313-326.

[2] M. CHALEYAT-MAUREL, D. MICHEL.- Conditions de Hörmander et filtrage. to appear

[3] N. KRYLOV, B.ROSOVSKII.- On the first integrals and Liouville equations for diffusion processes, in *Stochastic Differential Systems*, M. Arato, D. Vermes, A. Balakrishnan Eds, Lecture Notes in Control and Information Sciences Vol. 36, Springer (1981) pp. 117-125.

[4] H. KUNITA.- Stochastic partial differential equations connected with non-linear filtering. To appear in *Proc. CIME*, A. Moro Ed., Lecture Notes in Math.

[5] E. PARDOUX.- Stochastic PDEs and filtering of diffusion processes, *Stochastics* 3 (1979) pp. 127-167.

[6] E. PARDOUX.- The solution of the nonlinear filtering equation as a likelihood function. Proc. 20th IEEE Conference on Decision and control (1981) pp. 316-319.

[7] E. PARDOUX.- Equations du filtrage non linéaire, de la prédiction et du lissage. *Stochastics* 6 (1982) pp. 193-231.

[8] E. PARDOUX.- Equations of nonlinear filtering, and application to stochastic control with partial observation, to appear in *Proc.CIME*, A. Moro Ed., Lecture Notes in Math.

First Passage Times in Stochastic Models of Physical Systems and in Filtering Theory

by

Zeev Schuss(*)

School of Mathematical Sciences
Tel-Aviv University
Ramat-Aviv 69978, ISRAEL

Abstract

Langevin-type equations are used as models for diffusions in force fields in many physical situations. In particular, the mean first passage time of such a process over a potential barrier, or the time of transition from one stable state to another, is the physical quantity that determines chemical reactions rates (Arrhenius' law), diffusion tensors for atomic migration in crystals, ionic conductivity, the I - V characteristic in Josephson junction devices, relative stability or equilibrium and non-equilibrium steady states and so on. In filtering theory the phenomenon of cycle slip and threshold in phase locked loops can be modeled by a multi-stable system of differential equations driven by a random signal and by the noisy measurements with transitions between the stable states. An asymptotic method for the computation of mean first passage times and exit probabilities yields new formulas for various physical quantities. The method is based on singular perturbation and boundary layer theory and matched asymptotic expansions of solutions of boundary value problems for partial differential equations. It was recently developed by D. Ludwig [8], B. Matkowsky, Z. Schuss [1], [2], [3], [9], [11] and others.

---

*Partially supported by A.F.O.S.R. Grant No. AFOSR 75-3620A and by Israel Academy of Science Grant No. 7407.

1. First passage times in physical systems

In many cases fluctuations in physical systems can be described by a dynamical system driven by a relatively small white-noise-type random force. Thus the Langevin equation (or system of equations) [4]

$$\text{(1.1)} \qquad \begin{aligned} \dot{x} &= y \\ \dot{y} &= -\beta y - \nabla U(x) + \sqrt{2k\beta T}\,\dot{w} \end{aligned}$$

models the random motion of a particle in a force field, derived from a potential $U(x)$ with frictional force $-\beta y$, and a random driving force $\sqrt{2k\beta T}\,\dot{w}$. Here k is Boltzmann's constant, T is temperature and $\dot{w}$ is standard Gaussian white noise. The random force may represent thermal fluctuations, such as random collisions of the particle with the surrounding medium and so on. Equation (1.1) was used by Kramers [6] to model a chemical reaction. The function $U(x)$ represents the potential of the chemical bond and the noise term represents the random collisions of the particle with the surrounding medium. The particle is held in a potential well and has to overcome a potential barrier of height Q to break the chemical bond and enter the chemical reaction. The mean first passage time over the barrier determines the reaction rate [10] - [13]. A similar situation arises in describing atomic migration in crystals. In this case an interstitial particle is trapped in a potential well formed by the atoms of the crystallic lattice and jumps into an adjacent well as a result of the thermal vibrations of the lattice. The mean first passage time to the boundary of the well and the probability distribution of the directions of the jumps determine the diffusion tensor for atomic migration in crystals (cf., [7], [11] et. al.). Ionic conductivity in crystals can also be described by equation (1.1). The motion of a charged interstitial ion in an ionic or superionic crystal, subjected to a uniform electrostatic field can be described by (1.1) with

$$U(x) = -Ex + \mu(x)$$

where E is the uniform field and $\mu(x)$ is a periodic function, e.g.

$$\text{(1.2)} \qquad U(x) = -Ex - A\cos\omega x\,.$$

If $E/A\omega < 1$ the ion will be trapped in a stable equilibrium state. The thermal fluctuations will drive the ion out of the potential well and into the adjacent one. Thus ionic conductivity can be attributed to hopping ions. The mean first passage time determines the ionic conductivity by the relation

(1.3) $$\kappa \propto \frac{\partial}{\partial E} \frac{1}{\tau}$$

(cf. [10] - [12]).

A more sophisticated example is that of the Josephson junction [14]. Equation (1.1) with $U(x) = -Ix - \cos x$ represents the dynamics of the order paramater x, describing the junction. Here I is the (non-dimensional) current and β is a dissipation parameter determined by the capacitance, resistance and the Josephson frequency of the junction. The voltage across the junction is proportional to the mean "velocity" $\overline{\dot{x}}$ of the junction. The noisy term represents thermal fluctuations in the junction. The mechanical equivalent of the Josephson junction is the damped physical pendulum driven by constant torque I. The deterministic behavior of (1.1) depends on the ratio I/β. If $0 < I < 1$ and $I > I_{min}(\beta) \simeq \frac{4}{\pi}\beta$ (or $\beta < \beta_{max}(I) \simeq \frac{\Pi}{4} I$), (1.1) has two stable solutions if $T = 0$. There is a stable equilibrium solution $x = \arcsin I$, $\dot{x} = y = 0$, and there is a stable "running" solution for which y is a 2π-periodic function of x. For the first solution there is no voltage across the junction, as $x \equiv 0$, while the second solution S, given approximately by

(1.4) $$y = \frac{I}{\beta} + \frac{\beta}{I}\cos(x + \frac{\beta^2}{I}) - \frac{1}{4}(\frac{\beta}{I})^3 \cos 2x + 0(\frac{\beta}{I})^5$$

in the phase plane, yields $V \propto \overline{y} \cong \frac{I}{\beta}$. Thus at $T = 0$, (1.1) is a multistable system. In presence of thermal fluctuations there will be transitions from the stable.... equilibrium solution to the stable running solution S and v.v. at mean times $\overline{\tau}_e$ and $\overline{\tau}_s$ respectively. Thus the voltage on the noisy junction is given by

(1.5) $$V = \frac{\overline{\tau}_s\, I/\beta}{\overline{\tau}_e + \overline{\tau}_s}$$

2. Fluctuations and transitions in nonlinear oscillators and in filtering theory

The physical pendulum considered in Section 1 is an example of a nonlinear oscillator which has a stable limit cycle. The probability density of fluctuations about a stable limit cycle cannot be determined from Boltzmann's law

(2.1) $$p \propto e^{-E/kT}$$

where E is the energy of the system, because the system is not near equlibrium, so its energy diverges in time. Thus, for example, the energy of the physical pendulum is given by

$$E = \frac{y^2}{2} - Ix - \cos x \to -\infty \text{ as } x \to \infty$$

In such cases the stationary distribution of small fluctuations about a limit cycle can be found as follows. For a nonlinear oscillator given by

(2.2)
$$\dot{x} = y$$
$$\dot{y} = -\beta y f(x,y) - g(x,y) + \sqrt{2\beta kT}\, w$$

we seek a solution of the stationary Fokker-Plauck equation

(2.3)
$$\beta kT p_{yy} - y p_x + \{[\beta y f(x,y) + g(x,y)]p\}_y = 0$$

in the WKB form

(2.4)
$$p = \rho \exp(- W/kT).$$

The functions $W(x,y)$ and $\rho(x,y,T)$ are to be periodic on the limit cycle S and ρ is a regular function of T. Thus W plays a role similar to that of energy in the Boltzmann distribution. To find W we substitute (2.4) in (2.3) and obtain the Hamilton-Jacobi-type equation

(2.5)
$$\beta W_y^2 + y W_x - [\beta y f(x,y) + g(x,y)]W_y = 0.$$

At the same time, to leading order in T, the first term ρ_o in the expansion of ρ is the solution of the equation

(2.6)
$$\rho_{o_t} = -\beta(W_{yy} - yf(x,y))\rho_o + [g(x,y)\rho_o]_x$$

on the characteristics of (2.5) and ρ_o is periodic on S. We first show that the contours of constant W in phase space correspond to the limit cycles of a family of nonlinear oscillators into which (2.2) is embedded. To derive this result we will need to consider the following equations in phase space: the deterministic equations of motion of the oscillator (2.2); the parametric equations for the W-contours

(2.7)
$$\dot{x} = y$$
$$\dot{y} = -\beta y f(x,y) - g(x,y) + \beta W_y$$

and the parametric equations for the characteristic curves of W [5]

(2.8)
$$\dot{x} = y$$
$$\dot{y} = -\beta y f(x,y) - g(x,y) + 2\beta W_y$$
$$\dot{W}_x = [\beta y f_x(x,y) + g_x(x,y)]W_y$$
$$\dot{W}_y = -W_x + \beta[yf(s,y) + g(x,y)]_y W_y$$
$$\dot{W} = G W_y^2$$

We observe that on S

$$\dot{W} = -\beta W_y^2 \tag{2.9}$$

by (2.2) and (2.5). Hence, in order for W to be periodic on S, the right hand side of (2.9) must vanish identically on S, so that $W = \text{const.}$ on S, and consequently $\nabla W = 0$ on S. Next we imbed S in a family of limit cycle of (2.7) by showing that $Wy = yK$ on W-contours, where $K = K(W)$ is constant on each W-contour. To this end we consider the generalized Hamiltonian

$$H(x,y) = \frac{y^2}{2} + \int_{x_0}^{x} g(x,y)dx + \beta \int_{x_0}^{x} [yf(x,y) - Wy]dx \tag{2.10}$$

where the integral is a line integral along the W-contour that passes through the point (x,y) and x_0 is a fixed number. From (2.7) it is easy to see that on a characteristic curve

$$\dot{H} = \beta y W y. \tag{2.11}$$

Since the characteristics are transversal to the W-contours we obtain from (2.8) and (2.11)

$$\frac{dW}{dH} = \frac{Wy}{y} \tag{2.12}$$

so that $Wy = yK(W)$. Using this result we can rewrite (2.7) in the form

$$\dot{x} = y \tag{2.13}$$

$$\dot{y} = -\beta y(f(x,y) - K) - g(x,y)$$

Thus we have embedded the limit cycle S in a family of limit cycles, which are W-contours. For small values of K we have $W = \frac{1}{2} W_{kk}(0)K^2 + \ldots$ so that fluctuations about S are Gaussian in K. In the case of the physical pendulum K is the mean action A per cycle, so that the distribution of fluctuations about the running solution of the pendulum are Gaussian in action space. The fluctuations of the pendulum about a stable equilibrium are given by the Boltzmann distribution (2.1), but since

$$E \sim \omega A$$

(ω = frequency of the undamped oscillator), we see that the fluctuations are exponentially distributed in action space. Thus the effective potential for fluctuations of the pendulum is a linear function of the action in the basin of attraction of the stable equilibrium, and a quadratic function of the action in the basin of attraction of the running solution S

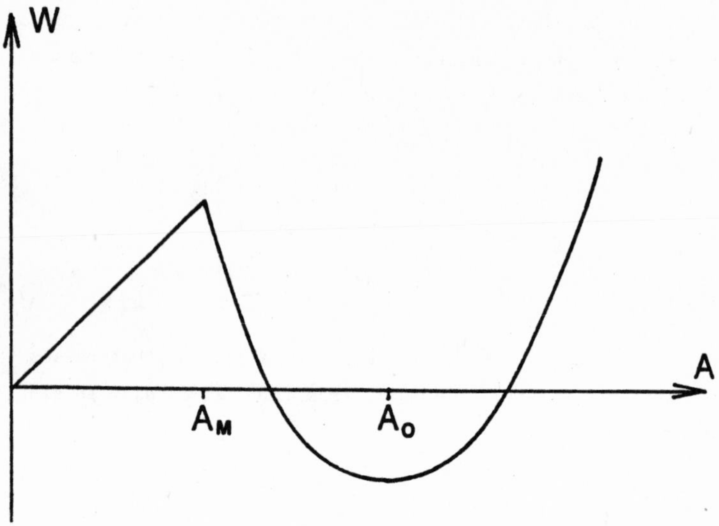

fig. 2.1 Effective potential in action space
for the pendulum

In a multiply stable system fluctuations cause spontaneous transitions from one stable state to another (cf. Section 1). It is known [11] that for small T in (2.2) the mean transition time $\bar{\tau}$ is given by

(2.14)
$$\bar{\tau} \propto e^{Q/kT}$$

where Q is the height of the potential barrier to be overcome. Equation (2.14) is valid for systems that can be described by a potential; however systems that have limit cycles cannot be described by a potential in state space. An effective potential for such systems is the function W = W(K). It is given by

(2.15)
$$W(K) = \int_0^K \frac{yK}{Ky} \, dK$$

where $y = y(K:\Omega)$ is the trajectory of the limit cycle of (2.13) (here Ω is an angular variable on the cycles; (K,Ω) is a system of "polar" coordinates near S). In the case of the physical pendulum (or the Josephson junctions) we have [1] by(1.4) and (2.13) and (2.15)

$$W(K) = \left(\frac{I}{G}\right)^2 \frac{K^2}{2(1 - K^2)} \tag{2.16}$$

for $K_c < K < 1$, where $K_c \simeq 1 - \frac{\pi I}{4\beta}$. The height of the potential barrier in this case is given by

$$W_c = W(K_c) \simeq \tfrac{1}{2}\left(\frac{I}{G} - \frac{4}{\pi}\right)^2 \tag{2.17}$$

Equation (2.16) can be written in the form

$$W(K) = \tfrac{1}{2}\,[A(K) - A(0)]^2 \tag{2.18}$$

Where the mean action $A(K)$ on the running solution of the corresponding equation (2.13) is given by

$$A = \frac{1}{2\pi}\int_0^{2\pi} y\,dx = \frac{I}{G(1 - K)} \quad . \tag{2.19}$$

The cases of the Van der Pol oscillator and the pendulum driven by shot noise are considered in [2]. In filtering theory first passage problems arise in the investigation of cycle slips in a phase-locked-loop (PLL). Thus, for example [3], if we denote by x and y the (properly scaled) frequency and phase estimation errors respectively, the following equations can be derived

$$(2.20) \qquad \begin{aligned} \dot{x} &= -\delta x - \sin y + \sqrt{\varepsilon}\,(\dot{w}_1 - \dot{w}_2) \\ \dot{y} &= x - \sin y - \sqrt{\varepsilon}\,\dot{w}_2 \end{aligned}$$

where δ is a positive constant, ε is a small.parameter, representing the noise to signal ratio, $\dot{w}_1$ and $\dot{w}_2$ are independent white noises. The system (2.20) has stable equilibrium points at $x = 0$, $y = 2\pi n$ $(n = 0,1,\dots)$. The domains of attraction of such equilibria are separated by curves through the saddle points $x = 0$, $y = (2n + 1)\pi$ (separatrices). A cycle slip occurs whenever the noise drives the trajectory of (2.20) across a separatrix. In this case several cycle slips can occur before relocking (at $x = o$, $y = 2\pi n$). This happens if several separatrices are crossed before relocking. The mean time of first passage to a separatrix determines the mean square frequency estimation error [3].

References

[1] Ben-Jacob, E., Bergman, D.J., Mathowsky, B.Y., and Schuss, Z. "The lifetime of non-equilibrium steady states," Phys. Rev. A. (to appear).

[2] ———. "Fluctuations and transitions in nonlinear oscillators," Proc. N.Y.A.S. (to appear).

[3] Bobrovsky, B.Z. and Schuss, Z. "Singular perturbation method for the computation of the mean first passage time in a nonlinear filter," SIAM J. Appl. Math. 42 (1982), 1, 174-187.

[4] Chandrasekhar, S. "Stochastic problems in physics and astronomy," Noise and Stochastic Processes (N. Wax, Ed.), Dover, N.Y., 1954.

[5] Courant, R. and Hilbert, D. Methods of Mathematical Physics, Vol. 2, Interscience, N.Y., 1962.

[6] Kramers, H.A. "Brownian motion in a field force and the diffusion model of chemical reactions," Physica 7 (1940), 284-304.

[7] Larsen, E. and Schuss, Z. "Diffusion tensor for atomic migration in crystals," Phys. Rev. B., 18 (1978), 5, 2050-2058.

[8] Ludwig, D. "Persistence of dynamical systems under random perturbations," SIAM Rev. 17 (1975), 4, 605-640.

[9] Matkowsky, B. and Schuss, Z. "The exit problem for randomly perturbed dynamical systems," SIAM J. Appl. Math. 33 (1977), 12, 365-382.

[10] ———. "Eigenvalues of the Fokker-Planck operator and the approach to equilibrium for diffusions in potential fields." SIAM J. Appl. Math. 40 (1981), 2, 242-254.

[11] Schuss, Z. Theory and Applications of Stochastic Differential Equations. Wiley, N.Y., 1980.

[12] ———. "Singular perturbations in stochastic differential equations of mathematical physics," SIAM Rev. 22 (1980), 2, 119-155.

[13] Schuss, Z. and Matkowsky, B. "The exit problem: a new approach to diffusion across potential barriers," SIAM J. Appl. Math. 36 (1979), 43, 604-623.

ADAPTIVE STOCHASTIC FILTERING PROBLEMS

- THE CONTINUOUS TIME CASE

J.H. van Schuppen
Mathematical Centre
P.O. Box 4079
1009 AB Amsterdam
The Netherlands

ABSTRACT

The adaptive stochastic filtering problem for Gaussian processes is considered. The selftuning-synthesis procedure is used to derive two algorithms for this problem. Almost sure convergence for the parameter estimate and the filtering error will be established. The convergence analysis is based on an almost-supermartingale convergence lemma that allows a stochastic Lyapunov like approach.

1. INTRODUCTION

The goal of this paper is to present some algorithms for a continuous-time adaptive stochastic filtering problem, and to establish almost sure convergence results for these algorithms.

What is the adaptive stochastic filtering problem? The adaptive stochastic filtering problem is to predict or filter a process when the parameter values of the dynamical system representing the process are not known. This problem is highly relevant for practical prediction and filtering problems in engineering and the social sciences.

Why should one consider the continuous-time version of this problem? In discrete-time the adaptive stochastic filtering problem has been investigated by many researchers. Hence the question why? Time is generally perceived to be continuous. In practice a continuous-time signal is sampled and the subsequent data processing is done in a discrete-time mode. One question then is what happens with the predictions when the sampling time gets smaller and smaller? To study these and related questions continuous-time algorithms must be derived, and their relationship with discrete-time algorithms investigated.

The questions that one would like to solve for the adaptive stochastic filtering problem are how to synthesize algorithms and how to evaluate the performance of these algorithms?

The selftuning synthesis procedure will be used in this paper. This procedure suggests first to solve the associated stochastic filtering problem, and secondly to estimate the values of the parameters of the filter system in an on-line fashion. A continuous-time parameter estimation algorithm is thus necessary. Although considerable attention has been given to off-line algorithms [1,2], on-line algorithms are scarce [3,7]. Below two new algorithms are presented.

As to the performance evaluation of the algorithms, the major question is the convergence of the parameter estimates and of the error in the filtering estimate. For these variables one may consider almost sure convergence and the asymptotic distribution. Below an almost sure convergence result for the given algorithms will be presented. This result is based on a convergence lemma that is of independent interest.

A brief outline of the paper follows. The problem formulation is given in section 2. Section 3 is devoted to the statement of the main results. The proofs of the results may be found elsewhere [14].

2. THE PROBLEM FORMULATION

The adaptive stochastic filtering problem is to predict or filter a stochastic process when the parameters of the distribution of this process are not known. The object of this section is to make this problem formulation precise.

Throughout this paper (Ω,F,P) denotes a complete probability space. Let $T = R$. The terminology of Dellacherie, Meyer [4,5] will be used.

Assume given an R-valued Gaussian process with stationary increments. From stochastic realization theory it is known [8] that under certain conditions this process has a minimal realization as the output of what will be called a *Gaussian system*:

$$dx_t = Ax_t dt + Bdv_t, \tag{1}$$

$$dy_t = Cx_t dt + \cdot dv_t, \tag{2}$$

where $y : \Omega \times T \to R$, $x : \Omega \times T \to R^n$, $v : \Omega \times T \to R^m$ is a standard Brownian motion process, $A \in R^{n\times n}$, $B \in R^{n\times m}$, $C \in R^{1\times n}$, $D \in R^{1\times m}$. The precise statement on the representation is that the distribution of the output y of this system is the same as that of the given process.

One may construct the asymptotic Kalman-Bucy filter for the system (1,2), say:

$$d\hat{x}_t = A\hat{x}_t dt + K(dy_t - C\hat{x}_t dt),$$

where $\hat{x}_t = E[x_t|F_t^y]$, $F_t^y = \sigma(\{y_s, \forall s \le t\})$.
This filter may be written as a Gaussian system

$$d\hat{x}_t = A\hat{x}_t dt + Kd\bar{v}_t, \tag{3}$$

$$dy_t = C\hat{x}_t dt + d\bar{v}_t, \tag{4}$$

where $\bar{v} : \Omega \times T \to R$ is the innovations process, which is a Brownian motion process, say with variance $\sigma^2.t$. It is a result of stochastic realization theory that the two relizations (1,2) and (3,4) are indistinguishable on the basis of information about y only. For adaptive stochastic filtering one may therefore limit attention to the realization (3,4). That realization has the additional advantage that it is suitable for prediction purposes. The fact that (1,2) is a minimal realization, and hence that (3,4) is a minimal realization, implies that the pair (A,C) is observable and that the spectrum of A is in $C^- := \{c \in C \mid Re(c) < 0\}$

2.1. PROBLEM. Assume given an R-valued Gaussian process with stationary increments having a minimal past-output based stochastic realization given by

$$d\hat{x}_t = A\hat{x}_t dt + K d\bar{v}_t, \tag{3'}$$

$$dy_t = C\hat{x}_t dt + d\bar{v}_t, \tag{4'}$$

$$\hat{z}_t := C\hat{x}_t, \tag{5}$$

with the properties mentioned above. Assume further thet the value of the dimension of this system and the value of σ^2 occurring in the variance of $\bar{v}$ are known, but that the values of the parameters A,K,C are not known. The *adaptive stochastic filtering problem* for the above defined Gaussian system is to recursively estimate $\hat{z}$ given y.

For the parameter estimation problem another representation of the Gaussian system (3,4) is needed. Such a representation is derived below. For notational convenience the time set is taken to be $T = R_+$ in the following.

2.2. PROPOSITION. *Given the Gaussian system* (3,4). *The two following representations describe the same relation between* $\bar{v}$ and $\hat{z}$.

a.

$$d\hat{x}_t = A\hat{x}_t dt + K d\bar{v}_t, \quad \hat{x}_0 = 0,$$

$$\hat{z}_t = C\hat{x}_t,$$

$$dy_t = C\hat{x}_t dt + d\bar{v}_t, \quad y_0 = 0.$$

b.

$$dh_t = F h_t dt + G_1 dy_t + G_2 (dy_t - h_t^T p dt), \quad h_o = 0, \tag{6}$$

$$\hat{z}_t = h_t^T p, \tag{7}$$

$$dy_t = h_t^T p dt + d\bar{v}_t, \quad y_0 = 0, \tag{8}$$

where $h : \Omega \times T \to R^{2n}$

$$h_t^T = (y_t^{(1)}, \ldots, y_t^{(n)}, \bar{v}_t^{(1)}, \ldots, \bar{v}_t^{(n)}), \tag{9}$$

$$y_t^{(1)} = y_t$$

$$y_t^{(i)} = \int_0^t y_s^{(i-1)} ds, \text{ for } i = 2,3,\ldots,n,$$

$p \in R^{2n}$ is related to A,K,C as indicated in the proof,

$$F_1 = \begin{pmatrix} 0 & \cdots & 0 \\ & & \vdots \\ I_{n-1} & & 0 \end{pmatrix} \in R^{n\times n}, \quad F = \begin{pmatrix} F_1 & 0 \\ 0 & F_1 \end{pmatrix} \in R^{2n\times 2n},$$

$$G_1 = e_1 \in R^{2n}, \; G_2 = e_{n+1} \in R^{2n},$$

where e_i *is the i-th unit vector.*

The proof of this result is given in [14].

3. THE MAIN RESULTS

In this section two algorithms are presented for the continuous-time adaptive stochastic filtering problem, and convergence results are provided. The proofs of these results may be found in [14].

In the following attention is restricted from the Gaussian system (3,4) or (6,8), to the auto regressive case described by

$$y_t = \sum_{i=1}^{n} a_i y_t^{(i+1)} + \bar{v}_t, \tag{10}$$

or

$$dy_t = h_t^T p\, dt + d\bar{v}_t, \; y_0 = 0, \tag{11}$$

where now $h : \Omega \times T \to R^n$, $p \in R^n$,

$$\begin{aligned} h_t^T &= (y_t^{(1)}, \ldots, y_t^{(n)}) \\ p^T &= (a_1, \ldots, a_n). \end{aligned} \tag{12}$$

Then

$$dh_t = \begin{pmatrix} a_1 \cdots & a_n \\ & 0 \\ I_{n-1} & \vdots \\ & 0 \end{pmatrix} h_t dt + \begin{pmatrix} 1 \\ 0 \\ \vdots \\ 0 \end{pmatrix} d\bar{v}_t, \; h_0 = 0, \tag{13}$$

and one concludes that asymptotically h is a stationary Gauss-Markov process. Since the interest here is in the stationary situtation, it will henceforth be assumed that h is a stationary Gauss-Markov process. By the stability of the Gaussian system the covariance function of h is integrable, thus h is an ergodic process [15, p. 69].

The first algorithm to be presented is based on the least-squares method.

3.1. DEFINITION. The adaptive stochastic filtering algorithm for the autoregressive representation based in the least-squares method is defined to be

$$\begin{aligned} d\hat{p}_t &= Q_t h_t \sigma^{-2}[dy_t - h_t^T \hat{p}_t dt]\; p_0 = 0, \\ dQ_t &= -Q_t h_t h_t^T Q_t \sigma^{-2} dt, \; Q_0 \in R^{n\times n}, \; Q_0 = Q_0^T > 0, \end{aligned}$$

$$\hat{\hat{z}}_t = h_t^T \hat{p}_t,$$

where $\hat{p} : \Omega \times T \to R^n$, $Q : \Omega \times T \to R^{n\times n}$, $\hat{\hat{z}} : \Omega \times T \to R$.
Here $\hat{\hat{z}}$ is the desired estimate of $\hat{z}$, and $\hat{p}$ is an estimate of the parameter p.

The algorithm of 3.1 may be derived as follows. One has the representation

$$dp_t = 0,\ p_0 = p$$
$$dy_t = h_t^T p_t dt + d\bar{v}t,\ y_0 = 0,$$

where it is now assumed that $\bar{v}$ is a Brownian motion process, $p : \Omega \times T \to R^n$, p is Gaussian $G(0,Q_0)$ and that p and $\bar{v}$ are independent objects. With (12) one concludes that $(h_t, F_t^y, t \in T)$ is adapted. The conditional Kalman-Bucy filter [9, 12.1] applied to the above representation then yields the algorithm given in 3.1.

3.2. THEOREM. *Given the stochastic dynamic system of* 2.1 *restricted to the autoregressive case as indicated above. Consider the adaptive stochastic filtering algorithm of* 3.1. *Under these conditions:*

a. $\quad as - \lim_{t\to\infty} \hat{p}_t = p;$

b. $\quad as - \lim_{t\to\infty} t^{-1} \int_0^t (\hat{z}_s - \hat{\hat{z}}_s)^2 ds = 0.$

The above result shows that under the stated conditions the difference between the filter estimate $\hat{z}$ with known parameters and the adaptive filter estimate goes to zero in the above stated sense. In addition the parameter estimate converges to the actual value.

One might conjecture that a result like 3.2 also holds if the restriction to the autoregressive case is relaxed to that of (6,8) and a recursive extended least-squares algorithm is used. An investigation indicates that this is unlikely. The reason why this is the case is not yet fully understood.

The second algorithm is related to that of Goodwin, Ramadge, and Caines [6], and that of Chen [3]. The latter also provides a continuous-time algorithm not only for the autoregressive case but also for the general case of 2.1.

3.3. DEFINITION. The adaptive stochastic filtering algorithm for the autoregressive representation (11, 12) based on the parameter estimation algorithm AML2 [6] is defined to be

$$d\hat{p}_t = h_t r_t^{-1} \sigma^{-2} [dy_t - h_t^T \hat{p}_t dt],\ \hat{p}_0 = 0,$$
$$dr_t = \sigma^{-2} h_t^T h_t dt,\ r_0 = 1,$$
$$\hat{\hat{z}}_t = h_t^T \hat{p}_t,$$

where $\hat{p} : \Omega \times T \to R^n$, $r : \Omega \times T \to R$, $\hat{\hat{z}} : \Omega \times T \to R$ and h is given in (12).

3.4. THEOREM. *Given the stochastic dynamic systems of 2.1. restricted to the autoregressive case as indicated above. Consider the adaptive stochastic filtering algorithm 3.3. Under these conditions*

$$\text{as} - \lim_{t\to\infty} t^{-1} \int_0^t (\hat{z}_s - \hat{\hat{z}}_s)^2 ds = 0.$$

In [3] a convergence result is given for the representation 2.1 with an algorithm that has the same structure as that of 3.3. There the convergence is obtained under an unnatural assumption. One possible reason for this assumption is that the second innovation process

$$d\bar{\bar{v}}_t = dy_t - \hat{h}_t^T \hat{p}_t dt$$

is directly used in $\hat{h}$ and not prefiltered.

The above convergence results for adaptive stochastic filtering are based on a convergence theorem to be provided below. As some of the other concepts and results of system identification the convergence theorem is also inspired by the statistics literature, in particular by the area of stochastic approximation. H. Robbins and D. Siegmund [11] have established a discrete-time convergence result for use in stochastic approximation theory. A simplified version of that result is given as an exercise in [10, II-4]. V. Solo [12, 13] has been the first to use this result in the system identification literature and since then it has become rather popular [6]. This popularity is due not only to the ease with which convergence results are proven but also to the formulation in terms of martingales which show up naturally in stochastic filtering problems. Below the continuous-time analog of [11,th.1.] is given.

A few words about notation follow. $(F_t, t \in T)$ denotes a σ-algebra family satisfying the usual conditions, A^+ is the set of increasing processes, M_{1uloc} the set of locally uniformly integrable martingales, and $\Delta x_t = x_t - x_{t-}$ is the jump of the process x at time $t \in T$.

3.5. THEOREM. *Let* $x : \Omega \times T \to R_+$, $a : \Omega \times T \to R_+$, $b : \Omega \times T \to R_+$, $e : \Omega \times T \to R_+$ *and* $m : \Omega \times T \to R$ *be stochastic processes. Assume that*

1. $x_0 : \Omega \to R_+$ *is* F_0 *measurable;*
2. $(a_t, F_t, t \in T) \in A^+$, $a_0 = 0$, $a_\infty < \infty$ a.s., *and for all* $t \in T$ $\Delta a_t \le c_1 \in R_+$,
   $(b_t, F_t, t \in T) \in A^+$, $b_0 = 0$;
3. $(e_t, F_t, t \in T)$ *is adapted and* $\int_0^\infty e_s ds < \infty$ a.s.;
4. $(m_t, F_t, t \in T) \in M_{1uloc}$, $m_0 = 0$;
5. *x satisfies the stochastic differential equation*

$$dx_t = e_t x_t dt + da_t - db_t + dm_t,\ x_0.$$

Then

a. $x_\infty := \text{as-}\lim_{t\to\infty} x_t$ *exists in* R_+, *hence* $x_\infty < \infty$ a.s.;

b. $b_\infty := \text{as-}\lim_{t\to\infty} b_t$ *exists and* $b_\infty < \infty$ a.s.

REFERENCES

1. A. Bagchi, Consistent estimates of parameters in continuous time systems, in: Analysis and Optimization of Stochastic Systems, O.L.R. Jacobs, M.H.A. Davis, M.A.H. Dempster, C.J. Harris, P.C. Parks., Academic Press, London, 1980, pp. 437-450.
2. A.V. Balakrishnan, Stochastic differential systems I, Filtering and Control, A function space approach, Lecture Notes in Economics and Mathematical Systems, volume 84, Springer-Verlag, Berlin, 1973.
3. Chen Han Fu, Quasi-least-squares identification and its strong consistency, Int. J. Control, 34 (1981), pp. 921-936.
4. C. Dellacherie, P.A. Meyer, Probabilités et Potentiel, Chapitres I à IV, Hermann, Paris, 1975.
5. C. Dellacherie, P.A. Meyer, Probabilités et Potentiel, Chapitres V à VIII, Theorie des Martingales, Hermann, Paris, 1980.
6. G.C. Goodwin, P.J. Ramadge, P.E. Caines, A globally convergent adaptive predictor, Automatica, 17 (1981), pp. 135-140.
7. Y.D. Landau, Adaptive control, Marcel Dekker, New York, 1979.
8. A. Lindquist, G. Picci, On the stochastic realization problem, SIAM J. Control and Optim., 17 (1969), pp. 365-389.
9. R.S. Liptser, A.N. Shiryayev, Statistics of random processes, I General theory, II Applications, Springer-Verlag, Berlin, 1977,1978.
10. J. Neveu, Martingales à temps discrets, Masson et Cie, Paris, 1972; english translation: Discrete parameter martingales, North-Holland, Amsterdam, 1975.
11. H. Robbins, D. Siegmund, A convergence theorem for non negative almost supermartingales and some applications, in 'Optimizing methods in statistics', J.S. Rustagi ed., Academic Press , New York, 1971, pp. 233-256.
12. V. Solo, The convergence of AML, IEEE Transactions on Automatic Control, 24 (1979), pp. 958-962.
13. V. Solo, The convergence of an instrumental-variable-like recursion, Automatica, 17 (1981), pp. 545-547.
14. J.H. van Schuppen, Convergence results for continuous-time adaptive stochastic filtering algorithms, preprint, 1982.
15. E. Wong, Stochastic processes in information and dynamical systems, McGraw-Hill, New York, 1971.

BETWEEN THE CHAPTERS: AN EDITOR'S NOTE

I think this is the right place to put a very recently discovered connection between filter and control. Unfortunately those who have contributed these results to the theory were unable to attend the conference. So I shall try to give a very rough description of this duality - as far as I understand it - , as I think that this link between filter and control might fertilize both theories. For the importance of these results in quantum physics see [2,3].

References:

[1] W.H.FLEMING, S.K.MITTER: Optimal control and nonlinear filtering for non-degenerate diffusion processes, preprint LCDS, Sept. 1981

[2] S.K.MITTER: On the analogy between mathematical problems in nonlinear filtering and quantum physics, Richerche di Automatica 10 (1979), 163-216

[3] S.K.MITTER: Non-linear filtering and stochastic mechanics, Stochastic Systems: The Mathematics of Filtering and Identification and Applications, D. Reichel Publ. Co (1981), M.Hazewinkel and J.C.Willems (eds), 479-503

In the last section we have seen that Zakai's equation for the partially observed system

$$(1) \qquad dx_t = f(x_t)dt + dw_t \qquad \text{(n-dim. signal)}$$

$$(2) \qquad dy_t = h(x_t)dt + d\tilde{w}_t \qquad \text{(1-dim. observation)}$$

with initial conditions x_0 and $y_0 = 0$, w and $\tilde{w}$ independent Brownian motions, is given by

$$(3) \qquad dq_t = A^* q_t dt + hq_t dy_t \ ,$$

where A^* is the adjoint of the generator of the signal process. After substituting

$$(4) \qquad q = \exp(yh)\cdot p$$

(3) becomes a linear p.d.e.

$$(5) \qquad \frac{\partial}{\partial t}p = \frac{1}{2}p_{xx} + fp_x + v\cdot p$$

with initial condition the distribution of $x(0)$.

Define now in an obvious notation the following abbreviations

$$(6) \qquad g = y\frac{\partial h}{\partial x} - f$$

and

$$(7) \qquad v = -\mathrm{div}(f - y\frac{\partial h}{\partial x}) - \frac{1}{2}h^2 - yAh + \frac{1}{2}y^2\frac{\partial h}{\partial x}\,\frac{\partial h}{\partial x} \quad .$$

The link between the filter problem and a control problem is now a log-transformation which is well known from a transformation of the heat equation in physics. Put

(8) $S = -\log p$

which gives us (for a positive solution of (5))

(9) $S_t = \frac{1}{2} S_{xx} + gS_x - \frac{1}{2} S_x S_x - v$

$S(x,0) = -\log p_o$

But this now is the dynamic programming equation of some control problem e.g. of the form [see next chapter]

(10) $d\xi_\tau = u(\xi_\tau,\tau)d\tau + dw_\tau \qquad o \leqq \tau \leqq t$

$\xi_o = x$

with criterion

(11) $J(x,t,u) = E_x \{{}_0\int^t L(\xi_\tau,t-\tau,u(\tau))d\tau + S_o(\xi_t)\}$

$L = \frac{1}{2}(u-g)'(u-g) - v$.

This establishes a control problem the dynamic programming equation of which is a transform of the Zakai equation of a partially observed system, and with this we have found a bridge to the next chapter:

STOCHASTIC DIFFERENTIAL SYSTEMS

PART III : STOCHASTIC CONTROL THEORY

ON PERTURBATION METHODS IN STOCHASTIC CONTROL

A. BENSOUSSAN (*)

We consider in this article some perturbation methods in stochastic control. The small coefficient ε is related to the variance of the noise. When ε is 0 the system is deterministic. Problems of this sort have been considered by W. Fleming [4] and he used the Bellman equation of Dynamic Programming to derive approximations results. We consider problems of a more general type, which cannot be easily dealt with, using D.P.

For this reason, we use a completely different approach, based upon perturbation arguments on the control problem itself. Besides its more general applicability, our method has several advantages. Firstly, regularity assumptions are limited to the optimal deterministic trajectory itself, and not to zones of regularity like in Fleming's approach. Moreover the method leads to new results, even in the case considered by Fleming and they may have a wider applicability in practice. A drawback of our method is that it is not easy to introduce constraints on the control. This drawback can be removed, but we shall not do it in this paper, to avoid too lenghty developments.

1. SETTING OF THE PROBLEM.

1.1. Notation. Assumptions.

Let $g_0(x,v)$, $g_1(x,v)$ such that :

(1.1) $$g_0 : R^n \times R^k \to R^n \quad , \quad g_1 : R^n \times R^k \to \mathcal{L}(R^n;R^n)$$

$$|g_0(x,v)| \leq \bar{g}_0 \, (1 + |x| + |v|)$$

$$|\text{tr}\; g_1 g_1^*| \leq \bar{g}_1(1 + |x|^2 + |v|^2)$$

(1.2) g_0 is C^3 in x,v, with Lipschitz third derivatives ; g_1 is C^2 in x,v with Lipschitz second derivatives. All derivatives of g_0,g_1 are bounded functions functions.

(*) University Paris-Dauphine and INRIA.

This work has been partly supported by DOE Office of Electric Energy Systems under contract 01-80, RA-50 154.

Let next f_0, f_1, h_0, h_1 such that :

$$(1.3)\qquad f_0,f_1 : R^n \times R^k \to R \quad , \quad h_0,h_1 : R^n \to R$$

$$|f_i(x,v)| \leq \bar{f}_i(1 + |x|^2 + |v|^2)$$

$$|h_i(x)| \leq \bar{h}_i(1 + |x|^2).$$

The functions f_0, h_0 are C^3 with Lipschitz third derivatives. The functions f_1,h_1 are C^2 with Lipschitz second derivatives. The second and third derivatives of f_0,h_0,f_1,h_1 are bounded functions.

Let $(\Omega,\mathcal{A},P,F^t,w(t))$ be a probability space equipped with a filtration F^t and a standard F^t Wiener process. An admissible control is a process adapted to F^t and with values in R^k. We do not consider constraints in the present framework.

For any given admissible control v(.) we can solve in the sense of Ito (strong sense) the stochastic differential equation

$$(1.4)\qquad \left| \begin{array}{l} dx_\varepsilon = g_0(x_\varepsilon,v)dt + \varepsilon\, g_1(x_\varepsilon,v)dw \\ x_\varepsilon(0) = x \end{array} \right.$$

This is possible by virtue of the regularity assumptions (1.1), (1.2). We next define the functional :

$$(1.5)\qquad J_\varepsilon(v(.)) = E \int_0^T [f_0(x_\varepsilon,v) + \varepsilon f_1(x_\varepsilon,v)]dt + $$
$$+ E\,(h_0(x_\varepsilon(T)) + \varepsilon\, h_1(x_\varepsilon(T))).$$

This functional is well defined by virtue of the assumptions (1.3).

Remark 1.1. : In the case $g_1=I$, $f_1=h_1=0$, we recover a problem studied by W. Fleming.

Remark 1.2. : We could also consider in (1.4) a deterministic perturbation, namely replace $g_1(x_\varepsilon,v)dw$ by $g_2(x_\varepsilon,v)dt + g_1(x_\varepsilon,v)dw$. The reader will easily adapt results of this paper to this case.

1.2. Necessary conditions.

Our objective is to study the behaviour of Inf $J_\varepsilon(v(.))$ as ε tends to 0. Although we shall not necessarily assume that this infimum is attained, let us state the necessary conditions of optimality, denoting by u an optimal control and by y the corresponding optimal trajectory. We introduce the Hamiltonians

$$(1.6) \qquad H_0(x,p,v) = f_0(x,v) + p \cdot g_0(x,v)$$

$$(1.7) \qquad H_1(x,r,v) = f_1(x,v) + \sum_j r_j g_{1j}(x,v)$$

where r denotes the matrix $(r_1,\dots,r_n)$. We have the relations :

$$(1.8) \qquad \begin{aligned} dy_\varepsilon &= g_0(y_\varepsilon,u_\varepsilon)dt + \varepsilon g_1(y_\varepsilon,u_\varepsilon)dw \\ y_\varepsilon(0) &= x \end{aligned}$$

$$(1.9) \qquad \begin{aligned} -dp_\varepsilon &= (g^*_{0x}(y_\varepsilon,u_\varepsilon)p_\varepsilon + \varepsilon \sum_j g_{1jx}(y_\varepsilon,u_\varepsilon)r_{j\varepsilon} + \\ &\quad + f_{0x}(y_\varepsilon,u_\varepsilon) + \varepsilon f_{1x}(y_\varepsilon,u_\varepsilon))dt - \sum_j r_{j\varepsilon}(t)dw_j(t) \\ p_\varepsilon(T) &= h_{0x}(y_\varepsilon(T)) + \varepsilon h_{1x}(y_\varepsilon(T)) \end{aligned}$$

$$(1.10) \qquad H_{0v}(y_\varepsilon,p_\varepsilon,u_\varepsilon) + \varepsilon H_{1v}(y_\varepsilon,r_\varepsilon,u_\varepsilon) = 0 \qquad \text{a.e.t, a.s.}$$

In (1.9), (1.10) the pair p_ε, r_ε represents the adjoint process arising in the stochastic maximum principle (see J.M. Bismut [2], A. Bensoussan [1]).

Setting $\varepsilon=0$, we obtain a deterministic control problem, namely

$$(1.11) \qquad \left| \begin{aligned} \frac{dx_0}{dt} &= g(x_0,v) \\ x_0(o) &= x \end{aligned} \right.$$

$$(1.12) \qquad J_0(v(.)) = \int_0^T f_0(x_0,v)dt + h_0(x_0(T))$$

to which corresponds the following necessary conditions of optimality (Pontryagin's Maximum principle).

$$(1.13) \qquad \begin{aligned} \dot{y}_0 &= g_0(y_0,u_0) \qquad\qquad y_0(0) = x \\ -\dot{p}_0 &= g^*_{0x}(y_0,u_0)p_0 + f_{0x}(y_0,u_0) \\ p_0(T) &= h_{0x}(y_0(T)) \\ H_{0v}&(y_0(t),p_0(t),u_0(t)) = 0 \end{aligned}$$

We shall formulate the next assumption as follows :

We shall assume that there exist $y_0(t), u_0(t), p_0(t)$ such that (1.13) holds and moreover :

(1.14) $\qquad H_{ovv}(x,p_0(t),v) \geq \beta_0 I \qquad \beta_0 > 0$

$\qquad \forall t \in [o,T] \ , \ \forall v,x$

(1.15) $\qquad H_{oxx} - H_{oxv} H_{ovv}^{-1} H_{ovx} \geq \beta_0 I$

with the same arguments as in (1.14)

(1.16) $\qquad h_{oxx}(x) \geq \beta_0 I.$

Remark 1.3 : It is important to notice that the assumptions (1.13), (1.14), (1.15), (1.16) imply that u_0 is necessarily an optimal control for the problem (1.11), (1.12) and that the optimal control is unique.

2. ASYMPTOTICS.

2.1. Formal expansion.

We can look for an expansion of the functions u_ε, y_ε, p_ε , r_ε as follows :

$$u_\varepsilon \sim u_0 + \varepsilon u_1 + \varepsilon^2 u_2 + \dots$$

$$y_\varepsilon \sim y_0 + \varepsilon y_1 + \varepsilon^2 y_2 + \dots$$

$$p_\varepsilon \sim p_0 + \varepsilon p_1 + \varepsilon^2 p_2 + \dots$$

$$r_\varepsilon \sim r_0 + \varepsilon r_1 + \varepsilon^2 r_2 + \dots$$

We first obtain (1.13) and next :

(2.1) $$dy_1 = (g_{ox} y_1 + g_{ov} u_1)dt + g_1 dw$$

$$y_1(o) = 0$$

(2.2) $$-dp_1 = (g_{ox}^* p_1 + H_{oxx} y_1 + H_{oxv} u_1 + H_{1x})dt - \sum_j r_{j1}\, dw_j$$

$$p_1(T) = h_{oxx}\, y_1(T) + h_{1x}$$

$$(2.3) \qquad H_{ovx}y_1 + g^*_{ov}\, p_1 + H_{ovv}\, u_1 + H_{1v} = 0$$

where all arguments are evaluated along the trajectory $y_0(t)$, $p_0(t)$, $u_0(t)$.

It is easy to check that (2.1) corresponds to the stochastic maximum principle applied to a linear quadratic problem where the state equation is given by (2.1) and the cost functional is given by :

$$(2.4) \qquad J_1(u_1) = \frac{1}{2} E \int_0^T (H_{oxx}y_1 \cdot y_1 + 2H_{oxv}u_1 \cdot y_1 + H_{ovv}u_1 \cdot u_1 +$$
$$+ 2H_{1x}\, y_1 + 2H_{1v}u_1)dt + \frac{1}{2} E\, (h_{oxx}y_1(T) \cdot y_1(T) +$$
$$+ 2\, h_{1x}\, y_1(T))$$

2.2. Solution of the linear quadratic problem.

Problem (2.1) (2.4) can be solved by standard methods relying on a Riccati equation. We consider the Riccati equation

$$(2.5) \qquad \dot{P}_0 + P_0(g_{ox} - g_{ov}H^{-1}_{ovv}\, H_{ovx}) +$$
$$+ (g^*_{ox} - H_{ovx}\, H^{-1}_{ovv}\, g^*_{ov})P_0 - P_0\, g_{ov}\, H^{-1}_{ovv}\, g^*_{ov}\, P_0 +$$
$$+ H_{oxx} - H_{oxv}\, H^{-1}_{ovv}\, H_{ovx} = 0$$
$$P_0(T) = h_{oxx}$$

and the linear equation

$$(2.6) \qquad \dot{\rho}_1 + (g^*_{ox} - P_0\, g_{ov}\, H^{-1}_{ovv}\, g^*_{ov})\rho_1 +$$
$$+ f_{1x} - H_{oxv}\, H^{-1}_{ovv}\, f_{1v} - P_0\, g_{ov}\, H^{-1}_{ovv}\, f_{1v} = 0$$
$$\rho_1(T) = h_{1x}$$

We then have :

$$(2.7) \qquad p_1(t) = P_0(t)\, y_1(t) + \rho_1(t)$$

$$(2.8) \qquad r_{j1}(t) = P_0(t) g_{1j}(y_0(t), u_0(t))$$

(2.9) $$u_1(t) = K_o(t)\, y_1(t) + \ell_o(t)$$

with the definitions :

(2.10) $$K_o(t) = -H_{ovv}^{-1}(H_{ovx} + g_{ov}^{*}\, P_o)$$

(2.11) $$\ell_o(t) = -H_{ovv}^{-1}(f_{1v} + g_{ov}^{*}\, \rho_1)$$

2.3. Approximation of the optimal control.

Considering $u_o + \varepsilon u_1$ as an approximation to u_ε , and using the feedback (2.9) we deduce that

$$u_o(t) + \varepsilon K_o(t) y_1(t) + \varepsilon\, \ell_o(t)$$

is an approximation to u_ε . In this expression we can replace $\varepsilon y_1(t)$ by $y_\varepsilon(t) - y_o(t)$ (which is a new approximation). Then we are naturally led to the following affine feedback :

$$u_o(t) + K_o(t)(y\ (t) - y_o(t)) + \varepsilon\, \ell_o(t).$$

We shall compare this approximate feedback to the optimal one, in the case studied by Fleming. Let us just notice at this level that this feedback can easily been implemented.

3. MAIN RESULTS OF APPROXIMATION.

3.1. Statement of the result.

Let us define the expression :

(3.1) $$L_1(u_1) = \frac{1}{6} E \int_0^T [H_{oxxx}(y_o,p_o,u_o) y_1 y_1 y_1 +$$

$$+ 3H_{oxxv}\, u_1 y_1 y_1 + 3H_{oxvv} u_1 u_1 y_1 + H_{ovvv}\, u_1 u_1 u_1 +$$

$$+ 3H_{oqxx}\, y_1 y_1 p_1 + 6H_{oqxv}\, u_1 y_1 p_1 + 3H_{oqvv} u_1 u_1 p_1]\, dt$$

$$+ \frac{1}{2} E \int_0^T [H_{1xx}(y_o,o,u_o) y_1 y_1 + 2H_{1xv} u_1 y_1 + H_{1vv}\, u_1 \cdot u_1$$

$$+ 2(H_{1r_jx}\, y_1 r_{j1} + H_{1r_jv}\, u_1 r_{j1})]\, dt +$$

$$+ E\, [\frac{1}{6}\, h_{oxxx}(y_o(T)) y_1(T) y_1(T) y_1(T) + \frac{1}{2}\, h_{1xx}(y_o(T)) y_1(T) y_1(T)]$$

This expression is well defined by virtue of the values of y_1, u_1, r_{j1} obtained above. Note that r_{j1} is deterministic. We have the following results :

Lemma 3.1. : The following estimate holds :

$$(3.2) \qquad |J_\varepsilon(u_0 + \varepsilon u_1) - J_0(u_0) - \varepsilon\left(\int_0^T f_1(y_0,u_0)dt + h_1(y_0(T))\right) - \varepsilon^2 J_1(u_1) - \varepsilon^3 L_1(u_1)| \leq C\varepsilon^4$$

□

Considering the feedback introduced in §2.3, namely :

$$(3.3) \qquad \mathcal{L}_\varepsilon(t;y) = u_0(t) + K_0(t)(y - y_0(t)) + \varepsilon \ell_o(t)$$

and the corresponding state equation :

$$(3.4) \qquad \begin{aligned} d\zeta_\varepsilon &= g_0(\zeta_\varepsilon, u_0 + K_0(\zeta_\varepsilon - y_o) + \varepsilon\ell_o)dt \\ &+ \varepsilon g_{1j}(\zeta_\varepsilon, u_0 + K_0(\zeta_\varepsilon - y_o) + \varepsilon\ell_o)dw_j \\ \zeta_\varepsilon(0) &= x \end{aligned}$$

and setting :

$$(3.5) \qquad v_\varepsilon(t) = u_0(t) + K_0(t)(\zeta_\varepsilon(t) - y_0(t)) + \varepsilon\ell_o(t)$$

we also have :

Lemma 3.2. : The following estimate holds :

$$(3.6) \qquad |J_\varepsilon(v_\varepsilon(.)) - J_\varepsilon(u_0 + \varepsilon u_1)| \leq C\varepsilon^4$$

□

Up to constants, controls v_ε and $u_0 + \varepsilon u_1$ yield the same accuracy, but of course v_ε derives from an easily implementable feedback, which is a very important advantage

Lemma 3.3. : Let u_ε be an admissible control such that

$$(3.7) \qquad J_\varepsilon(u_\varepsilon) \leq J_\varepsilon(u_0 + \varepsilon u_1)$$

and y_ε be the corresponding trajectory. Let p_ε, r_ε , be uniquely defined by (1.9). Then the following estimates hold

$$(3.8) \qquad E\int_0^T |u_\varepsilon(t) - u_0(t) - \varepsilon u_1(t)|^2\, dt \leq C\varepsilon^4$$

$$(3.9) \qquad E\,|y_\varepsilon(t) - y_0(t) - \varepsilon y_1(t)|^2 \le C\varepsilon^4$$

$$(3.10) \qquad E\int_0^T |p_\varepsilon(t) - p_0(t) - \varepsilon p_1(t)|^2\,dt \le C\varepsilon^4$$

$$(3.11) \qquad E\int_0^T |r_{j\varepsilon}(t) - \varepsilon r_{j1}(t)|^2 dt \le C\varepsilon^4$$

□

Lemma 3.4. : Let u_ε be as in Lemma 3.3, then one has

$$(3.12) \qquad |J_\varepsilon(u_\varepsilon) - J_0(u_0) - \varepsilon(\int_0^T f_1(y_0,u_0)dt + h_1(y_0(T)))$$

$$- \varepsilon^2\, J_1(u_1) - \varepsilon^3\, L_1(u_1)| \le C\varepsilon^4.$$

□

We can then state the main result of the paper

Theorem 3.1. : We assume (1.1), (1.2), (1.3) and (1.13), (1.14), (1.15), (1.16). Then we have :

$$(3.13) \qquad |\inf J_\varepsilon(v(.)) - J_\varepsilon(u_0 + \varepsilon u_1)| \le C\varepsilon^4$$

$$|\inf J_\varepsilon(v(.)) - J_\varepsilon(v_\varepsilon)| \le C\varepsilon^4$$

3.2. Some ideas of the proofs.

Setting :

$$\tilde{u}_\varepsilon = u_\varepsilon - u_0 - \varepsilon u_1$$

$$\tilde{y}_\varepsilon = y_\varepsilon - y_0 - \varepsilon y_1$$

the first step consists in deriving the following expression :

(3.14) $$J_\varepsilon(u_\varepsilon) = J_0(u_0) + \varepsilon\,[\int_0^T f_1(y_0,u_0)dt + h_1(y_0(T))]$$

$$+ E\int_0^T\int_0^1 \lambda[\int_0^1 (H_{oxx}(y_0 + \mu\lambda(\varepsilon y_1 + \tilde{y}_\varepsilon),u_0 + \mu\lambda(\varepsilon u_1 + \tilde{u}_\varepsilon))(\varepsilon y_1 + \tilde{y}_\varepsilon)^2 +$$

$$+ 2H_{ovx}(\quad)(\varepsilon y_1 + \tilde{y}_\varepsilon)(\varepsilon u_1 + \tilde{u}_\varepsilon) + H_{ovv}(\quad)(\varepsilon u_1 + \tilde{u}_\varepsilon)^2]\, d\mu d\lambda dt$$

$$+ \varepsilon E\int_0^T\int_0^1 [f_{1x}(y_0 + \lambda(\varepsilon y_1 + \tilde{y}_\varepsilon),u_0 + \lambda(\varepsilon u_1 + \tilde{u}_\varepsilon))(\varepsilon y_1 + \tilde{y}_\varepsilon) +$$

$$+ f_{1v}(\quad)(\varepsilon u_1 + \tilde{u}_\varepsilon)]\, d\lambda dt$$

$$+ E\int_0^1 \lambda[\int_0^1 h_{oxx}(y_0(T) + \mu\lambda(\varepsilon y_1(T) + \tilde{y}_\varepsilon(T)))(\varepsilon y_1(T) + \tilde{y}_\varepsilon(T))^2]\, d\mu d\lambda +$$

$$+ \varepsilon E\int_0^1 h_{1x}(y_0(T) + \lambda(\varepsilon y_1(T) + \tilde{y}_\varepsilon(T)))(\varepsilon y_1(T) + \tilde{y}_\varepsilon(T))d\lambda.$$

Using the assumptions (1.14), (1.15), (1.16), we have :

(3.15) $$E\int_0^T\int_0^1\int_0^1 [H_{oxx}(\quad)\tilde{y}_\varepsilon.\tilde{y}_\varepsilon + 2H_{ovx}(\quad)\tilde{y}_\varepsilon.\tilde{u}_\varepsilon + H_{ovv}(\quad)\tilde{u}_\varepsilon.\tilde{u}_\varepsilon]\, d\mu d\lambda dt$$

$$\geq \gamma_0\, E\int_0^T (|\tilde{y}_\varepsilon(t)|^2 + |\tilde{u}_\varepsilon + H_{ovv}^{-1} H_{ovx}\, \tilde{y}_\varepsilon|^2)dt$$

$$E\int_0^1\int_0^1 \lambda\, h_{oxx}(y_0(T))\tilde{y}_\varepsilon(T).\tilde{y}_\varepsilon(T) \geq \gamma_0\, E\, |\tilde{y}_\varepsilon(T)|^2$$

We also have the following estimates :

(3.16) $$E\int_0^T\int_0^1\int_0^1 \lambda[(H_{oxx}(y_0 + \mu\lambda(\varepsilon y_1 + \tilde{y}_\varepsilon),\, u_0 + \mu\lambda(\varepsilon u_1 + \tilde{u}_\varepsilon)) - H_{oxx}(y_0,u_0))\varepsilon y_1\tilde{y}_\varepsilon$$

$$+ (H_{ovx}(y_0 + \mu\lambda(\varepsilon y_1 + \tilde{y}_\varepsilon),\, u_0 + \mu\lambda(\varepsilon u_1 + \tilde{u}_\varepsilon)) - H_{ovx}(y_0,u_0))(\varepsilon y_1\tilde{u}_\varepsilon + \varepsilon u_1\tilde{y}_\varepsilon)$$

$$+ (H_{ovx}(y_0 + \mu\lambda(\varepsilon y_1 + \tilde{y}_\varepsilon),\, u_0 + \mu\lambda(\varepsilon u_1 + \tilde{u}_\varepsilon)) -$$

$$- H_{ovv}(y_0,u_0))\, \varepsilon u_1\, \tilde{u}_\varepsilon]\, d\mu d\lambda dt$$

$$\leq C\sqrt{\varepsilon}\, E\int_0^T (|\tilde{y}_\varepsilon(t)|^2 + |\tilde{u}_\varepsilon(t)|^2)dt + C\varepsilon^2(E\int_0^T(|\tilde{y}_\varepsilon(t)|^2 + |\tilde{u}_\varepsilon(t)|^2)dt)^{\frac{1}{2}}$$

$$(3.17) \qquad E \int_0^T (H_{oxx}(y_0,u_0)\varepsilon y_1\tilde{y}_\varepsilon + H_{ovx}(y_0,u_0)(\varepsilon y_1\,\tilde{u}_\varepsilon + \varepsilon u_1\tilde{y}_\varepsilon)$$

$$+ H_{ovv}(y_0,u_0)\;\varepsilon u_1\tilde{u}_\varepsilon + \varepsilon f_{1x}(y_0,u_0)\tilde{y}_\varepsilon + \varepsilon f_{1v}(y_0,u_0)\tilde{u}_\varepsilon)dt +$$

$$+ E(h_{oxx}(y_0(T))\varepsilon y_1(T)\tilde{y}_\varepsilon(T) + \varepsilon h_{1x}(y_0(T))\tilde{y}_\varepsilon(T)) =$$

$$= 2E\int_0^T\int_0^1\int_0^1 \lambda\,[p_1\cdot(g_{oxx}(y_0 + \mu\lambda(\varepsilon y_1 + \tilde{y}_\varepsilon),u_0 + \mu\lambda(\varepsilon u_1 + \tilde{u}_\varepsilon))(\varepsilon y_1 + \tilde{y}_\varepsilon)^2 +$$

$$+ 2\,g_{ovx}(\quad)(\varepsilon y_1 + \tilde{y}_\varepsilon)(\varepsilon u_1 + \tilde{u}_\varepsilon) + g_{ovv}(\quad)(\varepsilon u_1 + \tilde{u}_\varepsilon)^2]\;d\mu d\lambda dt$$

$$+ \varepsilon^2\;E\int_0^T\int_0^1 r_{j1}(g_{1jx}(y_0 + \lambda(\varepsilon y_1 + \tilde{y}_\varepsilon),\;u_0 + \lambda(\varepsilon u_1 + \tilde{u}_\varepsilon))(\varepsilon y_1 + \tilde{y}_\varepsilon) +$$

$$+ g_{1jv}(\quad)(\varepsilon u_1 + \tilde{u}_\varepsilon)d\lambda dt.$$

Treating other terms analogously and collecting them we obtain :

$$(3.18) \qquad J_\varepsilon(u_\varepsilon) \geq J_0(u_0) + \varepsilon[\int_0^T f_1(y_0,u_0)dt + h_1(y_0(T))]$$

$$+ \varepsilon^2\;J_1(u_1) + \varepsilon^3 L_1(u_1) + \gamma_1\;E\int_0^T(|\tilde{y}_\varepsilon|^2 + |\tilde{u}_\varepsilon|^2)dt$$

$$- \varepsilon^2\gamma_2(E\int_0^T(|\tilde{y}_\varepsilon(t)|^2 + |\tilde{u}_\varepsilon(t)|^2)dt)^{\frac{1}{2}}$$

$$+ \gamma_1\;E\;|\tilde{y}_\varepsilon(T)|^2 - \varepsilon^2\gamma_2(E|\tilde{y}_\varepsilon(T)|^2)^{\frac{1}{2}} - C\varepsilon^4$$

We next use (3.7) and the estimate (3.2) to deduce

$$E\int_0^T |\tilde{u}_\varepsilon(t)|^2 dt\;, \qquad E\int_0^T |\tilde{y}_\varepsilon(t)|^2 dt \leq C\varepsilon^4$$

Similar arguments are used to derive the remaining results.

4. DYNAMIC PROGRAMMING.

4.1. The Problem.

We shall specialize the preceding results to the case mentionned in Remark 1.1. To simplify the notation we write

$$g_0 = g,\; f_0 = f,\; h_0 = h.$$

We thus consider the following problem :

(4.1) $$\begin{cases} dx = g(x,v)dt + \varepsilon dw \\ x(s) = x \end{cases}$$

(4.2) $$J_{\varepsilon,x,s}(v(.)) = E\left[\int_s^T f(x_\varepsilon,v)dt + h(x_\varepsilon(T))\right]$$

and we set :

(4.3) $$\Phi_\varepsilon(x,s) = \operatorname*{Inf}_{v(.)} J_{\varepsilon,x,s}(v(.))$$

which is solution of Bellman equation :

(4.4) $$\frac{\partial\Phi^\varepsilon}{\partial t} + \frac{\varepsilon^2}{2}\Delta\Phi^\varepsilon + \operatorname*{Inf}_v [f(x,v) + D\phi^\varepsilon . g(x,v)] = 0$$

$$\Phi^\varepsilon(x,T) = h(x)$$

The limit problem is defined by :

(4.5) $$\frac{\partial\Phi^0}{\partial t} + \operatorname*{Inf}_v [f(x,v) + D\Phi^0 . g(x,v)] = 0$$

$$\Phi^0(x,T) = h(x)$$

4.2. Assumptions.

We define a zone of strong regularity of (4.5) as an open set Q of $R^n \times [o,T]$ of the form

$$Q = \{x,s \mid s \in (T_0,T),\ x \in \Gamma(s)\}$$

where

(4.6) the derivatives Φ^o_t, Φ^0_x, Φ^0_{tx}, Φ^0_{xx}, Φ^0_{txx}, Φ^0_{xxx} exist and are continuous ; Φ^0 has a quadratic growth

(4.7) The minimization in (4.5) is attained in a unique point $\bar{v}(x,t)$, where $\bar{v}$ is continuously differentiable in x, and $\bar{v}_x$ is bounded ; $\bar{v}$ has a linear growth.

In a zone of strong regularity, one can thus solve locally the differential equation

$$\frac{dy_0}{dt} = g(y_0, \bar{v}(y_0,t)) \tag{4.8}$$
$$y_0(s) = x$$

We set :

$$u_0(t) = \bar{v}(y_0(t),t) \tag{4.9}$$

$$p_0(t) = D\Phi^0(y_0(t),t). \tag{4.10}$$

We shall assume that :

(4.11) (x,s) belongs to a zone of strong regularity, and that the trajectory (4.8) belongs to the zone of strong regularity up to time T. Moreover we assume that (1.14), (1.15), (1.16) hold for $p_0(t)$ defined by (4.10)

If (4.11) holds, we may define

$$\Phi^1(x,s) = \frac{1}{2}\int_s^T \Delta\Phi^0(y_0(t),t)dt \tag{4.12}$$

which is solution of :

$$\left| \begin{array}{l} \dfrac{\partial\Phi^1}{\partial t} + \dfrac{\partial\Phi^1}{\partial x} \cdot g(x,\bar{v}(x,t)) + \dfrac{1}{2}\Delta\Phi^0 = 0 \\ \Phi^1(x,T) = 0 \end{array} \right. \tag{4.13}$$

4.3. The convergence result.

We can state the following :

Theorem 4.1. : We assume (4.11). Then we have :

$$|\Phi^\varepsilon(x,s) - \Phi^0(x,s) - \varepsilon^2\Phi^1(x,s)| \leq C\varepsilon^4 \tag{4.14}$$

where the constant may depend on the point (x,s).

□

The proof follows from the estimate (3.12), noticing that in the present situation

$$J_1(u_1) = \Phi^1(x,s) \tag{4.15}$$

$$L_1(u_1) = 0 \tag{4.16}$$

and the regularity is used to establish (4.15). It is worth to notice the following relations (at least if (4.11) holds).

(4.17) $$P_0(t) = \Phi^o_{xx}(y_0(t),t)$$

(4.18) $$K_0(t) = \bar{v}_x(y_0(t),t)$$

Remark 4.1. : In the present case the feedback (3.3) reduces to :

$$\mathcal{L}_\varepsilon(t;y) = u_0(t) + K_0(t)(y - y_0(t))$$

and from (4.18)

$$\mathcal{L}_\varepsilon(t;y) = u_0(t) + \bar{v}_x(y_0(t),t)(y - y_0(t)).$$

Therefore, it is just the linearization of $\bar{v}(y,t)$ around the trajectory $y_0(t)$. W. Fleming had proven that using the optimal deterministic feedback in the stochastic control problem one gets an order of approximation in ε^4. He has also mentionned that it was possible to use a linearization technique for computational purposes. We have shown here that the order of approximation is not affected by the linearization. Moreover one has an other characterization of $\bar{v}_x(y_0(t),t)$ which does require the knowledge of the optimal feedback $\bar{v}(x,t)$. This is perhaps the most important point for practical purposes.

Remark 4.2. : From the precedings considerations, it is clear that it is not easy to take care of constraints on the control. The knowledge of the feedback $\bar{v}(x,t)$ when possible is preferable in that respect.

Remark 4.3. : Our treatment has analogies with a general approach to regular perturbations in optimal control due to J. Cruz [3].

REFERENCES.

[1] A. BENSOUSSAN, Lectures on Stochastic Control, CIME Course on Stochastic Control and Filtering, Cortona, July 1981.

[2] J.M. BISMUT, An introductory Approach to Duality in Optimal Stochastic Control, SIAM Rev., vol. 20, n°1, Jan. 1978.

[3] J.B. CRUZ, Feedback Systems, MacGraw Hill, 1972.

[4] W.H. FLEMING, Stochastic Control for Small Noise Intensities, SIAM Cont.9, pp. 473-517 (1971).

A CONTROL PROBLEM IN A MANIFOLD

WITH NONSMOOTH BOUNDARY

R. BOEL
Lab. for System Dynamics
State University of Ghent
Ghent, Belgium

M. KOHLMANN
Institut für Angewandte Mathematik
Universität Bonn
Bonn, BRD

ABSTRACT

Diffusions in manifolds with non-smooth boundaries (e.g. in $R_+^r \times R^{n-r}$*) often arise as approximations to queueing networks. We consider the following stochastic differential equation :*

$$dx_t = f(x_t)dt + \sigma(x_t)dw_t + \sum_{i=1}^{r} \alpha^i(x_t)d\ell_t^i + \sum_{i=1}^{r} \beta^i(x_t)dm_t^i$$

where w_t *is a Brownian motion,* ℓ_t^i *the local time at* $\{x^i = 0\}$ *and* m_t^i *a Brownian motion at the time scale* ℓ_t^i.

We first prove existence of weak solutions to this s.d.e. A Girsanov theorem modifying the drift f *inside the domain and the average angle of reflection* α^i *on the boundary* $\{x^i = 0\}$ *will be proved.*

Using this theorem we can allow f *and* α^i *to depend on a full information control law* $u(t,\omega)$. *A reasonable cost structure*

$$J(u) = E\left[\int_0^T c(x_s,u_s)ds + \sum_{i=1}^{r} \int_0^T d^i(x_s, u_s)d\ell_s^i \right]$$

then assigns extra cost for hitting the boundaries $\{x_i = 0\}$. *So, on the one hand we can control the behaviour around the boundary via the average angle of reflection* $\alpha^i(x,u)$*, on the other hand we penalize the system for being close to this boundary. This leads to a well-posed optimal control problem, with dynamic programming type optimality conditions and existence theorems under fairly standard - but strict - assumptions.*

1. INTRODUCTION

It is well known that the analysis of queueing networks - and in particular their transient behaviour - is very difficult, because the problem is highly non-linear. The non-linearity is due to the fact that the transition probabilities change when the queue is empty or when the waiting room is full. If the arrival rate is high, and if we look at the system at a time scale such that many events occur in a short time, then via the central limit theorem the queue-length process becomes continuous and locally Gaussian. This means that we obtain a reflected multi-dimensional Ornstein-Uhlenbeck process as approximation to the queue length process.

The most general statement of this diffusion approximation is due to Reiman [11], but earlier results were obtained by Gaver [4], Kobayashi [8], Harrison [6] and others. Reiman proceeds as follows. Consider a network of n ./G/1 queues, with respective queue length $Q_t^i \geq 0$. Note that we do not write GI/G/1 since the arrival process depends on departures at other queues and can therefore not be independently specified. In general it will not even be a renewal process. As shown in fig.1 we can specify an external arrival process with average arrival rate λ^i at queue i (λ^i = 1/average interarrival time), average service time $1/\mu^i$ and transition probabilities $p_{ij} = P(dQ_t^j = 1/dQ_t^i = -1)$ = P(job moves to queue j immediately after leaving queue i). The total average arrival rate

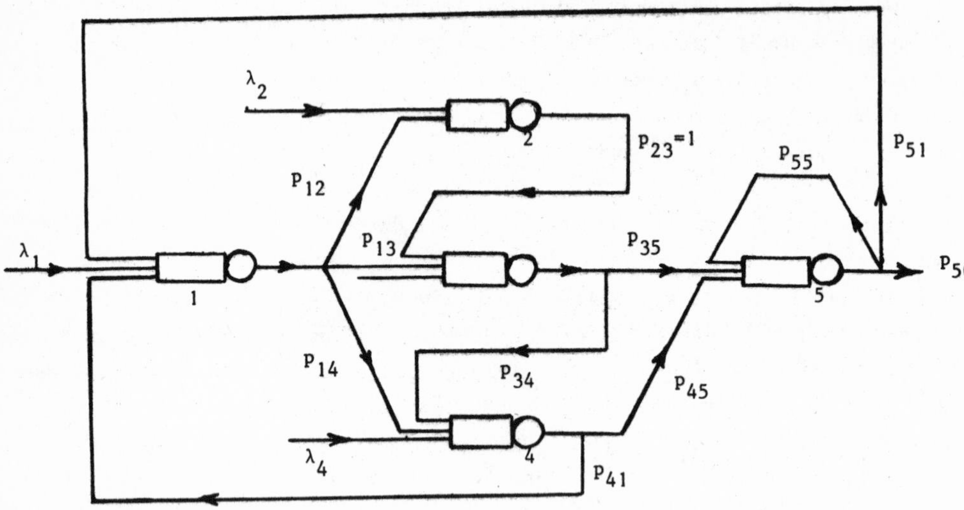

fig.1

at queue i is then given by γ^i specified by the set of linear equations

$$\gamma^i = \lambda^i + \sum_j \gamma^j p_{ji}$$

If now we consider a sequence of such queueing processes Q^n_t, with routing matrices $(P^n)_{ij} = p^n_{ij}$, with external arrival rates λ^{in}, service rates γ^{in}, total arrival rates λ^{in}, and traffic intensities $\rho^{in} (= \frac{\gamma^{in}}{\mu^{in}})$, if all $\rho^{in} \to 1$ equally fast, then at the time scale $nt(1-\rho^{in} \sim \frac{1}{\sqrt{n}})$ the system undergoes a lot of changes in a short time. A central limit theorem then shows that $\frac{1}{\sqrt{n}} . Q^n_{nt}$ approaches a continuous and locally Gaussian process (in the sense of distributions). In other words, the distribution of $\frac{1}{\sqrt{n}} Q^n_{nt}$ is approximated by that of the multidimensional reflected Ornstein- Uhlenbeck process Z_t :

$$dZ_t = F.Z_t.dt + \Sigma.dW_t + (I-P^T)d\ell_t \qquad (1)$$

where W_t is a Brownian motion, while $\ell_t = (\ell^1_t,\dots,\ell^n_t)^T$ is the smallest non-decreasing adapted process such that $Z^i_t \geq 0$, $\forall i$, $\forall t$. We will see later that ℓ^i_t can be interpreted as a local time, viz. a measure of the time Z_t spends on the boundary $\{Z^i = 0\}$. The columns $e_i - p_i$ of $I - P^T$ can be interpreted as directions of reflection at $\{Z^i_t = 0\}$. The weak convergence argument used by Reiman [11] specifies the differential operator:

$$Af(x) = \lim_{t \downarrow 0} \frac{1}{t} [E(f(X_t) - f(x)) | X_o = x]$$

for continuously differentiable functions f which are 0 in the neighbourhood of a corner $\{x^i = 0, x^j = 0; i \neq j\}$. For applications, such as recursive estimation, we have to derive from A the forward operator A^* of Z_t (via partial integration). Clearly we will not be able to specify the boundary conditions of A^* in a corner if all test functions f are zero in its neighbourhood. This is one reason for trying a different approach.

A second limitation of the approach of Reiman for both filtering and control applications is the fact that the drift F and the diffusion Σ and $(I-P^T)$, the directions of reflection, are fixed. In a controlled queue these will depend on Q_t, the state. Thus we would like to study weak solutions of the stochastic differential equation (2) (i.e. we want to find a probability space (Ω, F, P_u) such that a mapping $X_t(\omega)$ exists which is a.s. P_u continuous and satisfies (2)) :

$$dx_t = f(x_t,u_t)dt + \sigma(x_t)dw_t + \sum_{i=1}^r \alpha^i(x_t,u_t)d\ell^i_t \qquad (2)$$

where $f(x)$ replaces F.X, $\sigma(x)$ replaces the constant Σ, w_t is still an n-dimensional Brownian motion, $\alpha^i(x)$ represents the direction of reflection and ℓ_t^i is again the local time of x_t^i. ℓ_t^i is a continuous, non-decreasing, non-differentiable process with probability 1. This is a consequence of the fact that if a diffusion process hits a boundary at a given time t, it hits this boundary infinitely often in any neighbourhood $(t,t+\varepsilon)$. Thus we determine the angle of reflection with probability one, and hence the probability measure P_u and P_v corresponding to controls u and v, giving different directions of reflection $\alpha^i(x,u) \neq \alpha^i(x,v)$ for some x and i, can be distinguished. This means $P_u \perp\!\!\!\perp P_v$, i.e. they are singular.

To describe the control problem we want to use the Girsanov approach i.e. a change in control means a change in probability measure. This requires P_u and P_v to be absolutely continuous. It will be shown in chapter 2 that this can be achieved by adding randomness to the reflections :

$$dx_t = f(x_t,u_t)dt + \sigma(x_t)dw_t + \sum_{i=1}^{r} \alpha^i(x_t,u_t)d\ell_t^i + \sum_{i=1}^{r} \beta^i(x^t)d\tilde{w}^i_{\ell_t^i} \qquad (3)$$

where $\tilde{w}_t^i$ are Brownian motions independent of each other and of w_t. Hence $\tilde{w}^i_{\ell_t^i}$ is a martingale with quadratic variation $\langle w^i_{\ell^i} \rangle_t = \ell_t^i$. In § 2 we will also derive the likelihood ration $\frac{dP_u}{dP_v}$ corresponding to these changes. This is probably the most useful new result in this paper, both for control and filtering applications.

Given the requirements expressed above, it is reasonable to try to extend the weak convergence results of Stroock and Varadhan [12] to the case where the boundary has corners. In general the region could be defined as $G = \{x\varepsilon R^n : \phi_i(x) > 0\}$, an open set with boundaries $\partial_i G = \{x\varepsilon R^n : \phi_i(x) = 0\}$ where $\phi_i : R^n \to R$ are continuous functions. We will treat the existence of weak solutions to (3) in chapter 2, for the special case $G = R_+^2 \times R$. This simplifies the notation without hiding any essential difficulties. We will emphasize the boundary conditions, and especially the behaviour at the corner, of this process since experience has indicated that the usefulness of diffusion approximations for analysing queues depends heavily on choosing the right boundary conditions (Gelenbe-Mitrani [5]) . It turns out that the Stroock-Varadhan approach uses as an intermediary result the case of sticky Brownian motion, which has been suggested as a better approximation for queueing networks , to model the fact that when a queue hits a boundary it remains there for an exponentially distributed time (see Gelembe-Mitrani [5] and for a quite sophisticated analysis Harrison and Lemoine [6]). This stickyness makes ℓ_t^i stochastically larger since the process x_t "sticks" to the boundary longer i.e. $\{x_t^i = 0\}$ becomes a larger set.

As Stroock and Varadhan define it, the stickyness is measured via $\rho^i(x)$:

$$\int_0^t 1_{\partial_i G}(x_s)ds = \int_0^t \rho^i(x_s)d\ell_s^i$$

In § 2 we first consider $\rho^i(x) \equiv 1$, then $\rho^i(x) \geqslant \varepsilon > 0$ and then take the weak limit for $\rho_n^i(x) \downarrow 0$. This leads to the non-sticky boundary condition which has been implicitly assumed earlier. Note that sticky boundary conditions are quite complicated to state. Hence in this paper applications to optimal control will be for the case $\rho^i \equiv 0$ only.

For each F_t-adapted control law $u(t,\omega)$, chapter 2 will have shown existence of a probability space (Ω, F, P_u) which solves (3). Thus we control the average drift inside G and the average angle of reflection on each boundary $\partial_i G$. When one thinks of a queueing network with finite buffers at some nodes, it is important for the throughput, to avoid blocking (i.e. keep $Q_t^i \ll B^i$ away from the full buffer), to keep service stations active (i.e. avoid empty queues and unused equipment) and to keep the waiting times small which is often equivalent to Q_t^i small. This last requirement can be modeled as a cost $\int_0^T c(x_s,u_s)ds$ in the diffusion approximation case. The first two requirements impose a cost for hitting the boundary, which in the diffusion approximation is modeled by a cost running at the rate of the local time : $\int_0^T d^i(x_s, u_s)d\ell_s^i$ (Note that ℓ_t^i increases only when $x_t^i = 0$, and ℓ_t^i measures the amount of time spent on the boundary $\{x^i = 0\}$).

The optimal control problem is then : find an admissible control law u, such that

$$J(u) = E_u\left[\int_0^T c(x_s, u_s)ds + \sum_{i=1}^r \int_0^T d^i(x_s, u_s)d\ell_s^i + G(x_s)\right] \qquad (4)$$

is minimized. Here E_u is expectation corresponding to (Ω, F, P_u) where P_u is a solution to (3) for coefficients $(f(x,u), \sigma(x), \alpha^i(x,u), \beta^i(x), \rho^i(x) \equiv 0)$. Dynamic programming arguments can then be used to show that the value functions

$$V(t,x) = \inf_{u \text{ adm.}} E_u\left[\int_t^T c(x_s, u_s)ds + \sum_{i=1}^r \int_t^T d^i(x_s, u_s)d\ell_s^i \mid x_t = x\right]$$

is well defined. The Bellman-Hamilton-Jacobi equation (5) then gives a necessary and sufficient condition for optimality :

$$
\left.\begin{aligned}
0 &= \inf_{u\in V} \Big[\frac{\partial V}{\partial x_j}(t,x) + \sum_{j=1}^{n} f_j(x,u)\frac{\partial V}{\partial x_j}(t,x) + c(x,u) \\
&\qquad + \frac{1}{2}\sum_{j,\ell} (\sigma(x)\sigma^T(x))_{j\ell}\frac{\partial^2 V}{\partial x_j \partial x_\ell}(t,x)\Big] \\
&\text{with boundary conditions :} \\
0 &= \inf_{u\in V} \Big[\sum_{j=1}^{n} \alpha_j^i (x,u)\frac{\partial V}{\partial x_j}(t,x) + d^i(x,u) \\
&\qquad + \frac{1}{2}\sum_{j,\ell} (\beta^i(x)\beta^{iT}(s))_{j\ell}\frac{\partial^2 V}{\partial x_j \partial x_\ell}(t,x)\Big] \\
&\qquad\qquad \text{for } x^i = 0,\ i = 1,\dots,r \\
V(T,x) &= G(x)
\end{aligned}\right\} \quad (5)
$$

Only the boundary condition involving a minimization is new, but expected given that the average angle of reflection has to be decided by the controller at the boundary, and given that a non-negligible cost (in Lebesgue's sense) is incurred there.

It should be remarked that some optimal control results depend on the martingale representation property and the fact that (P_u, x_t) is strong Markov and Feller. All these properties depend on the uniqueness of the solution (Ω, F, P_u) corresponding to coefficients $(f, \sigma, \alpha^i, \beta^i)$. The proofs of uniqueness in Stroock-Varadhan [12] are long and tedious, involving bounds for partial differential operators (appendix, chapter 7 of [12]) Because we have corners and diffusion on the boundaries we will require more smoothness of σ, α^i, β^i than in [12], but we have not yet worked out the details. In chapter 4 the optimal control results will be stated assuming uniqueness explicitly whenever necessary.

The results of this paper can also be used to derive an optimal recursive nonlinear estimator (e.g. Mortensen-Zakai filter) when the process (P_u, x_t) is the signal which is observed in white noise (Pardoux [10]) or acts as the rate of an observed point process. This last case will be treated in a forthcoming paper. In the case of sticky boundaries one also has a conditional density on the boundary, and a non zero conditional probability of being on the boundary. Then a partial observation optimal control problem can be formulated.

The optimal control problem was justified in the preceding paragraphs for queueing networks. In fact, in every large system which is controlled hierarchically, a higher level controller will become active whenever the state reaches certain boundaries (e.g. safety levels) and special decisions will then be taken as well as extra costs incurred. The present model gives a good description for such control problems.

Finally we want to emphasize a difference in our approach, compared to Stroock-Varadhan [12] . Both for control and for recursive estimation applications it is essential to use stochastic calculus results. We will therefore observe as soon as possible that x_t is a semi-martingale and then use freely results of modern theory of stochastic processes (Meyer [8]).

2. EXISTENCE OF A DIFFUSION PROCESS x_t IN $R_+^2 \times R$, WITH PARAMETER $(f(x), \sigma(x), \alpha^i(x), \beta^i(x), \rho^i(x))$

In this chapter we outline the construction, via a weak limit, of a probability measure solving the following equation :

$$\begin{pmatrix} dx_t^1 \\ dx_t^2 \\ dx_t^3 \end{pmatrix} = \begin{pmatrix} f_1(x_t) \\ f_2(x_t) \\ f_3(x_t) \end{pmatrix} dt + \begin{pmatrix} \sigma_{11}(x_t) & \sigma_{12}(x_t) & \sigma_{13}(x_t) \\ \sigma_{12}(x_t) & \sigma_{22}(x_t) & \sigma_{23}(x_t) \\ \sigma_{13}(x_t) & \sigma_{23}(x_t) & \sigma_{33}(x_t) \end{pmatrix} \begin{pmatrix} dw_t^1 \\ dw_t^2 \\ dw_t^3 \end{pmatrix}$$

$$+ \begin{pmatrix} 1 & \alpha_1^2(x_t) \\ \alpha_2^1(x_t) & 1 \\ \alpha_3^1(x_t) & \alpha_3^2(x_t) \end{pmatrix} \begin{pmatrix} d\ell_t^1 \\ d\ell_t^2 \end{pmatrix} + \begin{pmatrix} 0 & 0 \\ \beta_{21}^1(x_t) & \beta_{22}^1(x_t) \\ \beta_{22}^1(x_t) & \beta_{32}^1(x_t) \end{pmatrix} \begin{pmatrix} d\tilde{w}^1_{1,\ell_t^1} \\ d\tilde{w}^1_{2,\ell_t^1} \end{pmatrix}$$

$$+ \begin{pmatrix} \beta_{11}^2(x_t) & \beta_{12}^2(x_t) \\ 0 & 0 \\ \beta_{13}^2(x_t) & \beta_{33}^2(x_t) \end{pmatrix} \begin{pmatrix} d\tilde{w}^2_{1,\ell_t^2} \\ d\tilde{w}^2_{3,\ell_t^2} \end{pmatrix} \qquad (3')$$

or

$$dx_t = f(x_t)dt + \sigma(x_t)dw_t + \sum_{i=1}^{2} \alpha^i(x_t)d\ell_t^i + \sum_{i=1}^{2} \beta^i(x_t)d\tilde{w}^i_{\ell_t^i}$$

where w_t, $\tilde{w}_t^1$, $\tilde{w}_t^2$ are independent standard Brownian motions (resp.3 and 2 dimensional) while ℓ_t^i is the local time of x_t^i, i.e.

$$\int_0^t 1(x_s^i=0).ds = \int_0^t \rho^i(x_s)d\ell_s^i \qquad P - a.s. \qquad (3'')$$

We assume :

$f(x)$: $R^2_+ \times R \to R^3$, measurable, bounded.

$\sigma(x)$: $R^2_+ \times R \to R^{3\times 3}$, continuous, bounded, positive-definite matrix, which is the unique positive definite square root of the local variance $\Sigma(x)=\sigma(x)\sigma^T(x)$.

$\alpha^i(x)$: $R_+ \times R \to R^3$, measurable, bounded, with $\alpha^i_i(x) = 1$ (defined on $\partial_i G = \{x \in R^3 : x^i = 0\}$).

$\beta^i(x)$: $\partial_i G \to R^{2\times 3}$, continuous, bounded, with $\beta^i_{ik}(x) = 0$. The matrix $\tilde{\beta}^i$ left over by eliminating the zero row i is positive definite.

$\rho^i(x)$: $\partial_i G \to R_+$, bounded, Lipschitz-continuous.

We make the following extra assumptions :

A1. $\beta^i(x) \downarrow 0$ at corners, i.e. whenever a sequence x_n is such that $x^i_n \downarrow 0$, $x^j_n \downarrow 0$ $(i,j \in \{1,2\}$, then $\beta^i(x_n) \downarrow 0$ and $\beta^j(x_n) \downarrow 0$.

A2. $\rho^i(x)$ and $\rho^j(x)$ agree in their common corner, i.e. $\rho^i(x) = \rho^j(x)$, for $x \in \partial_i G \cap \partial_j G$.

Remark : The last property (3"), which holds a.s. P for all sample paths, is essential to uniquely specify ℓ^i_t. If $\rho^i(x) \equiv 0$, then the boundary is non-sticky and ℓ^i_t is a.e. non-differentiable w.r.t.Lebesgues measure. Then we will be able to show :

$$\ell^i_t = \lim_{\varepsilon \downarrow 0} \varepsilon.\#\{\text{downcrossing of } x^i_t \text{ from } \varepsilon \text{ to } 0\}$$

$$= \lim_{n\to\infty} n. \int_0^t 1(x^i_s < \tfrac{1}{n}).ds$$

The case of sticky boundaries, $\rho^i(x) > 0$, is more complicated. As constructed in this paper $\frac{d\ell^i_t}{dt} = \frac{1(x^i_t=0)}{\rho^i(x_t)}$ a.e. However Harrison and Lemoine [7] construct a diffusion process with local time ℓ^i_t which is a.e. non-differentiable by changing the time scale when $x^i_t=0$. Further analysis is necessary to decide which definition is most useful for applications.

Consider a probability space (Ω, F, P) and sample paths $x_t(\omega)$. We say that (Ω, F, P) *solves the submartingale problem* for $(f, \sigma, \alpha^i,\beta^i, \rho^i)$ if :

Si) $P(x_t \in \overline{G}) = P(x_t \in \overline{R_+^2 \times R}) = 1$

Sii) $P(x_{t_o} = x_o) = 1$

Siii) for any $g \in C_o^{1,2}([0,\infty) \times R_+^2 \times R)$ such that $\rho^i(x).\frac{\partial g}{\partial t}(t,x) + J_i g(t,x) \geq 0$

$\forall x \in \partial_i G$, it holds that

$g(t,x_t) - g(t_o,x_o) - \int_{t_o}^{t} 1_G(x_s)(\frac{\partial g}{\partial s}(s,x_s) + Lg(s,x_s))ds$ is a (P, F_t) submartingale where $F_t = \sigma(x_s, t_o \leq s \leq t)$

$$Lg(t,x) = \sum_{j=1}^{3} f_j(x)\frac{\partial g}{\partial x^j}(t,x) + \frac{1}{2}\sum_{j,\ell=1}^{3}(\sigma(x)\sigma^T(x))_{j\ell}\frac{\partial^2 g}{\partial x_j \partial x_\ell}(t,x) \qquad (6)$$

$$J_i g(t,x) = \sum_{j=1}^{3}\alpha_j^i(x)\frac{\partial g}{\partial x_j}(t,x) + \frac{1}{2}\sum_{j,\ell=1}^{3}(\beta^i(x)\beta^{iT}(x))_{j\ell}\frac{\partial^2 g}{\partial x_j \partial x_\ell}(t,x) \qquad (7)$$

for $x \in \partial_i G$

With the above assumptions and notation we can state and prove most theorems of chapter 2 of Stroock and Varadhan [12] .

Theorem 2.1 (thm.2.4 of [12]). If (Ω, F, P) solves the submartingale problem, then there exists a unique vector of continuous, non-decreasing, F_t-adapted processes ℓ_t^i :

$$\ell_o^i = 0 \qquad \text{a.s.}$$

$$E\,\ell_t^i < \infty \qquad \forall\, t$$

$$\ell_t^i = \int_0^t 1_{\partial_i G}(x_s)d\ell_s^i$$

and for all $g \in C_o^{1,2}([0,\infty) \times R_+^2 \times R)$ satisfying Siii) we have :

$$M_g^t = g(t,x_t) - g(t_o,x_o) - \int_{t_o}^{t} 1_G(x_s)Lg(s,x_s)ds$$

$$- \sum_{i=1}^{2}\int_{t_o}^{t}(\rho^i(x_s)\frac{\partial g}{\partial s}(s,x_s) + J_i g(s,x_s))d\ell_s^i$$

is a (P, F_t)-martingale.

Proof. The proofs of lemma 2.2 to 2.5 of [12] are adapted to the case with corners ($\phi(x) \geq 0$ replaced by $x^1 \geq 0$, $x^2 \geq 0$). By defining $\overline{g}(t;x) = g(t,x)+\sum_{i=1}^{2}\alpha_i\eta_i(x)x^i$

analogously to the definition in [12] , we obtain Lemma 2.5.

$$g(t,x_t) - \int_0^t Lg(s,x_s)ds = \ell_t^g + \text{martingale}$$

where the unique, increasing, F_t-adapted process $d\ell_t^g$ can be written as

$$d\ell_t^g = \sum_{i=1}^{2} k_i(t,x_t)d\ell_t^i = \sum_{i=1}^{2} (\rho^i(x_t) \frac{\partial g}{\partial t}(t,x_t) + J_i g(t,x_t).d\ell_t^i$$

for every neighbourhhood V of $x \in \partial_1 G \cup \partial_2 G$, away from the corner. At the corner A1 and A2 guarantee that $k_i(t,x)$ is continuous. Therefore the argument also works in an open neighbourhood V of a corner.

Theorem 2.2 (12, thm.2.5). $(\Omega, F, P),(F_t)$ solves the submartingale problem corresponding to $(f, \sigma, \alpha^i, \beta^i, \rho^i)$ iff there exist continuous non-decreasing F_t-adapted ℓ_t^1 and ℓ_t^2 s.t.

i) $\ell_o^i = 0$ a.s.

$$\forall \lambda^i > 0, \forall t > 0 \; : \; E\, e^{\lambda^1 \ell_t^1 + \lambda^2 \ell_t^2} < \infty$$

ii) $\ell_t^i = \int_0^t 1_{\{x_i=0\}}(x_s).d\ell_s^i$

iii) $\forall \lambda^1, \lambda^2, \; \forall \theta \in R^3$:

$$Z_{\lambda,\theta}(t) = \exp(\theta^T(x_t - x_o) - \frac{1}{2}\int_0^t 1_G(x_s)\theta^T\sigma(x_s)\sigma^T(x_s)\theta ds - \int_0^t 1_G(x_s)\theta^T f(x_s)ds$$
$$- \sum_{i=1}^{2} \int_0^t 1_{\partial_i G}(x_s)\theta^T\alpha^i(x_s)d\ell_s^i$$
$$- \frac{1}{2}\sum_{i=1}^{2} \int_0^t 1_{\partial_i G}(x_s)\theta^T\beta^i(x_s)\beta^{iT}(x_s)\theta d\ell_s^i)$$
$$.\exp \sum_{i=1}^{2} \lambda^i(\int_0^t 1_{\partial_i G}(x_s)ds - \int_0^t \rho^i(x_s)d\ell_s^i)$$

is a (P, F_t)-martingale.

Moreover if P is such a solution, then (ℓ_t^1, ℓ_t^2) are uniquely determined, P.a.s., by the condition :

$$x_t^i - \int_0^t 1_G(x_s) f_i(x_s) ds - \int_0^t \sum_{j=1} \alpha_i^j(x_s) d\ell_s^j$$

is a (P, F_t) martingale.

The following properties also hold :

$$\int_0^t 1_{\partial_i G}(x_s) dx_s = \sum_{\substack{j=1 \\ j\neq i}}^{2} \int_0^t 1_{\partial_i G}(x_s) \alpha_i^j(x^s) d\ell_s^i \qquad P.a.s.$$

$$\int_0^t 1_{\partial_i G}(x_s) ds = \int_0^t \rho^i(x_s) d\ell_s^i \qquad P.a.s.$$

Proof : immediately as in [12] .

The last theorem is important because it shows us how to use the (P, F_t)-semi-martingale x_t in stochastic calculus. Of course we only have a weak solution and we cannot specify a priori the space (Ω, F) and filtration (F_t) as would have been the case with a strong solution along the lines of Watanabe [14] .

Since x_t is a semi-martingale, we can use theorem 3.2 of van Schuppen and Wong [13] to define absolutely continuous changes of the probability measure. This gives a Girsanov type theorem which modifies the drift f and the average angles of reflection α^i. From assumption (ii) of the following theorem it is clear that this change of the average angle of reflection is possible only because $\tilde{\beta}^i(x)\tilde{\beta}^{iT}(x)$ is invertible, i.e. the randomness in the angle of refection is (n-1)-dimensional, non-degenerate ($\tilde{\beta}^i$ is β^i with the 0-row i deleted). This requires $\tilde{\beta}^i$ strictly positive definite. Since $\tilde{\beta}^i \downarrow 0$ at each corner, we must assume $\|\alpha^i(x)-\alpha^{io}(x)\| \downarrow 0$ in a corner, faster than β^i. Some reflection about what can be statistically measured from one sample, will explain the above comment. One will also find then that indeed equivalence of the probability measures $(P \sim P_o)$ requires that $\sigma(x)$, $\beta^i(x)$ and $\rho^i(x)$ remain unchanged.

Theorem 2.3. Given a weak solution (Ω, F, P_o), (F_t) to the submartingale problem corresponding to $(f^o(x), \sigma(x), \alpha^{io}(x), \beta^i(x), \rho^i(x))$ and given a different set of parameters $(f(x), \sigma(x), \alpha^i(x), \beta^i(x), \rho^i(x))$, both sets of parameters satisfying all the above assumptions as well as :

i) $\quad E_o \int_0^t \|(f(x_s)-f^o(x_s))(\sigma(x_s)\sigma^T(x_s))^{-1}\|^{1+\varepsilon}.ds < \infty \qquad \forall t$

ii) $E_0 \int_0^t \|(\alpha^i(x_s)-\alpha^{io}(x_s))(\beta^i(x_s)\beta^{iT}(x_s))^{-1}\|^{1+\varepsilon}.d\ell^i_s < \infty \quad \forall i, \forall t$

iii) $\|(\alpha^i(x_n)-\alpha^{io}(x_n))(\beta^i(x_n)\beta^{iT}(x_n))^{-1}\| \downarrow 0$ if $x^i_n \downarrow 0$, $x^j_n \downarrow 0$, $j \neq i$

Then we can define, for all finite t, the Radon-Nikodym derivative :

$$L_t = \left.\frac{dP}{dP_o}\right|_{F_t}$$

$$= \exp\left[\int_0^t (f(x_s)-f^o(x_s))(\sigma(x_s)\sigma^T(x_s))^{-1}\sigma(x_s)dw_s - \frac{1}{2}\int_0^t \|(f(x_s)-f^o(x_s))(\sigma(x_s)\sigma^T(x_s))^{-\frac{1}{2}}\|^2.ds\right]$$

$$.\exp\sum_{i=1}^{2}\left[\int_0^t (\alpha^i(x_s)-\alpha^{io}(x_s))(\beta^i(x_s)\beta^{iT}(x_s))^{-1}\beta^i(x_s)dw_{\ell^i_s}^i - \frac{1}{2}\int_0^t \|(\alpha^i(x_s)-\alpha^{io}(x_s))(\beta^i(x_s)\beta^{iT}(x_s))^{-\frac{1}{2}}\|^2.d\ell^i_s\right] \qquad (8)$$

where L_t is a (P_o, F_t)-martingale. Then (Ω, F, P) is a probability space which solves the submartingale problem for the parameters $(f, \sigma, \alpha^i, \beta^i, \rho^i)$.

Proof : One has to show that

$$x_t - \int_0^t f(x_s)ds - \sum_{i=1}^{2}\int_0^t \alpha^i(x_s)d\ell^i_s$$

is a (P, F_t)-martingale. This is equivalent to

$$L_t(x_t - \int_0^t f(x_s)ds - \sum_{i=1}^{2}\int_0^t \alpha^i(x_s)d\ell^i_s)$$

being a (P_o, F_t)-martingale, which can be obtained using stochastic calculus. The fact that the sample path property

$$\int_0^t 1_{\partial_i G}(x_s)ds = \int_0^t \rho^i(x_s)d\ell^i_s$$

holds a.s.P if it holds a.s.P_o follows from the equivalence of P and P_o, under the assumptions on $f, f^o, \sigma, \alpha^i, \alpha^{io}, \rho^i$.

From the preceding theorems it follows that we only have to prove existence of a solution to the equation :

$$\begin{cases} dx_t^i = dw_t^i + d\ell_t^i + \beta_{11}^2(x_t).d\tilde{w}^2_{1\ell_{2t}} \\ dx_t^2 = dw_t^2 + d\ell_t^2 + \beta_{22}^1(x_t)d\tilde{w}^1_{2\ell_{1t}} \\ dx_t^3 = dw_t^3 + \beta_{33}^1(x_t)d\tilde{w}^1_{3\ell_{1t}} + \beta_{33}^2(x_t)d\tilde{w}^2_{3\ell_{2t}} \end{cases} \tag{9}$$

We will construct a probability space (Ω, F, P) via a sequence of approximating Markovian probability spaces (x_t^n, P_n), with (Ω, F) the Skhorokhod embedding. P_n is defined as follows : if $x^n = x_o$ is an interior point of $R_+^2 \times R$, then let $T_1 = \inf \{s > 0 : (w_s^1, w_s^2, w_s^3) \in \partial_1 G \cup \partial_2 G\}$ be the first time Brownian motion hits the boundary $x^1=0$ or $x^2=0$. Then define $P_n\big|_{F_{T_1}} = P^o_{x_o}\big|_{F_{T_1}}$ where $P^o_{x_o}$ is Brownian motion starting in x_o at t_o. If $x^n_{T_1} \in \partial_1 G$, $x^n_{T_1} \notin \partial_2 G$, then under P_n, x_t^n remains at $x^n_{T_1}$ for an exponentially distributed time T_2-T_1 with mean ε_n. At T_2, x_t jumps to $x_{T_2} = (\varepsilon_n x^n_{2,T_2} + \sqrt{\varepsilon}_n z_2.\beta_{11}(x^n_{T_1}))^+, (x^n_{3,T_1} + \sqrt{\varepsilon}_n.z_3.\beta_{33}(x^n_{T_1}))^+)$ where z_2, z_3 are i.i.d. random variables taking the values $+1,0,-1$ w.p. $\frac{1}{4}$, $\frac{1}{2}$, $\frac{1}{4}$ resp. This defines $P_n\big|_{F_{T_2}}$ if $x^n_{T_1} \in \partial_1 G$. The analogous definition of $P_n\big|_{F_{T_2}}$ if $x^n_{T_1} \in \partial_2 G$ is obvious. Note that $x^+ = \max(0,x)$. It prevents $x^n_{T_2}$ from leaving $R_+^2 \times R$. From $x^n_{T_2}$ we define $P_n\big|_{F_{T_3}}$, as a Wiener measure if $x^n_{T_2}$ is an interior point or as another "sticking to the boundary for an exponentially distributed time". Since $T_n \uparrow \infty$ a.s. $P_n\big|_{F_{T_n}}$, the above construction defines inductively a probability measure P_n on (Ω, F). The trajectories are piecewise continuous processes in $(D,\mathcal{D})$. Weak compactness, for the Skorokhod metric on D of the sequence P_n, for $\varepsilon_n \downarrow 0$, will follow (Billingsley [2, § 8 and 15]) if we can show

$$E_n[(x^n_{t_3} - x^n_{t_2})^2 (x^n_{t_2} - x^n_{t_1})^2] \leq C(t_3 - t_1)^2$$

for all $t_1 \leq t_2 \leq t_3$ (here E_n denotes expection under measure P_n). Since x_t^n behaves, depending on the time interval, as a Wiener or compound Poisson process with bounded jumps ($\leq \sup_{x,i} [1 + \beta^i_{ij}(x)\sqrt{\varepsilon}_n]$) this is a straightforward calculation, using

$$E_n[(x^n_{t_3} - x^n_{t_2})^2 \mid x^n_{t_2}] = E_n(\langle x^n\rangle_{t_3} - \langle x^n\rangle_{t_2} \mid x^n_{t_2})$$

$$\leq C_1(t_3 - t_2)$$

It follows then that the weak limit P_o of (P_n) exists and moreover that P_o is concentrated on $C([0,\infty) \times R^2_+ \times R)$.

We must now show that P_o solves the submartingale problem. Under P_n we have for all $f \in C^{1,2}_o([0,\infty) \times R^2_+ \times R)$:

$$f(t,x^n_t) - \int_0^t \frac{\partial}{\partial s} f(s,x^n_s)ds - \int_0^t 1_G(x^n_s).\Delta f(s,x^n_s)ds$$

$$- \sum_{i=1}^{2} \sum_{T_i \ni x^n_{T_i} \in \partial_i G} J^i_{\varepsilon_n} f(T_i, x^n_{T_i}).\varepsilon_n$$

is a (P_n, F_t)-martingale. Here

$$J^1_\varepsilon f(t,x) = \frac{1}{16\,\varepsilon}[f(t,(\varepsilon,x_2+\beta_{22}(x)\sqrt{\varepsilon}, x_3+\beta_{33}(x)\sqrt{\varepsilon}))$$

$$+ 2f(t,(\varepsilon,x_2, x_3+\beta_{33}(x)\sqrt{\varepsilon}))$$

$$+ f(t,(\varepsilon, x_2-\beta_{22}(x)\sqrt{\varepsilon}, x_3+\beta_{33}(x)\sqrt{\varepsilon})) + \ldots\ldots$$

$$- 16f(t,(0, x_2,x_3))]$$

for $x = (0, x_1, x_2) \in \partial_1 G$; $J^2_\varepsilon f$ is similarly defined. When $\varepsilon_n \downarrow 0$ then $J^i_\varepsilon f(t,x) \to J^i f(t,x)$ (using the fact that f is twice differentiable).

Consider : $G_{i\varepsilon} = \{y \in G: |y-\partial_i G| \leq \varepsilon\}$. As in lemma 3.2 of [12] one finds that

$$E_n \int_s^t 1_{G_{1\varepsilon} \cup G_{2\varepsilon}}(x^n_s)ds \leq 2\,C_T.\varepsilon$$

for $s \leq t \leq T$.

Hence for all $f \in C^{1,2}_o([0,\infty), \overline{R^2_+ \times R})$ such that

$$\frac{\partial f}{\partial t}(t,x) + J_i f(t,x) \geq 0$$

for $x \in \partial_i G$, then

$$f(t,x_t) - \frac{\partial f}{\partial t}(t,x_t) - \int_0^t 1_G(x_s).\Delta f(x_s)ds$$

is a (P_o, F_t)-submartingale.

This solves the submartingale problem for the case $\rho^i(x) \equiv 1$. All problems $\rho^i(x) > 0$ can be derived from it via a rescaling of time and space. The case of non-sticky boundaries, $\rho^i(x) = 0$ is obtained as in lemma 3.3 of [12] by considering a sequence $\rho^i_n(x) \downarrow 0$ as $n \to \infty$, and approximating sequences (corresponding to $\varepsilon_k \downarrow 0$) P_{k,ρ_n}. The diagonal sequence P_{n,ρ_n} is again weakly compact and has a limit P_o solving the submartingale for $\rho^i(x) = 0$.

Summarizing the above, and combining it with the earlier theorems of this chapter, we obtain :

Theorem 2.4 : If f, σ, α^i, β^i, ρ^i satisfy all the conditions of theorems 2.1, 2.2 and 2.3, then there exists a solution to the corresponding submartingale problem. Equivalently there exists a *weak solution* (x_t, ℓ_t) to the stochastic differential equation

$$dx_t = f(x_t)dt + \sigma(x_t)dw_t + \sum_{i=1}^{r} \alpha^i(x_t)d\ell^i_t + \sum_{i=1}^{r} \beta^i(x_t)d\tilde{w}^i_{\ell^i_t}$$

with the extra conditions

ℓ^i_t continuous, non-decreasing, $\ell_o = 0$

$$\int_0^t 1_{\partial_i G}(x_s).d\ell^i_s = \ell^i_t$$

$$\int_0^t 1_{\partial_i G}(x_s)ds = \int_0^t \rho^i(x^s)d\ell^i_s.$$

Remark : For applications it is useful to know that (x_t, P) is a strong Feller process. This would follow if we had shown that P is the unique solution to the submartingale problem, a result which is in itself important to allow us to use martingale representation results. Under reasonable continuity assumptions on f, σ, α^i, β^i and ρ^i this will follow by proving results analogous to those of Stroock-Varadhan [12, last part of chapter 2, and chapters 4,5 and 7] . We have not yet carried out this lengthy verification.

3. OPTIMAL CONTROL OF A DIFFUSION PROCESS WITH BOUNDARIES, INCLUDING BOUNDARY CONTROL

Consider a weak solution (Ω, F, P_u), (F_t) to the stochastic differential equation ($x_t \in R_+^r \times R^{n-r}$, $t \in [0, T]$)

$$dx_t = f(x_t, u_t)dt + \sigma(x_t)dt + \sum_{i=1}^{r} \alpha^i(x_t, u_t)d\ell_t^i + \sum_{i=1}^{r} \beta^i(x_t)d\tilde{w}^i_{\ell_t^i} \tag{10}$$

with the extra conditions

$$\ell_t^i \text{ continuous, non-decreasing, } \ell_o^i = 0$$

$$\int_0^t 1_{\partial_i G}(x_s)d\ell_s^i = \ell_t^i \tag{11}$$

$$\int_0^t 1_{\partial_i G}(x_s)ds = \int_0^t \rho^i(x_s)d\ell_s^i$$

where $u_t(\omega) : \Omega \to U$(a compact set) is an F_t-adapted function, called an admissible control law. Assume that (for each fixed $u \in U$) $f(., u)$, σ, $\alpha^i(., u)$, β^i and ρ^i satisfy the assumption of § 2. We also assume that the weak solution P_u corresponding to each admissible control law $u(u \in U)$ is absolutely continuous w.r.t. a reference measure P_o, solving the submartingale problem for $(f^o(x), \sigma(x), \alpha^{io}(x), \beta^i(x), \rho^i(x))$. The Girsanov functional $\Lambda_t(u) = \left.\frac{dP_u}{dP_o}\right|_{F_t}$ can be written according to theorem 2.3. It is useful to know that it satisfies the exponential equation :

$$d\Lambda_t(u) = \Lambda_t(u) \left[(f(x_t, u_t) - f^o(x_t))\sigma(x_t)dw_t + \sum_{i=1}^{r} (\alpha^i(x_t, u_t) - \alpha^{io}(x_t))\beta^i(x_t)d\tilde{w}_{\ell_t^i} \right]$$

To each admissible control law u, we associate a cost

$$J(u) = E_u \left[\int_0^T c(x_s, u_s)ds + \sum_{i=1}^{r} \int_0^T d^i(x_s, u_s)d\ell_s^i + G(x_T) \right] \tag{12}$$

where c, d^i, G are non-negative,measurable functions of x and u. Let U_t^T be the class of admissible control laws u(s), restricted to $t \leqslant s \in T$, which can be used after using control law $\{u(s), 0 \leqslant s \leqslant t\} = u_o^t$. We assume that U_t^T is independent of u_o^t, but it may depend on x_t. The value function is defined as follows :

$$V(t,x) = \inf_{u_t^T \in U_t^T} J(t, x, u_t^T)$$

where $J(t, x, u_t^T)$ is the future cost when using control law $u_t^T = \{u_s, t \leq s \leq T\}$:

$$J(t, x, u_t^T) = E_u [\int_t^T c(x_s, u_s)ds + \sum_{i=1}^r \int_t^T d^i(x_s, u_s)d\ell_s^i + G(x_T) | x_t = x]$$

independent of u_o^t

$$= E_o [\Lambda(u_t^T)(\int_t^T c(x_s, u_s)ds + \sum_{i=1}^r \int_t^T d^i(x_s, u_s)d\ell_s^i + G(x_T)) | x_t=x, u_t^T]$$

where $\Lambda(u_t^T) = \dfrac{\Lambda_T(u)}{\Lambda_t(u)} = \dfrac{\Lambda_T(u_o^T)}{\Lambda_T(u_o^t)}$

This problem can be cast in the abstract form necessary to apply the results of Boel and Kohlman [3] by introducing the control martingales

$$n_{j,t}^u = \int_0^t [(f(x_s, u_s) - f^o(x_s))(\sigma(x_s)\sigma^T(x_s))^{-1})\sigma(x_s)]_j dw_s^j$$

$$+ \sum_{i=1}^r \int_0^t [(\alpha^i(x_s, u_s) - \alpha^{io}(x_s))(\beta^i(x_s)\beta^{iT}(x_s))^{-1}\beta^i(x_s)]_j d\tilde{w}_{j,\ell_t^i}^i$$

We also rewrite the cost as the conditional expectation of a final cost, by introducing independent martingales:

w_t^o Brownian motion, $< w^o >_t = t$

$\tilde{w}_{i,t}^o$ Brownian motion, $< \tilde{w}_i^o >_{\ell_t^i} = \ell_t^i$

$$< w^o, \tilde{w}_i^o >_t = 0, \quad < \tilde{w}_{i\ell^i}^o, \tilde{w}_{j\ell^j}^o >_t = \delta_{ij}\ell_t^i$$

This requires extending the probability space (Ω, F, P_o) to $(\tilde{\Omega}, \tilde{F}, \tilde{P}_o)$ as in [3] and Beneš [1]. If on the probability space $(\tilde{\Omega}, \tilde{F})$ we define the Radon-Nikodym derivative :

$$\left.\frac{d\tilde{P}_u}{d\tilde{P}_0}\right|_{\tilde{F}_t} = \tilde{\Lambda}_t(u) =$$

$$= \Lambda_t(u)\exp\Big[\int_0^t c(x_s, u_s)dw_s^o - \frac{1}{2}\int_0^t c^2(x_s, u_s)ds + \sum_{i=1}^r \int_0^t d^i(x_s,u_s)d\tilde{w}^o_{\ell_s^i} - \frac{1}{2}\sum_{i=1}^r \int_0^t d^{i^2}(x_s, u_s)d\ell_s^i\Big]$$

then on $(\tilde{\Omega}, \tilde{F}, \tilde{P}_u)$ the stochastic differential equation (10), (11) is a.s. $\tilde{P}_u$-satisfied. Under $\tilde{P}_u$ the cost can be written as :

$$J(x, u) = E_u(Y_T^{n,u})$$

for the martingale :

$$Y_t^{n,u} = w_t^o + \sum_{i=1}^r \tilde{w}^o_{i,\ell_t^i} + E_u(G(x_T)|\tilde{F}_t)$$

Note that $Y_t^{n,u}$ is independent of u if the final cost $G(x_T) \equiv 0$.

We can now apply theorem 3.3.3 of [3] to obtain a necessary and sufficient condition for optimality :

Theorem 3.1 : Assume (Ω, F, P_u), (F_t) is the unique solution to the submartingale problem corresponding to $(f(x,u), \sigma(x), \alpha^i(x,u), \beta^i(x), \rho^i(x))$(equivalently, it is the unique weak solution to (10), (11)), then for all admissible control laws u ($\in$ U), for all F_t-stopping times τ and for all $h \geq 0$:

$$\begin{aligned} V(\tau,x_\tau) &\leq E_u\Big[\int_\tau^{\tau+h} c(x_s, u_s)ds + \sum_{i=1}^r \int_\tau^{\tau+h} d^i(x_s, u_s)d\ell_s^i + V(\tau+h,x_{\tau+h})|F_\tau\Big] \\ V(T,x) &= G(x) \end{aligned} \tag{13}$$

with equality in (13) holding if and only if u is optimal.

Remark : The dependence on u is clarified if one rewrites the right-hand side of (13) as :

$$E_{u_\tau^{\tau+h}}\{\Lambda^{\tau+h}(u_\tau^{\tau+h})\ [\ \int_\tau^{\tau+h} c(x_s,\ u_s)ds + \sum_{i=1}^{r} \int_\tau^{\tau+h} d^i(x_s,\ u_s)d\ell_s^i + V(\tau+h,\ x_{\tau+h}) |\ F_\tau\ ]$$

In order to give a local optimality, it is necessary to first derive the differential generator of the process $(x_t,\ P_u)$. Consider $g(x)$, $g \in C_o^2(\overline{R_+^r \times R^{n-r}})$.

Then by theorem 2.2 stochastic calculus can be applied to the case $\rho^i(x) \equiv 0$.

$$g(x_{t+dt}) = g(x_t) + \nabla g(x_t).dx_t + \frac{1}{2}\sum_{j,\ell=1}^{n} \frac{\partial^2 g}{\partial x_j \partial x_\ell}(x_t)d < x^j,\ x^\ell >_t$$

$$= g(x_t) + \sum_{j=1}^{n} \frac{\partial g}{\partial x_j}(x_t).[f_j(x_t,\ u_t)dt + \sum_{i=1}^{r} \alpha_j^i(x_t,\ u_t)d\ell_t^i$$

$$+ \sum_{\ell=1}^{n} \sigma_{j\ell}(x_t)dw_{\ell,t} + \sum_{i=1}^{r}\sum_{\ell=1}^{n} \beta_j^i(x\ )d\tilde{w}^i_{\ell,\ell_t^i}] + \frac{1}{2}\sum_{j,\ell=1} \frac{\partial^2 g}{\partial x_j \partial x_\ell}(x_t)$$

$$[\ (\sigma(x_t)\sigma^T(x_t))_{j\ell}dt + \sum_{i=1}^{r}(\beta^i(x_t)\beta^{iT}(x_t))_{j\ell}d\ell_t^i]$$

Hence for $g \in C_o^2(\overline{R_+^r \times R^{n-r}})$ such that

$$J_g^i(x) = \alpha^i(x)\nabla g(x) + \frac{1}{2}\sum_{j,\ell=1}^{n} (\beta^i(x)\beta^{iT}(x))_{j\ell}\frac{\partial^2 g}{\partial x_j \partial x_\ell}(x) = 0$$

for $x \in \partial_i G$, we can take the limit :

$$A^u g(x) = \lim_{dt \downarrow 0} \frac{1}{dt} E_u\ [g(x_{t+dt}) - g(x_t) |\ x_t = x]$$

$$= f^T(x)\nabla g(x) + \frac{1}{2}\sum_{j,\ell=1} (\sigma(x)\sigma^T(x))_{j\ell}\cdot\frac{\partial^2 g}{\partial x_j \partial x_\ell}(x)$$

because the other terms are martingales. Note that we also used the fact that ℓ_t^i is singular w.r.t. Lebesgues measure dt and that ℓ_t^i increases only when $x \in \partial_i G$ (i.e. the second term of $\frac{g(x_{t+dt})-g(x_t)}{dt}$ can be written as $1_{\partial_i G}(x_t) \cdot J^i g(x_t).\frac{d\ell_t^i}{dt}$ which diverges unless $J^i g(x).1_{\partial_i G}(x) \equiv 0$).

Summarizing we obtain :

<u>Lemma 3.2 :</u> Let $T_t^u g(x) = E_u(g(x_t)\,|x_o = x)$ for $g \in \mathcal{D}(A^u) = \{g \in C_o^2(\overline{R_+^r \times R^{n-r}})$ $: \forall\ x \in \partial_i G : J_i g(x)=0\}$ and $A_g^u(x)=f^T(x,u)\nabla g(x) + \frac{1}{2}\sum_{j,\ell} (\sigma(x)\sigma^T(x))_{j\ell}\frac{\partial^2 g}{\partial x_j \partial x_\ell}(x)$.

Then $(A^u, \mathcal{D}(A^u))$ is the differential generator of T_t^u, corresponding to $(f(x,u), \sigma(x), \alpha^i(x,u), \beta(x), 0)$.

Remark : The above backward operator A^u (the differential generator) suffices for dynamic programming applications. In filtering applications, we will need the dual forward operator. This is easily obtained via partial integration, assuming densities $p^u(t,x)$ exist.

Under very strong assumptions we can now state a local optimality criterion.

Theorem 3.3. : Given the optimal control problem of this chapter, assume that $(\Omega, \mathcal{F}, P_u)$, $(\mathcal{F}_t)$ is the unique weak solution to (10), (11). Assume that the value function $V \in C_0^{1,2}([0, T], \overline{R_+^r \times R^{n-r}})$. Then

$$0 = \inf_{u \in U} \{ \frac{\partial V}{\partial t}(t,x) + \sum_{j=1}^{n} f_j(x,u) \frac{\partial V}{\partial x_j}(t,x)$$

$$+ \frac{1}{2} \sum_{j,\ell=1}^{n} (\sigma(x)\sigma^T(x))_{j\ell} \cdot \frac{\partial^2 V}{\partial x_j \partial x_\ell}(t,x)$$

$$+ c(x,u) \}$$

subject to the boundary conditions :

$$0 = \inf_{u \in U} \{ \sum_{j=1} \alpha_j^i(x,u) \frac{\partial V}{\partial x_j}(t,x)$$

$$+ \frac{1}{2} \sum_{j,\ell=1}^{n} (\beta^i(x)\beta^{iT}(x))_{j\ell} \frac{\partial^2 V}{\partial x_j \partial x_\ell}(t,x)$$

$$+ d^i(x,u) \} \qquad \forall x \in \partial_i G$$

$$V(T,x) = G(x)$$

A control law $u_t^*(\omega)$ is optimal if and only if

$$\frac{\partial V}{\partial t}(t,x_t) + A^{u_t^*} V(t, x_t) + c(x_t, u_t^*) = 0 \qquad \text{on} \quad (x_t > 0)$$

$$J^i_{u_t^*} V(t,x_t) + d^i(x_t, u_t^*) = 0 \qquad \text{on} \quad (x_t^i = 0)$$

for almost all t (i.e. u_t^* has to achieve the infimum a.e.)

Proof: Directly from theorem 3.1 and lemma 3.2. Indeed, apply thm. 3.1 to $\tau = t$, $h = dt$:

$$0 \leqslant E_{\substack{t+dt \\ u_t}} [V(t+dt, x^u_{t+dt}) - V(t,x_t) + c(x_t, u_t)dt + \sum_{i=1}^{r} d^i(x_t, u_t)d\ell^i_t]$$

with equality holding iff u_t is optimal. Then apply lemma 3.2 or stochastic differentiation, and divide by dt to get formally the required result. The limit interchanges are justified because all terms are bounded ($V \in C^{1,2}_o$ has derivatives with compact support).

Remark : As in theorem 4.3.1 of Boel&Kohlmann [3] we could have stated a stochastic maximum principle, but since we do not have an explicit procedure to calculate the dual variable $g^*(t,x)$ (or even $g^u(t,x)$) the above form of the optimality criterion was chosen .

Remark : As explained in Meyer [9, appendix] the differentiability conditions on V can be weakened probably to : V is the difference of two convex functions. The derivatives, in the Ito-Meyer differentiation rule, are then defined in the sense of distributions. Both the notations and the rigorous justification of the interchanges of limits become too complicated. Especially at corners the differentiability conditions of theorem 3.3 are much too strong, and the suggested improvement would be very useful. □

Finally we give an example of an existence result for optimal, full information controls of the type of this chapter :

Theorem 3.4 : Assume $\mathcal{D}(n^u) = \{\Lambda_T(u) : u \in U\}$ is strongly uniformly integrable and weakly closed in L, and if $G(x)=0$ (no final cost) and if for some $\gamma > 1$:

$$E\left(|Y_T|^{\frac{\gamma}{\gamma-1}} \cdot 1_{\{|Y_T|^{\frac{\gamma}{\gamma-1}} > N\}}\right) \xrightarrow[N\to\infty]{} 0$$

then there exists an optimal control.

Proof : Since $G(x) = 0$ implies that Y_T is independent of the control law u, the result follows from theorem 3.2.2 of [3] . The conditions of the theorem will be satisfied (remembering U compact) if $f(x, u)$, $\alpha^i(x,u)$: continuous in x, u, $c(x,u)$, $d^i(x,u)$: bounded, continuous in x and u.

4. CONCLUSIONS

This paper has shown that abstract results on stochastic optimal control can be applied to a model which was chosen, not to fit the abstract theory, but on the basis of the literature for a specific application, queueing networks. Of course the results become complicated, and some very strong conditions on the value function have to be imposed. Some of these conditions (non-sticky boundary, full information) will be removed in a future paper.

REFERENCES

1. V.E. Beneš : Existence of optimal stochastic control laws, *SIAM J. Control, 9* (1971), p. 446-472.
2. P. Billingsley : *Convergence of probability measures*, Wiley, New York, 1968.
3. R. Boel and M. Kohlmann : Semi-martingale models of stochastic optimal control, with applications to double martingales, *SIAM J. Control and Optimization, 18* (1980), p.511-533.
4. D.P. Gaver and G.S. Shedler : Multiprogramming system performance via diffusion approximations, IBM Research report RJ - 938(1971), Yorktown Heights, N.Y.
5. E. Gelenbe and I. Mitrani : *Analysis and Synthesis of Computer Systems*, Academic Press, London, 1980.
6. J.M. Harrison : The diffusion approximation for tandem queues in heavy traffic, *Adv. Applied Prob., 10*(1978), p.886-905
7. J.M. Harrison and A.J. Lemoine : Sticky Brownian motions as the limit of storage processes, *J. Appl. Prob. 18*(1981), p. 216-226.
8. H. Kobayashi : Application of the diffusion approximation to queueing networks, parts I and II : *J.A.C.M., 21* (1974), p. 316-328; p. 459-469.
9. P.A. Meyer : Un cours sur les intégrales stochastiques, Séminaire de Probabilités X, Lecture Notes in Mathematics, vol. 511, Springer Verlag, Berlin, 1976.
10. E. Pardoux : Characterization of the density of the conditional law in the filtering of a diffusion with boundary, Recent Developments in Statistics, J.R. Barra etal, eds., North Holland, 1977.
11. M. Reiman : Queueing networks in heavy traffic, Ph.D. Dissertation, Dept. of Operations Research, Stanford University, 1977.
12. D.W. Stroock and S.R.S. Varadhan : Diffusion processes with boundary conditions, *Comm. Pure and Appl. Math, 24* (1971), p. 147-225.
13. J. van Schuppen and E. Wong : Transformation of local martingales under a change of law, *Ann. Probability 2* (1974), p. 879-888.
14. S. Watanabe : On stochastic differential equations for multi-dimensional diffusion processes with boundary conditions, *J.Math. Kyoto University, 11* (1971), p. 169-180.

SOME RECENT RESULTS ON THE CONTROL OF PARTIALLY OBSERVABLE STOCHASTIC SYSTEMS

N.Christopeit
Institut für Ökonometrie
und Operations Research
Universität Bonn
Adenauerallee 24-42
53 Bonn
West Germany

M. Kohlmann
Institut für Angewandte
Mathematik
Universität Bonn
Wegelerstr. 6
53 Bonn
West Germany

1. Introduction

We shall be concerned with systems whose evolution is described by a stochastic differential equation

(1.1) $\quad dx_t = f(t,\{x_s,s\leq t\},u(t,\{y_s,s\leq t\}))dt + \sigma(t,\{x_s,s\leq t\})dw_t$,

(w_t) a Brownian motion. By choosing an appropriate control u the controller wants to influence the system in such a way as to minimize a certain criterion

(1.2) $\quad J[u] = E\Phi(\{x_s,y_s,u_s,s\leq T\})$.

The interpretation to be given to (1.1) is then the following. It is supposed that the state (x_t) of the system is itself not directly amenable to observation, but rather some process (y_t) - the observation process - which is related to the state in some way. It is then natural to require that the control action taken at time t depends only on the information $\{y_s,s\leq t\}$ available at time t. One may try the following approach to solve this problem. We estimate the current state on the basis of the available information and restrict ourselves to controls depending only on this estimate. If it can be shown that in this class there exists an optimal control and that it is at least as good as any other control depending on the observations, then we have indeed found a solution of our problem. To begin with, let us have a brief look at a situation where this approach via the "separated control problem" works pretty well, namely the linear case.

Let the system dynamics be described by the linear stochastic differential equation

(1.3) $dx_t = [Ax_t + Bu_t]dt + Cdw_t,$

(1.4) $dy_t = Dx_t + v_t$

with Gaussian initial values, where (w_t) and (v_t) are independent standard Brownian motions. Then, if the control process is treated as a random parameter which is nonanticipative with respect to (w_t), the system (1.3), (1.4) has a unique strong solution (x_t,y_t). If now the control u is adapted to the observation σ-fields

$$\mathcal{Y}_t^u = \sigma\{y_s, s \leq t\},$$

then the process (x_t,y_t) is conditionally Gaussian, which means in particular that the conditional distribution of x_t given $\mathcal{Y}_t^u$ is Gaussian with mean $\hat{x}_t$ and covariance R_t satisfying the stochastic differential equation

(1.5) $d\hat{x}_t = [A\hat{x}_t + Bu_t]dt + F_t d\nu_t,$

where $F_t = R_t D'$ and R_t is given as solution of the Riccati-equation

$$R_t = AR + RA' - RD'DR + CC',$$

$$R_o = \mathrm{Cov}(x_o) .$$

ν_t is the innovation process, which is given by

$$\nu_t = y_t - \int_o^t D\hat{x}_s ds .$$

(ν_t) is a Wiener process with respect to (y_t). Let n(t,x) denote the normal density with mean O and covariance R_t. Then, letting $\ell \equiv 0$ for simplicity,

$$J[u] = E\{E\{g(x_T)/\mathcal{Y}_T^u\}\}$$

$$= E\int g(x)n(T,x - \hat{x}_T)dx$$

(1.6) $$= E\hat{g}(\hat{x}_T)$$

$$= \hat{J}[u]$$

with the new cost functional

$$\hat{g}(x) = \int g(y)n(T,y - x)dy .$$

If, moreover, $\mathcal{Y}_t^u = \mathcal{Y}_t^o$ for all t, where $\mathcal{Y}_t^o = \sigma\{y_s^o, s \leq t\}$ and (y_t^o) is the observation process obtained from (1.3), (1.4) for $u \equiv 0$, then the innovation processes (ν_t) coincide (with (ν_t^o)) for all such u. Hence, for the class $\mathcal{U}$ of controls satisfying

(1.7) (u_t) is adapted to (Y^o_t), and $Y^u_t = Y^o_t$ for all t,

we obtain a neat well-defined stochastic control problem, in which the evolution of the state is described by (1.5) and (1.6) is the criterion to be minimized. The important point is, that it is a problem in which the state is completely observable. (1.5)-(1.7) is called the separated problem. Our interest in this problem stems from the fact, that

$$J[u] = \hat{J}[u]$$

for any $u \in \mathcal{U}$. Hence, any optimal control for the separated problem will at the same time be optimal for the original (partially observable) problem of minimizing J[u] subject to (1.3), (1.4) in the class $\mathcal{U}$. The problem becomes then one of characterizing the class $\mathcal{U}$. The famous result of Wonham ([19],[31]) is that $\mathcal{U}$ contains all controls obtained (by inserting the actual observation process) from Lipschitz feedback function of past observations. From this result it follows in particular that, if there is a Lipschitz Markov feedback control $u_t = \upsilon(t,\hat{x}_t)$ minimizing $\hat{J}[u]$ s.t. (1.5), then u_t is in $\mathcal{U}$ and hence optimal for both the separated and the partially observable problem. This applies in particular to the case of the linear regulator where $\ell(x,u)$ and g(x) are quadratic functionals in x and u. Hence, in such cases, the conditional mean is a sufficient statistic. One may wish to admit any feedback function of past observations. In this case, things become more involved, since in general there will only exist a weak solution to (1.3), (1.4). For the formulation of the separated problem and its solution some results have been obtained in [2], [8], [12], [29].

In all of these cases, a sufficient statistic for the controls is given by the conditional mean, which is due to the linearity in x of the dynamics equation. In the nonlinear case, one cannot hope to obtain such a simple finite-dimensional statistic. Some interesting extensions where this remains possible have recently been given by Beneš [1]. In general, however, one will have to consider the whole conditional distribution of x_t given the past of (y_t) as sufficient statistic.

2. Review of nonlinear filtering.

Let us briefly recall some basic facts from nonlinear filter theory. Let $(x_t, \mathcal{F}_t)$ be a (homogeneous, for simplicity) Markov process (defined on some probability space $(\Omega, \mathcal{F}, P)$) with transition function $P(t,x,A)$ and extended generator L, i.e.

$$T_t\phi(x) = \phi(x) + \int_0^t T_u L\phi(x)\,du$$

for all $\phi \in \mathcal{D}(L)$, where T_t denotes the semigroup associated with (x_t):

$$T_t\phi(x) = \int \phi(y) P(t,x,dy) .$$

The observation process (y_t) is supposed to be given by

$$y_t = \int_0^t h(x_s)\,ds + w_t ,$$

where h is such that the integral exists. For simplicity we assume that (x_t) and (w_t) are independent. Let $\mathcal{Y}_t = \sigma\{y_s, s \leq t\}$ denote the observation σ-fields and

$$\pi_t(\phi) = E\{\phi(x_t) | \mathcal{Y}_t\} .$$

Then, if

$$E\phi(x_t)^2 < \infty, \quad 0 \leq t \leq T,$$

$$E\int_0^T h(x_t)^2 dt < \infty , \quad E\int_0^T |\phi(x_t)h(x_t)|^2 \,dt < \infty ,$$

the evolution of $\pi_t(\phi)$ is described by the stochastic differential equation

$$(2.1) \qquad d\pi_t(\phi) = \pi_t(L\phi)dt + [\pi_t(\phi h) - \pi_t(\phi)\pi_t(h)]d\nu_t$$

for all $\phi \in \mathcal{D}(L)$ (cf. [21], chapter 8.6), in which (ν_t) is the innovation process

$$\nu_t = y_t - \int_0^t \pi_s(h)\,ds .$$

It is well known (cf. [14], [21], [25]) that $(\nu_t, \mathcal{Y}_t)$ is a Wiener process. In particular, the increments $\nu_t - \nu_s$ are independent of $\mathcal{Y}_s$, conveying the idea that $\nu_t - \nu_s$ represents the "new information" concerning the signal process $h(x_t)$ available from the observations between times s and t. For the case of particular interest to us when (x_t) is generated by the stochastic differential equation

$$dx_t = f(x_t)dt + \sigma(x_t)dv_t ,$$

with (v_t) a Brownian motion independent of (w_t) and f,σ satisfying the Ito conditions, thus guaranteeing existence of a unique strong solution, then the filter equation (2.1) holds with

$$(2.2) \qquad L = \sum_{i=1}^{n} f^i(x) \frac{\partial}{\partial x_i} + \frac{1}{2} \sum_{i,j=1}^{u} a^{ij}(x) \frac{\partial^2}{\partial x_i \partial x_j} ,$$

$$a = \sigma\sigma' .$$

For reasons of stochastic modelling and in particular for the purposes of stochastic control it turns out more convenient to work with some unnormalized version of π_t. This is defined as

$$\sigma_t(\phi) = E_o\{\phi(x_t)\zeta_t | Y_t\} ,$$

where

$$\zeta_t = \exp[\int_o^T h(x_s)dy_s - \frac{1}{2}\int_o^T h(x_s)^2 ds]$$

and P_o is the probability measure given by

$$dP_o = \exp[-\int_o^T h(x_s)dy_s + \frac{1}{2}\int_o^T h^2(x_s)ds]dP .$$

Under P_o, (y_s) is a standard Brownian motion independent of (x_t), and (x_t) has the same distribution as under P. From the filtering equation it may be deduced that $\sigma_t(\phi)$ satisfies the stochastic DE

$$d\sigma_t(\phi) = \sigma_t(L\phi)dt + \sigma_t(h\phi)dy_t .$$

This is Zakai's equation (cf. [14], [32]).

Taking a robust version of $\sigma_t^y(\phi)$, it can be shown that, for each fixed $y \in C[0,T]$, there exist measures Λ_t^y such that

$$\sigma_t^y(\phi) = \langle\phi,\Lambda_t^y\rangle, \qquad \phi \in C_b,$$

where C_b denotes the space of bounded continuous functions on $\mathbb{R}^n$.

Λ_t^y is the unnormalized conditional distribution. If a density $q(t,x)$ (depending on the paths y, of course) exists, then it satisfies the stochastic partial differential equation

$$(2.3) \qquad dq(t,x) = L^*q(t,x)dt + h(x)q(t,x)dy_t \ .$$

For the sakes of optimal stochastic control, the assumptions made - especially the independence of (x_t) and (w_t) - turn out too restrictive. Actually, the filter equation and the Zakai equation continue to hold if the state dynamics are given by

$$(2.4) \qquad dx_t = f(x_t,\alpha_t)dt + \sigma(x_t)dv_t \ ,$$

where (v_t) is independent of (w_t) and (α_t) is a nonanticipative process w.r. to (v_t) such that there exists a solution (x_t,y_t) to (2.4), (1.2) (under some probability measure $P = P^\alpha$) for which $\mathcal{F}_t^{x,w}$ is independent of the future increments of (w_t). The decisive point is that under P^α, for $\phi \in C_b^2$, the martingale

$$(2.5) \qquad M_t(\phi) = \phi(x_t) - \phi(x_o) - \int_o^t L^\alpha\phi(x_s)ds = \int_o^t \sigma(x_s)f(x_s,\alpha_s)dv_s$$

is orthogonal to (w_t), i.e.

$$< M(\phi),w > \equiv 0$$

(cf. [21]). Thereby, in (2.5), L^α is the operator obtained by applying Ito's formula to $\phi(x_t)$, i.e.

$$(2.6) \qquad L^\alpha\phi(x_t) = f(x_t,\alpha_t)\phi'(x_t) + \frac{1}{2}a(x_t)\phi''(x_t) \ .$$

In control applications, we will typically have $\alpha_t = u(t,\{y_s,s\leq t\})$. For the case of correlated noises, see [16] and [23].

A different description of the unnormalized conditional distribution is the Kallianpur-Striebel formula which we write down in the form needed later, assuming $\sigma \equiv 1$ for simplicity:

$$\Lambda_t(A) = \int 1_{[\xi_t\in A]} \exp\left\{ \int_o^t f(\xi_s,\alpha_s)d\xi_s - \frac{1}{2}\int_o^t |f(\xi_s,\alpha_s)|^2 ds \right.$$

$$+ \int_0^t h(\xi_s)dy_s - \frac{1}{2}\int_0^t |h(\xi_s)|^2 ds \Big\} \mu(d\xi) \ ,$$

where μ is the distribution measure of x_o+v. The stochastic integral w.r. to y is only defined for a set of full Wiener measure. But integrating by parts (i.e. applying the Ito-formula to $y_t h(\xi)$) we obtain an expression of the form

$$(2.7) \qquad \Lambda_t(A) = \int 1_{[\xi_t \in A]} \exp[\Phi_t(\{\xi_s, y_s, s \leq t\}, \alpha_t)]\mu(d\xi) \ ,$$

which is now well defined for all paths (y_s). For the specific form of the functional Φ_t cf. [3].

3. Wide-sense admissible controls and the separated nonlinear control problem.

We shall consider a system whose state and observation process are governed by the stochastic differential equations

$$(3.1)\qquad dx_t = [f_o(x_t,y_t) + f_1(x_t,y_t)u_t]dt + \sigma(x_t,y_t)dw_t ,$$

$$(3.2)\qquad dy_t = h(x_t)dt + dv_t .$$

(v_t) and (w_t) are independent standard Brownian motions. The control process (u_t) takes values in some convex compact set $\mathcal{U}$. x_o has initial distribution μ and $y_o = 0$. Our aim is to minimize the performance criterion

$$(3.3)\qquad J = E\{\int_o^T \ell(x_t,u_t)dt + g(x_T)\} .$$

A natural class of admissible controls would be the class of all measurable processes $u = (u_t)$ taking values in $\mathcal{U}$ and adapted to the observation σ-fields $\mathcal{Y}_t = \sigma\{y_s, s\leq t\}$. This is usually called the strict sense version of the problem. Unfortunately, all efforts to solve this problem in its whole generality have so far been without success. If we are looking for a way of doing something, nevertheless, we find ourselves facing the following dilemma: either to enlarge the class of admissible controls, or to restrict it. The first way leads to the concept of randomized controls and is the subject of this section. The second way will be dealt with in the next section.

We make the following assumptions:

(A1) $f_o, f_1, \sigma \in C_b$ and Lipschitz in x (uniformly in y) .

(A2) $h \in C_b^2$.

Here $C_b^k(E)$ denote the space of bounded functions on E whose derivatives up to order k are bounded and continuous. In order to define the class of admissible controls, consider the canonical sample space

$$\Omega = \underbrace{C[0,T] \times C[0,T]}_{\Omega_1} \times \underbrace{C[0,T] \times L_2([0,T],\mathcal{U})}_{\Omega_2}$$

with elements $\omega = (x,w,y,u)$. The control-observation space Ω_2 will be endowed with the (metric) topology of uniform convergence on C and weak convergence on L_2. The corresponding canonical (= Borel) filtration is given by

$$G_t = Y_t \times U_t ,$$

where

$$Y_t = \sigma\{y_s, s \leq t\} \text{ and } U_t = \sigma\{\int_0^s u_\theta d\theta, s \leq t\} .$$

The following notion of admissible controls is adopted in [18], [7].

Definition 3.1. An admissible (wide-sense) control is a probability measure π *on* (Ω_2, G_T) *such that* (y_t) *is a Wiener process with respect to* (G_t) *and* π.

Hence, in addition to requiring that the projection $(y,u) \to y$ maps π onto Wiener measure, this definition requires that $\int_0^t h_\theta d\theta$ is independent of $y_s - y_t$, $t \leq s \leq T$. Let $\mathcal{A}_w$ denote the class of wide-sense admissible controls.

Definition 3.2. $\pi \in \mathcal{A}_w$ *is a strict-sense admissible control if there exists a* $(Y_T - U_T)$*-measurable function* $\upsilon : C[0,T] \to L_1([0,T]; \mathcal{U})$ *such that for every* G_T*-measurable function* $\psi \geq 0$

$$\int \psi(y,u)d\pi = \int \psi(y,\upsilon(y))d\mu ,$$

where μ *is Wiener measure on* Y_T.

In other words: for strict-sense admissible controls the conditional probability

$$\pi^y(du) = \delta_{\upsilon(y)}(du).$$

That this definition of strictly admissible controls indeed coincides with the intuitive concept mentioned at the beginning of this section comes from the fact that to every strictly-admissible π there exists a causal function $\gamma : [0,T] \times C \to \mathcal{U}$ such that $u_t = \gamma(t,y)$, and vice-versa. Denote the strict-sense admissible controls by $\mathcal{A}_s$. Then:

(a) $\mathcal{A}_s$ is dense in $\mathcal{A}_w$ (for the topology of weak convergence of probability measures).

(b) Every strict-sense control is an extreme point of $\mathcal{A}_w$.

Unfortunately, the converse is not true, as the example of Varadhan (based on the Cirelson counter example) shows (cf. [18]).

Given an observation and control $(y,u) \in \Omega_2$, assumption (A1) implies that there is a pathwise unique solution to (3.1) corresponding to initial value $x_o = \xi$ with unique law $\bar{P}_\xi^{y,u}$ on Ω_1. In other words: under $\bar{P}_\xi^{y,u}$, the coordinate process (w_t) is a standard Brownian motion and (x_t) satisfies the stochastic differential equation (3.1). The fundamental result is

Proposition 3.1. $\bar{P}_\xi^{y,u}$ *depends continuously on* (ξ,y,u) (*cf.* [18], [30]).

Given $\pi \in \mathcal{A}_w$ and a fixed initial distribution μ of x_o, we define on the canonical sample space the probability measures

$$P_\pi^o(dx,dw,dy,du) = \bar{P}_\xi^{y,u}(dx,dw)\,\pi(dy,du)\,\mu(d\xi) \ .$$

Then, under P_π^o, (w_t) and (y_t) are independent standard Brownian motions. Finally, we introduce the probability measures

$$dP_\pi = \zeta_T dP_\pi^o \tag{3.4}$$

with

$$\zeta_T = \exp\left[\int_o^T h(x_s)dy_s - \frac{1}{2}\int_o^T h(x_s)^2 ds\right] . \tag{3.5}$$

Then, under P_π, (3.1), (3.2) hold (for the coordinate processes) with $x_o \sim \mu$ and with

$$v_t = y_t - \int_o^t h(x_s)ds$$

a standard Brownian motion independent of (w_t). The assumptions concerning the drifts f_o, f_1 may be weakened somewhat to

(A1') f_o, f_1 are bounded and continuous, and $a = \sigma\sigma'$ satisfies

$$0 < \lambda|\theta|^2 \le \langle\theta, a\theta\rangle \le \Lambda|\theta|^2 < \infty ,$$

$$|a(\xi,\eta) - a(\xi',\eta')| \le \delta(|(\xi-\eta,\xi'-\eta')|)$$

(locally suffices).

Then, for fixed (ξ,y,u), a solution to (3.1) can be defined as the (unique) solution of the martingale problem associated with initial value ξ, drift $f_o(x,y_t) + f_1(x,y_t)u_t$ and diffusion $a(x,y_t)$, and the continuity of $(\xi,y,u) \to \bar{P}_\xi^{y,u}$ continues to hold (cf. [18], [30]).

Given $\pi \in \mathcal{A}_w$, the criterion may now be written as the function space integral

$$(3.6) \qquad J[\pi] = \int [\int_0^T \ell(x_t,u_t)dt + g(x_T)]dP_\pi \ .$$

Proposition 3.2. *The set $\mathcal{A}_w$ is convex and sequentially compact in the topology of weak convergence of probability measures.*

Tightness follows since every $\pi \in \mathcal{A}_w$ projects onto Wiener measure under $(y,u) \to y$ and the second component takes values in the compact space $L_2([0,T];\mathcal{U})$ (w.r. to weak topology). Closedness and convexity are straightforward.

Proposition 3.3. *Under "appropriate conditions", $J[\pi]$ is lower semi-continuous on $\mathcal{A}_w$.*

Since a l.s.c. function adopts its infiminum on compacts, we finally arrive at

Theorem 3.1. *Under the assumptions made there exists an optimal control $\pi \in \mathcal{A}_w$.*

Since $\mathcal{A}_s$ is dense in $\mathcal{A}_w$, an immediate further consequence of Proposition 3.3. is

Theorem 3.2. $\inf_{\pi\in\mathcal{A}_s} J[\pi] = \inf_{\pi\in\mathcal{A}_w} J[\pi]$.

"Appropriate conditions" are either of the following:

(A3) $\ell \equiv 0$, $g \geq 0$ and continuous.

(A4) $\ell \geq 0$, $g \geq 0$, continuous; $\ell(x,u)$ convex in u.

$\sigma \in C_b^1$ with bounded inverse.

The initial distribution μ has density $p_0(\xi)$.

(A5) $\ell \geq 0$, $g \geq 0$, continuous, $\ell(x,u)$ convex in u.

(A3) and (A4) are used in [18], (A5) in [7]. The derivation of Proposition 3.3 and Theorem 3.1 may be based on the following results. The approach is similar to the one in [7], yet somewhat more transparent.

Lemma 3.1. The class $\mathcal{P} = \{P_\pi, \pi \in \mathcal{A}_w\}$ *is tight.*

Proof. It suffices to show that the measures P_π^x, P_π^w, etc. corresponding to the projections $(x,w,y,u) \to x$, etc. are tight.

As to P_π^x, note first that for fixed (y,u) (x_t) is a solution to (3.1) with driving noise (w_t). Hence, under $\bar{P}_\mu^{y,u}(dx,dw) = \int \bar{P}_\xi^{y,u}(dx,dw)\mu(d\xi)$,

$$\bar{E}_\mu^{y,u}(x_t-x_s)^4 \le 16\,\bar{E}_\mu^{y,u}\Big[\Big(\int_s^t (\|f_0\| + \|f_1\|\sup_{v\in\mathcal{U}}|v|)dr\Big)^4 + \Big(\int_s^t \sigma(x_r,y_r)dw_r\Big)^4\Big]$$

$$\le K'[(t-s)^4 + (t-s)^2]$$

$$\le K(t-s)^2 .$$

This implies

$$E_\pi^x|x_t-x_s|^4 = \int \pi(dy,du) \int |x_t-x_s|^4 \bar{P}_\mu^{y,u}(dx,dw) \le K(t-s)^2 ,$$

hence, by the moment condition (cf. Theorem in [6]), $\{P_\pi^x, \pi \in \mathcal{A}_w\}$ is tight.

P_π^w is Wiener measure for each $\pi \in \mathcal{A}_w$.

Under P_π,

$$y_t - y_s = \int_s^t h(x_r)dr + (v_t - v_s)$$

for some Brownian motion (v_t). Hence

$$E_\pi^y|y_t-y_s|^4 \le L'[(t-s)^4 + (t-s)^2] \le L(t-s)^2 ,$$

showing that $\{P_\pi^y, \pi \in \mathcal{A}_w\}$ is tight.

Finally, $\{P_\pi^u, \pi \in \mathcal{A}_w\}$ is tight since (u_t) takes values in the compact space $L_2([0,T];\mathcal{U})$.

Lemma 3.2. $\mathcal{P}$ *is weakly sequentially compact.*

Proof. In view of Proposition 3.2, it suffices to show that the mapping $\mathcal{A}_w \to \mathcal{P}$ defined by $\pi \to P_\pi$ is (weakly sequentially) continuous.

Let $\pi_n \to \pi$ weakly. We shall show that, for every bounded continuous $\phi(x,w,y,u)$,

$$\int \phi \, dP_{\pi^n} \to \int \phi \, dP_\pi \ .$$

Since $\mathcal{P}$ is tight, we may suppose that ϕ has compact support and hence is uniformly continuous. Since

$$\int \phi \, dP_{\pi^n} = \int_{\Omega_2} \pi_n(dy,du) \int \phi(x,w,y,u) \bar{P}_\xi^{y,u}(dx,dw)\mu(d\xi) \ ,$$

the assertion will be proved if

$$\psi(y,u) = \int \phi(x,w,y,u) \bar{P}_\xi^{y,u}(dx,dw)\mu(d\xi)$$

is continuous. To show this, let $(y_n,u_n) \to (y,u)$. Writing $\bar{P}_\xi^n = \bar{P}_\xi^{y_n,u_n}$ and $\phi_n(x,w) = \phi(x,w,y_n,u_n)$, we find (from the uniform continuity) that $\phi_u(x,w)$ converges uniformly (in (x,w)) to $\phi(x,w,y,u)$. Hence

$$|\psi(y_n,u_n) - \psi(y,u)| \leq \int |\phi_n(x,w) - \phi(x,w,y,u)| \bar{P}_\xi^n(dx,dw)\mu(d\xi)$$

$$+ \left| \int \phi(x,w,y,u) \bar{P}_\xi^n(dx,dw)\mu(d\xi) - \int \phi(x,w,y,u) \bar{P}_\xi^{y,u}(dx,dw)\mu(d\xi) \right| \ .$$

The first term can be made arbitrarily small by the uniform convergence of the integrands, while the second converges to 0 by the continuous dependence of $\bar{P}_\xi^{y,u}$ on (y,u).

Let now $\Phi(x,w,y,u)$ be any function $\Omega_1 \times \Omega_2 \to \mathbb{R}$ satisfying

(A6) $\Phi \geq 0$ and l.s.c.

Then, for any sequence $P_{\pi_n} \to P_\pi$

$$\int \Phi \, dP_\pi \leq \varliminf_{n\to\infty} \int \Phi \, dP_{\pi^n} \ .$$

Hence we have

Theorem 3.3. Under assumptions (A1), (A2) *and* (A6) *there exists in* $\mathcal{A}_w$ *an optimal control for the criterion* $J[\pi] = E_\pi \Phi$.

A somewhat different definition of wide-sense admissible controls has been given in [20]. The dynamics are somewhat more general, replacing linearity in u by a Roxin condition and allowing σ and h to depend on

y_t, but for simplicity, we shall continue to refer to (3.1), (3.2) as the description of the dynamical system. An admissible randomized control is then any adapted stochastic process $u = (u_t)$ taking values in $\mathcal{U}$, which is defined on some filtered probability space $(\Omega, F, (F_t), P)$ carrying independent Brownian motions $w = (w_t)$, $v = (v_t)$, a random variable ζ with values in some metric space and adapted processes $x = (x_t)$, $y = (y_t)$ such that (3.1), (3.2) are satisfied and

(i) $u(t,\omega)$ is $B_{[0,T]} \times F^{y\zeta}$ - measurable;

(ii) $\{(u_s, y_s), s \leq t\}$ and $\{y_s - y_t, s > t\}$ are independent with respect to the measure Q which makes (y_t) a Brownian motion (obtained by a Girsanov transformation);

(iii) (u,y) and w are independent under Q;

(iv) y and ζ are independent under Q;

(v) (x,w,v) and ζ are conditionally independent given (y,u).

The random variable ζ represents the randomization of the controls.

It can be shown that in fact every randomized control can be canonically identified with a wide-sense control in the sense of [13]. Existence of an optimal randomized control is then proven by techniques involving weak convergence of measures and invariance theorems, the crucial point being the construction of the random variable ζ corresponding to the limiting process.

Recalling the definition of the probability measure P_π^0 (cf. (3.4), (3.5)), we may write the criterion (3.6) as

$$J[\pi] = E_\pi^0\{\int_0^T \zeta_t \ell(x_t, u_t)dt + \zeta_T g(x_T)\}$$

$$= E_\pi^0\{\int_0^T E_\pi^0[\zeta_t \ell(x_t, u_t) | G_t]dt + E_\pi^0[\zeta_T g(x_T) | G_T]\}$$

$$= \int_{\Omega_2} \pi(dy, du)\{\int_0^T \bar{E}^{y,u}[\zeta_t \ell(x_t, u_t)]dt + \bar{E}^{y,u}\zeta_T g(x_T)\}$$

$$(3.7) \qquad = \int_{\Omega_2} \pi(dy,du)\{\int_0^T <\ell(\cdot,u_t),\Lambda_t^{y,u}>dt + <g,\Lambda_T^{y,u}>\},$$

where $\bar{P}^{y,u}(A) = \int \bar{P}_\xi^{\,y,u}(A)\mu(d\xi)$ (μ the fixed initial distribution) and $\Lambda_t^{y,u}$ is the pathwise version of the conditional distribution of x_t given the past G_t of the observation process y and the control process u. In [18], existence and continuous dependence of $\Lambda_t^{y,u}$ on (y,u) and the initial distribution μ is shown. Moreover, it is shown that $\Lambda_t^{y,u}$ satisfies the Zakai equation

$$(3.8) \qquad d<\phi,\Lambda_t> = <L_t^{y,u}\phi,\Lambda_t>dt + <h\phi,\Lambda_t>dy_t$$

for all $\phi \in C_b^2$, with

$$L_t^{y,u} = \sum_{i=1}^d f^i(x,y_t,u_t)\,\frac{\partial}{\partial x_i} + \frac{1}{2}\sum_{i,j=1}^d a^{ij}(x,y_t)\,\frac{\partial^2}{\partial x_i \partial x_j} \quad .$$

It should be noted that these results do not directly follow from the corresponding results in nonlinear filtering theory, but can be derived along similar lines (cf. [18],[28]). (3.7), (3.8) is called the "separated" control problem. In the separated problem, we regard the unnormalized conditional distribution $\Lambda_t = \Lambda_t^{y,u}$ as state with initial value $\Lambda_0 = \mu$, and (3.7) is the criterion to be minimized. If Λ_t possesses a density q(t,x), then, as in nonlinear filtering, q(t,x) obeys Zakai's stochastic partial differential equation. Equivalently, one may work with the gauge-transformed density

$$\tilde{q}(t,x) = e^{-y_t h(x)}\,q(t,x) \ ,$$

whose evolution is described by the partial differential equation (2.7). For strictly admissible controls, the separated problem may be derived directly from the corresponding results in nonlinear filtering. As state may either be taken the unnormalized conditional distribution or the unnormalized conditional density, both satisfying the corresponding Zakai equation. One would like to obtain controls depending only through the conditional density or the conditional distribution on the observations. In this direction, approaches have been made by Mortensen, Beneš/Karatzas ([26], [4]), who use a dynamic programming approach to obtain a verification theorem for "Markov" controls (w.r.t. the conditional density). The crucial point is, however, to obtain a solution of the Zakai equation (2.3) (for the unnormalized conditional density) which possesses the "right" measurability properties (i.e. q(t,x) is adapted to $(\mathcal{Y}_t)$) when in the operator $L = L^{\alpha}$ (cf. (2.6)) a "Markov"-feedback control (of the unnormalized conditional density) $\alpha_t = \nu(t,p(t,x))$ is substituted. A considerable step towards the solution of this problem has been made in a recent paper by Beneš and Karatzas (cf.[3]) who consider the corresponding Kallianpur-Striebel formula instead of the Zakai equation. In this formulation, the problem of finding a version of the unnormalized conditional distribution adapted to the observation presents itself in the form of a fixed point problem for an operator acting on measures. Namely, for controls depending on the observations only through the current unnormalized conditional distribution, i.e. $\alpha_t = \nu(t,\Lambda_t)$ in (2.7), we are looking for a solution of (2.7) which is adapted to $(\mathcal{Y}_t)$. Or, if we define the space C of continuous functions from [0,T] into the space of finite measures on x-space (endowed with a certain metric) and introduce, for fixed $y \in C[0,T]$, the operator $T = T^y : C \to C$ defined by $(T\Lambda)_t(A) =$ rhs of (2.7) with $\alpha_t = \nu(t,\Lambda_t)$, then the problem to be solved is equivalent to finding a solution $\Lambda = (\Lambda_t)$ of $T\Lambda = \Lambda$ which is adapted to $(\mathcal{Y}_t)$. Under continuity assumptions about the feedback control law ν Beneš and Karatzas are able to show that there does indeed exist a desired version of the unnormalized conditional distribution.

Starting from the Zakai PDE for the unnormalized conditional density, a different approach has been taken by Bensoussan [5] who derives a stochastic maximum principle and an adjoint process, thus making rigorous an idea of Kwakernaak [24]. Then dynamic programming methods are used to characterize the value function of the control problem as maximum element of a certain set of functions. This approach is very much near the idea introduced by Nisio [27] into control theory under complete observation. Nisio's approach to characterize the value function as the envelope of a family of semigroups was adapted to control problems with partial observation by Fleming [17] and, under a different point of view, by Davis and Kohlmann [13].

4. Some results for strictly admissible controls.

Strictly admissible controls have been considered in [9] and [22]. For state dynamics described by

$$(4.1) \qquad dx_t = f(x_t,u_t)dt + \sigma(x_t,u_t)dw_t$$

and criterion (3.3), it is supposed that the observation process is given by

$$(4.2) \qquad y_t = \pi(x_t) \ ,$$

the projection on some subvector of x_t, or, more generally, by

$$(4.3) \qquad y_t = \phi[x_s, s \leq t] \ ,$$

where ϕ is some continuous $C^1[0,T]$-valued function of the past of the state process. Note that the case where the observation is described by (3.2) is included in both (4.1) and (4.2) if we take as state $\tilde{x}_t = (x_t, y_t)$. Since we cannot expect to find an optimal control in the whole class of strictly admissible controls (as defined earlier), we have to restrict admissibility to certain subclasses. The approach taken in [22] and [9], [10], consists in imposing tightness conditions on the time dependence and equicontinuity conditions on the dependence on observations of the controls admitted. More precisely, an admissible control will be any function $u\colon [0,T] \times C^1[0,T] \to \mathcal{U}$ (compact) such that

(i) u is measurable with respect to the Borel σ-field on $[0,T] \times C^1[0,T]$ (implying $u(\cdot,y) \in L^1[0,T]$);

(ii) $\|u(\cdot,y) - u(\cdot,\bar{y})\|_t \leq r_{\bar{y}}(\sup_{0 \leq s \leq t} |y_s - \bar{y}_s|)$,
where $r_y(\alpha) \searrow 0$ as $\alpha \searrow 0$ (and $\|\cdot\|_t = L_1$-norm on $[0,t]$);

(iii) $\|u_h(\cdot,y) - u(\cdot,y)\| \leq r_y(h)$,
where, for $v \in L_1[0,T]$,

$$v_h(t) = \frac{1}{2h} \int_{t-h}^{t+h} v(s)ds.$$

(iv) There exists a probability space (Ω, F, P) carrying a Wiener process (w_t, F_t) such that the equation

$$x_t = x_o + \int_o^t f(x_s,u(s,\phi(x)))ds + \int_o^t \sigma(x_s,u(s,\phi(x)))dw_s$$

(4.4)

$$= x_o + F(t) + G(t) \qquad \text{(for further reference)}$$

has a solution (x_t) adapted to $(\mathcal{F}_t)$.

Denote this class of admissible controls by $\mathcal{A}_r$. Finally assume

(A7) f,σ,ℓ,g are continuous; $\ell \geq 0$, $g \geq 0$; f and σ satisfy a linear growth condition (uniformly in u).

First, as a consequence of (ii), there exists an admissible modification of $u(t,y)$ that is adapted to the natural filtration on $C^1[0,T]$ and leads to the same solution of (4.4). The main reason to require (ii) and (iii) is, however, that the set

$$\mathcal{A} = \{U : U(y)(\cdot) = u(\cdot,y) \text{ for } u \text{ satisfying (i) - (iii)}\} \subset C(C^1[0,T]; L_1[0,T])$$

is sequentially compact with respect to pointwise convergence on $C^1[0,T]$. Actually, it follows from (ii) that convergence holds in the somewhat stronger sense

(4.5) $\qquad \|U_n(y_n) - U(y)\|_{L_1} \to 0 \quad \text{if} \quad \|y_n - y\| \to 0$

for all y. To establish sequential compactness, note that by virtue of (iii) for every y the set

$$\{U(y),\ U \in \mathcal{A}\}$$

is tight in $L_1[0,T]$, and by virtue of (ii) $\mathcal{A}$ is equicontinuous. By Arzela-Ascoli's theorem then, $\mathcal{A}$ is relatively compact in $C(C^1[0,T],L_1[0,T])$ w.r.t. the topology of pointwise convergence (even w.r.t. uniform convergence on compacts). But since the topology of pointwise convergence satisfies the first countability axiom, it follows that $\mathcal{A}$ is relatively sequentially compact. (Note that (2.5) does then imply sequential compactness w.r.t. uniform convergence on compacts.) Closedness is easily established.

In order to show existence of an optimal admissible control we proceed as follows. Let u^N be a minimizing sequence of admissible controls with corresponding solutions x^N and Wiener processes w^N, i.e.

$$x_t^N = x_o + F^N(t) + G^N(t) \ ,$$

where the random functions F^N and G^N are defined as in (4.4). Then it follows from (A7) that the random functions

$$\Phi^N = (x^N, F^N, G^N)$$

are tight in C^{3n} (cf. [11]). Let $\Phi^{N'}$ be a subsequence converging weakly to some Φ. By Skorokhod's embedding theorem, we can find random functions $\tilde{\Phi}^{N'}$, all defined on the same probability space $(\tilde{\Omega}, \tilde{\mathcal{F}}, \tilde{P})$, such that

$$\tilde{\Phi}^{N'} \sim \Phi^{N'} \qquad \text{(stochastic equivalence)}$$

and

$$\tilde{\Phi}^{N'} \to \tilde{\Phi} \qquad \text{a.e. (in the topology of } C^{3n}) \ .$$

Moreover, we may suppose that N' is chosen in such a way that

$$\|U^{N'}(y^{N'}) - U(y)\|_{L_1} \to \quad \text{if} \quad \|y^{N'} - y\| \to 0$$

for all $y \in C^1[0,T]$, with $U \in \mathcal{A}$. It can then be established (cf. [9]-[11], [22]) that

$$\tilde{x}_t^{N'} = \tilde{x}_o + \int_o^t f(\tilde{x}_s^{N'}, U^{N'}(\phi(\tilde{x}^{N'}))(s))ds + $$
$$+ \int_o^t \sigma(\tilde{x}_s^{N'}, U^{N'}(\phi(\tilde{x}^{N'}))(s))d\tilde{w}_s^{N'}$$

and

$$\tilde{x}_t = \tilde{x}_o + \int_o^t f(\tilde{x}_s, U(\phi(\tilde{x}))(s))ds + \int_o^t \sigma(\tilde{x}_s, U(\phi(\tilde{x}))(s))d\tilde{w}_s$$

with Wiener processes $(\tilde{w}_t^{N'}, \tilde{\mathcal{F}}_t^{N'})$, $(\tilde{w}_t, \tilde{\mathcal{F}}_t)$, with respect to which $(\tilde{x}_t^{N'})$ and $(\tilde{x}_t)$, respectively, are nonanticipative. Hence

$$u(t,y) = U(y)(t)$$

is indeed an admissible control. Moreover,

$$\tilde{E}\{\int_o^T \ell(\tilde{x}_s, u(s, \phi(\tilde{x})))ds + g(\tilde{x}_T)\}$$
$$\leq \underline{\lim}_{N' \to \infty} \tilde{E}\{\int_o^T \ell(\tilde{x}_s^{N'}, U^{N'}(\phi(\tilde{x}^{N'}))(s))ds + g(\tilde{x}_T^{N'})\}$$

$$= \lim_{N\to\infty} E\{\int_0^T \ell(x_s^N, u^N(\phi(x^N))(s))ds + g(x_T^N)\} ,$$

i.e. u is optimal.

Hence:

<u>*Theorem 4.1*</u>. *Under* (A7) *there exists an optimal control in the class* $\mathcal{A}_r$.

If the drift and cost functional can be decomposed as

$$(4.6) \qquad f(x,u) = f_o(x) + f_1(x)f_2(u), \ \ell(x,u) = \ell_o(x) + \ell_1(x)\ell_2(u) ,$$

where f_2, ℓ_2 satisfy Roxin's condition:

(4.7) The set

$$\mathcal{V} = \{(f_2(u), \ell_2(u)) : u \in \mathcal{U}\}$$

is compact and convex, and σ is independent of u;

then, defining

$$(4.8) \qquad U(y)(t) = (\int_0^t f_2(u(s,y))ds, \int_0^t \ell_2(u(s,y))ds) ,$$

we require

$$(ii') \qquad \|U(y) - U(\bar{y})\|_t \leq r_{\bar{y}}(\sup_{0\leq s\leq t} |y_s - \bar{y}_s|) .$$

As is shown in [10], [22], corresponding to each such U we may find, using McShane and Warfield's implicit function theorem, a measurable function $\tilde{u}:[0,T] \times C^1[0,T] \to \mathcal{U}$, adapted to the natural filtration on C^1, such that (4.8) holds with u replaced by $\tilde{u}$. Let $\mathcal{A}_r'$ denote the class of controls satisfying (i), (ii') and)iv). Consider now the set

$$\mathcal{A}' = \{U \text{ defind by (4.6) with } u:[0,T] \times C^1[0,T] \to \mathcal{U} \text{ measurable, } U \text{ satisfies (ii')}\}.$$

For each y, the set

$$\{U(y), U \in \mathcal{A}'\}$$

is then automatically tight (in some $C^m[0,T]$) and it may be shown (as with $\mathcal{A}$) that $\mathcal{A}'$ is relatively compact in the same sense as $\mathcal{A}$. Closedness is established again with the help of McShanes and Warfield's implicit function theorem. The existence of an optimal admissible control may then be shown along the same lines as above (cf. [10]).

Theorem 4.2. If (4.6) *and* (4.7) *hold, then (under* (A7)), *there exists an optimal control in* $\mathcal{A}'_r$.

If the observation process is given by (3.1) and f and σ satisfy the Ito conditions, things turn out much simpler. In this case, using a Girsanov measure transformation (depending on the controls), we may consider the observation process to be in the same for all u and treat the controls as random parameters, which are compact in some L_1-space. For each u, we have then a unique strong solution x^u, which depends continuously on u (cf. [20]).

Let us finally have a brief look at some models in which the existence of an optimal control is ensured by assuming special information patterns (cf. [11], [15]). Thus we assume that observations can only be made at certain fixed times

$$0 \leq t_1 < \ldots < t_p \leq T$$

and that at these times, it can only be observed if some function

$$(4.8) \qquad \tilde{y}_j = \tilde{\phi}_j(x_{t_1},\ldots,x_{t_j}) \in \mathbb{R}^m , \quad j=1,\ldots,p ,$$

falls into a certain cell D_i, where $\{D_i\}$ is a finite or countable decomposition of $\mathbb{R}^m$ into disjoint sets. It is then convenient to take a point z_i from each D_i and define the observation space $Y = \{z_i\}$ as well as the observation functions

$$y_j = \phi_j(x_{t_1},\ldots,x_{t_j}) = z_i \quad \text{if} \quad \tilde{\phi}_j(x_{t_1},\ldots,x_{t_j}) \in D_i .$$

As an example, we may take as D_i the unit cubes

$$[\frac{k_1}{N}, \frac{k_1+1}{N}) \times \ldots \times [\frac{k_m}{N}, \frac{k_m+1}{N}), \quad k_1,\ldots,k_m \in \mathbb{Z} ,$$

(for fixed N) and as z_i all vectors $(k_1/N,\ldots,k_m/N)$. An admissible control is now defined as any measurable function

$$u:[0,T] \times C[0,T] \to \mathcal{U}$$

which at any time $t \in [t_j,t_{j+1})$ depends only on y_j, i.e.

$$u(t,x) = \mathcal{V}_j(t,\phi_j(x_{t_1},\ldots,x_{t_j})), \quad t \in [t_j,t_{j+1}) ,$$

for some measurable $\mathcal{V}_j:[t_j,t_{j+1}) \times Y \to \mathcal{U}$. Note that the case where $u(t_i)$ may depend on $(y_1,\ldots,y_j)$ at time $t \in [t_j,t_{j+1})$ can easily be reduced to the former one by introducing the augmented observation space Y^p and the augmented observation functions $Y_j = (y_1,\ldots,y_j)$. In [15], the admissible control may also depend on some observable subvector of the current state x_t. In [11], the observation space may also be continuous, i.e. it may be supposed that the $\tilde{y}_j$ in (4.8) can be measured exactly and

$$u(t,x) = \mathcal{U}_j(t,\tilde{\phi}_j(x_{t_1},\ldots,x_{t_j})), \quad t \in [t_j,t_{j+1}) ,$$

for $\mathcal{U}_j:[t_j,t_{j+1}) \times \mathbb{R}^m \to \mathcal{U}$. Let $\mathcal{A}_d$ denote any of these classes of admissible controls. Under appropriate conditions on the drift and the diffusion function in (4.1) there exists a solution to the dynamic equation (4.1), defined - for instance - via the Girsanov measure transformation. If, in addition, the drift and the integral cost (in 3.3)) factor as in (4.5) such that the extended velocity set is compact and convex (cf. (4.6)), then it can be shown:

Theorem 4.3. In any of the classes $\mathcal{A}_d$ there exists an optimal control.

The proof in [15] (for countable observation space and Markov-controls) uses the concept of relaxed controls, which take values in the space of probability measures on $\mathcal{U}$, thereby depending in a nonanticipative way on the observations. In this class, an optimal control always exists, and under Roxin's condition it can be represented as an ordinary admissible control.

The approach in [11] (for finite or continuous observation space) employs the technique based on weak convergence of probability measures and the construction of stochastically equivalent processes (Skorokhod embedding), which was explained above. The crucial point is to show that the limit control process has the "right" measurability properties, i.e. depends in a nonanticipative way on the observations. As a byproduct of both approaches it follows that every strictly admissible control u which, at time t, depends on the whole past (up to time t) of certain components (y_t) of the state process (x_t) can be approximated by controls $u^n \in \mathcal{A}_d$ for any of the three classes $\mathcal{A}_d$ (in the sense that the values of the criterion (3.3) corresponding to the u^n converge to the value corresponding to u).

REFERENCES.

[1] V.E. Beneš, Exact finite dimensional filters for certain diffusions with nonlinear drift. To appear in Stochastics.

[2] V.E. Beneš and I. Karatzas, Examples of optimal control for partially observable systems: Comparison, classical and martingale methods. Stochastics 5 (1981) 43-64.

[3] V.E. Beneš and I. Karatzas, Filtering of diffusions controlled through their conditional measures. To appear in Stochastics.

[4] V.E. Beneš and I. Karatzas, On the relation of Zakai's and Mortensen's equations. To appear in SIAM J. Control and Optimization.

[5] A. Bensoussan, Maximum principle and dynamic programming approaches of the optimal control of partially observed diffusions. Reprint, University Paris Dauphine, 1982.

[6] P. Billingsley, Convergence of Probability Measures. Wiley, New York, 1968.

[7] J.M. Bismut, Partially observed diffusions and their control. SIAM J. Control and Optimization 20 (1982) 302-309.

[8] N. Christopeit and K. Helmes, Optimal control for a class of partially observable systems. To appear in Stochastics.

[9] N. Christopeit, Existence of optimal controls under partial observation. Z. Wahrscheinlichkeitstheorie und Verw. Gebiete 51 (1980) 201-213.

[10] N. Christopeit, A note on the existence of optimal stochastic control for partially observed systems. Preprint, Bonn, 1980.

[11] N. Christopeit, Optimal stochastic control with special information patterns. SIAM J. Control and Optimization 18 (1980) 559-575.

[12] M.H.A. Davis, The separation principle in stochastic control via Girsanov solutions. SIAM J. Control and Optimization 14 (1976) 176-188.

[13] M.H.A. Davis and M. Kohlmann, On the nonlinear semigroup of stochastic control under partial observations. Preprint, Bonn 1981.

[14] M.H.A. Davis and S.I. Marcus, An introduction to nonlinear filtering. In: Stochastic Systems, M. Hazewinkel and J.C. Willems (eds.), University of Rotterdam, 1981.

[15] R.J. Elliott and M. Kohlmann, On the existence of optimal partially observed controls. To appear in SIAM J. Control and Optimization.

[16] R.J. Elliott and M. Kohlmann, Robust filtering for correlated multidimensional observations. Math. Z. 178 (1981) 559-578.

[17] W.H. Fleming, Nonlinear semigroup for controlled partially observed diffusions. SIAM J. Control and Optimization 20 (1982) 286-301.

[18] W.H. Fleming and E. Pardoux, Optimal control for partially observed diffusions. SIAM J. Control and Optimization 20 (1982) 261-285.

[19] W.H. Fleming and R.W. Rishel, Deterministic and Stochastic Optimal Control. Springer-Verlag, New York, 1975.

[20] N.G. Haussmann, On the existence of optimal controls for partially observed diffusions. SIAM J. Control and Optimization 20 (1982) 385-407.

[21] G. Kallianpur, Stochastic Filtering Theory. Springer-Verlag, New York 1980.

[22] M. Kohlmann, Existence of optimal controls for a partially observed semimartingale control problem. To appear in Stochastic Proc. Appl. .

[23] M. Kohlmann, Robust filtering for systems with correlation between signal and observation. IFIP-ISI Conference Proc. . Bangalore 1982.

[24] H. Kwakernaak, A minimum principle for stochastic control problems with output feedback. Systems and Control Letters 1 (1981).

[25] R.S. Liptser and A.N. Shiryayev, Statistics of Random Processes I. Springer-Verlag, New York, 1977.

[26] R.E. Mortensen, Stochastic optimal control with noisy observations. Int. J. Control 4 (1966) 455-464.

[27] M. Nisio, On stochastic optimal controls and envelope of Markovian semi-groups. Stochastic Differential Equations (ed. K. Ito), Wiley, New York 1978.

[28] E. Pardoux, Nonlinear filtering, prediction and smoothing. In: Stochastic Systems, M. Hazewinkel and J.C. Willems (eds), University of Rotterdam, 1981.

[29] J. Ruzicka, On the separation principle with bounded coefficients. Appl. Math. Optimization 3 (1977) 243-261.

[30] D.W. Strook and S.R.S. Varadhan, Multidimensional diffusion processes. Springer-Verlag, New York, 1979.

[31] W.M. Wonham, On the separation theorem of stochastic control. SIAM J. Control 6 (1968) 312-326.

[32] M. Zakai, On the optimal filtering of diffusion processes. Z. Wahrscheinlichkeitstheorie und Verw. Gebiete 11 (1969) 230-243.

Optimal Controls for Partially Observed Stochastic Systems using Nonstandard Analysis

By Nigel J. Cutland

Dept. of Pure Mathematics, University of Hull, England

§1. INTRODUCTION

In this paper we describe informally a new approach to optimal stochastic control using techniques from nonstandard analysis. This has been applied to partially observed systems with information obtained at discrete times, generalising results of Elliott and Kohlmann [5]. Our systems take the following form.

State dynamics

$$(1.1)\qquad \begin{cases} dx_t = f(t,(x_s)_{s\le t},u_t)\,dt + g(t,(x_s)_{s\le t})\,db_t & (0 \le t \le 1) \\ x_o = 0 \end{cases}$$

(Conditions on f, g are given later; b is a Brownian motion)

Observations Times $0 \le t_1 < t_2 < \ldots < t_p < 1$ are fixed; for $x \in C[0,1]^d$ we observe

$$y(x) = (y_1(x), y_2(x), \ldots y_p(x))$$

where $y_i(x)$ is the observation made at time t_i; y_i is a fixed measurable function, depending on $x\restriction t_i$, with values lying in a fixed countable observation space $\mathcal{Y}$.

Controls The set $\mathcal{U}$ of (ordinary) admissible controls consists of measurable functions $u : [0,1] \times \mathcal{Y}^p \to A$, where A is a compact subset of Euclidean space; $u(t,y)$ depends only on $(y_1,\ldots,y_m)$, where $t_m \le t < t_{m+1}$.

We also consider relaxed controls v, where A is replaced by the space of Borel measures on A; the set of admissible relaxed controls is denoted by $\mathcal{V}$.

Cost For simplicity we assume a terminal cost

$$J(u) = E(c(x_1))$$

where c is bounded, measurable.

Conditions on the drift and diffusion.

(a) f is jointly measurable, and $f(t,x,\cdot)$ is continuous, all t, x;

(b) g is jointly measurable, nonsingular, and $g(t,\cdot)$ is uniformly Lipschitz;

(c) $f(t,x,a), g(t,x)$ depend only on $x\restriction t = \{x_s : s \le t\}$;

(d) f, g, $g^{-1}f$ satisfy linear growth conditions (e.g. $\|f(t,x,a)\| \le k(1+\|x\|)$).

In [5] systems similar to the above are considered, but having Markovian dynamics: the drift and diffusion depend only on x_t. The methods used are essentially Markovian. The systems in [5] possess an additional feature not discussed here: the control u_t is allowed to depend on a projection of x_t (as well as the observations).

In the paper [4] , to appear, the following is proved:

1.2 Theorem For the system described above:

(a) $\inf_{u \in \mathcal{U}} J(u) = \inf_{v \in \mathcal{V}} J(v)$,

(b) there is an optimal relaxed control,

(c) if $f(t,x,a) = f_1(t,x)h(t,a)$ and for each t, $h(t,A)$ is convex, there is an optimal ordinary control.

In the rest of this present paper we will indicate how nonstandard methods are used in [4]; this involves sketching the basics of nonstandard anlysis (§2), the construction of Brownian motion and the Itô integral due to Anderson (§3), the non-standard approach to the solution of stochastic differential equations (§4), and finally the application to stochastic control theory (§5). We give references where full details about each of these topics may be found. In the conclusion of this paper (§6) we mention another partially observed optimal control problem that yields to an extension of these methods.

§2. NONSTANDARD ANALYSIS Nonstandard analysis begins by extending the reals $\mathbb{R}$ to a set of hyperreals ${}^*\mathbb{R} \supsetneq \mathbb{R}$, which contains both infinite and infinitesimal elements. The following picture indicates how to think of this extension.

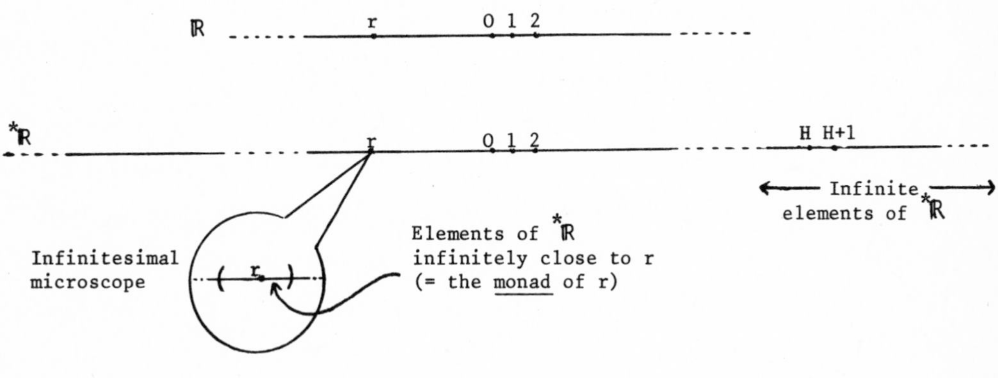

One way to construct ${}^*\mathbb{R}$ is to set ${}^*\mathbb{R} = \mathbb{R}^{\mathbb{N}}/D$ where D is a non-principal ultrafilter on $\mathbb{N}$. It is best for practical purposes to assume ${}^*\mathbb{R}$ as given axiomatically as in [6] or [7] . The essential feature of ${}^*\mathbb{R}$ is that it has 'the same properties as $\mathbb{R}$' (in a sense that can be made precise). In particular, ${}^*\mathbb{R}$ is an ordered field. With ${}^*\mathbb{R}$ given(or constructed) we make the following definitions.

Definition For $x,y \in {}^*\mathbb{R}$

(a) x is infinite if $|x| > n$, all $n \in \mathbb{N}$

(b) x is infinitesimal if $|x| < \frac{1}{n}$, all $n \in \mathbb{N}$

(c) $x \simeq y$ (x is infinitely close to y) if $x - y$ is infinitesimal.

The next lemma is very important:

Standard Part Lemma If $x \in {}^*\mathbb{R}$ is finite there is a unique $r \in \mathbb{R}$ with $x \simeq r$. We say r is the standard part of x, written ${}^o x$.

In order to discuss continuity, we need the following notion , which is the non-standard counterpart of standard continuity.

Definition A function $F : {}^*[0,1] \to {}^*\mathbb{R}$ is S-continuous if for all $t_1,t_2 \in {}^*[0,1]$ $(= \{x \in {}^*\mathbb{R}: 0 \le x \le 1\})$

$$t_1 \simeq t_2 \Rightarrow F(t_1) \simeq F(t_2).$$

The standard part lemma has the following counterpart for S-continuous functions.

Lemma If F is S-continuous and internal (see below) and F(0) is finite there is a unique $f \in C[0,1]$ such that

$$f({}^o t) = {}^o(F(t)) \qquad \text{all } t \in [0,1] .$$

We write $f = {}^o F$; this means that the following diagram commutes:

$$\begin{array}{ccc} {}^*[0,1] & \xrightarrow{\quad F \quad} & {}^*\mathbb{R} \\ \downarrow{\scriptstyle o} & & \downarrow{\scriptstyle o} \\ [0,1] & \xrightarrow[\quad f \quad]{} & \mathbb{R} \end{array}$$

Internal objects To do 'higher' mathematics we need not only $\mathbb{R}$ but subsets of $\mathbb{R}$, functions, sets of functions, etc., etc. An adequate setting for most higher mathematics is the superstructure $V(\mathbb{R})$ defined by $V_o(\mathbb{R}) = \mathbb{R}$; $V_{n+1}(\mathbb{R}) = V_n(\mathbb{R}) \cup \mathcal{P}(V_n(\mathbb{R}))$; $V(\mathbb{R}) = \bigcup_{n \in \mathbb{N}} V_n(\mathbb{R})$. ($\mathcal{P}$ denotes the power set.)

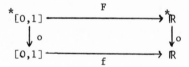

Similarly, for nonstandard mathematics, we have $V({}^*\mathbb{R}) = \bigcup_{n\in\mathbb{N}} V_n({}^*\mathbb{R})$. A mapping ${}^* : V(\mathbb{R}) \to V({}^*\mathbb{R})$ is constructed (or postulated, if proceeding axiomatically), so that every standard object A has a nonstandard extension ${}^*A$. An object B is internal if $B \in {}^*A$ for some A; other objects in $V({}^*\mathbb{R})$ are external . The way to picture this is as follows.

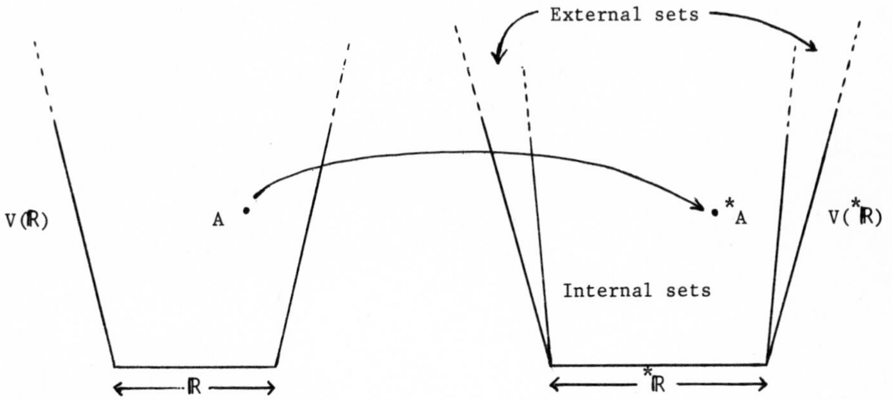

The distinction between internal and external objects is fundamental, because the internal objects inherit all the properties of the standard objects in $V(\mathbb{R})$.

Futher details about the superstructure approach and the properties of the mapping * may be found in [2], [7] or [11]. With this brief sketch we proceed to explain how nonstandard methods are used in stochastic analysis.

§3 BROWNIAN MOTION AND ITÔ INTEGRATION ON [0,1] Anderson [1] gave a very intuitive construction of Brownian motion and Itô integration essentially as follows (in 1-dim).

Fix an infinite $H \in {}^*\mathbb{N}$; let $\Delta t = H^{-1} \approx 0$. Let $T = \{0,\Delta t,2\Delta t,\ldots,H\Delta t = 1\}$, a hyperfinite time line.

Let Ω be the (internal) set of all internal functions $\omega : T \to {}^*\mathbb{R}$ such that (i) $\omega(0) = 0$, (ii) $\omega(t + \Delta t) - \omega(t) = \pm\sqrt{\Delta t}$. Then Ω is hyperfinite, with internal cardinality $2^H = |\Omega|$ (using $|\cdot|$ to denote cardinality of a set).

Define the counting measure P on internal subsets $A \subseteq \Omega$ by

$$P(A) = {}^\circ(|A|/|\Omega|) \in [0,1] .$$

It was shown by Loeb ([9],[10]) that there is a unique extension of P to a σ -additive probability measure on $F = \sigma(\{\text{internal subsets of } \Omega\})$. Denoting the extension by P we thus have a standard probability space (Ω, F, P), a hyperfinite Loeb space. Now Anderson [1] showed

3.1 Theorem Define $B : \Omega \times T \to {}^*\mathbb{R}$ by

$$B(\omega,t) = \omega(t).$$

Then (a) $B(\omega,\cdot)$ is S-continuous for a.a. $\omega \in \Omega$;

(b) the continuous process b on Ω defined by

$$b = {}^oB \qquad \text{(a.s. in } \Omega)$$

is a Brownian motion.

A natural filtration $(F_t)_{t\in[0,1]}$ on Ω is given by $F_t = \{A : A \text{ is closed under } \sim_t\}$, where $\omega \sim_t \omega'$ if $\omega(s) = \omega'(s)$ all $s \lesssim t$. Now suppose that $f(\omega,t)$ is bounded, measurable and F_t-adapted. Then we have

3.2 Theorem (Anderson [1]) There is a bounded internal function $F : \Omega \times T \to {}^*\mathbb{R}$ (a lifting of f) such that

(a) for a.a. (ω,t) with respect to the Loeb counting measure on $\Omega \times T$

$${}^oF(\omega,t) = f(\omega,{}^ot).$$

(i.e. for a.a. (ω,t) the following diagram commutes:

$$\begin{array}{ccc} \Omega \times T & \xrightarrow{\quad F \quad} & {}^*\mathbb{R} \\ \downarrow{\scriptstyle o} & & \downarrow{\scriptstyle o} \\ \Omega \times [0,1] & \xrightarrow[\quad f \quad]{} & \mathbb{R} \end{array}$$

(c) For any such F

$$\int_0^t f\,db_s = {}^o\Big(\sum_{s<t} F(\omega,s)\Delta B_s(\omega)\Big),$$

where $\Delta B_s(\omega) = B(\omega,s+\Delta t) - B(\omega,s)$.

§4 STOCHASTIC DIFFERENTIAL EQUATIONS In [8] Keisler initiated the study of stochastic differential equations on Loeb spaces, and showed how to solve equations of the form

$$dx_t = f(t,x_t)dt + g(t,x_t)db_t$$

under very general conditions on f, g, and with b prescribed. Keisler's methods were extended in [3] to equations of the form (1.1) (without the control u_t).

The procedure (in 1-dim) is roughly as follows. Take suitable non-anticipating liftings (as in (3.2)) F, G of f, g and define an internal process $X : \Omega \times T \to {}^*\mathbb{R}$ by

$$\begin{cases} X(\omega,0) = 0 \\ X(\omega,t+\Delta t) = X(\omega,t) + F(t,X(\omega,\cdot))\Delta t + G(t,X(\omega,\cdot))\Delta B_t(\omega) \, . \end{cases}$$

It is fairly routine to show that $X(\omega,)$ is a.s. S-continuous, so we may define the continuous process $x = {}^oX$ a.s. Then we show that x is a solution to the equations; this involves calculation of a Radon-Nikodym derivative ρ of the measure induced by the process X, to show that $X(\omega,\cdot)$ a.s. avoids the set where F behaves badly. Not surprisingly ρ is given by the usual Girsanov formula.Notice that a solution is obtained on the given space with a prescribed Brownian motion.

<u>§5 OPTIMAL CONTROL</u> We now return to the controlled system described in §1 , and for simplicity of exposition restrict ourselves to 1 dimension. Take suitable liftings F, G, Y, C of the fixed functions f, g, y, c of the system. Suppose that u is an admissible control, and let U be a lifting of u. Then we can define an internal process X^U by

$$(5.1) \qquad \begin{cases} X^U(0) = 0 \\ X^U(t+\Delta t) = X^U(t) + F(t,X^U(\cdot),U(t,Y(X^U(\cdot))))\Delta t + G(t,X^U(\cdot))\Delta B_t \, . \end{cases}$$

As in §4, X^U is a.s. S-continuous, and the process $x^u = {}^oX^U$ is a solution to (1.1). Moreover

$$J(u) = {}^o\bar{E}(C(X^U(1))) = {}^oJ(U), \text{ say,}$$

where $\bar{E}$ denotes the internal counting expectation.

Now let $J_o = \inf \{J(u) : u \in \mathcal{U}\}$ and take (u_n) with $|J(u_n) - J_o| < n^{-1}$. Choose liftings U_n of u_n; so for all finite n

$$(5.2) \qquad |J(U_n) - J_o| < n^{-1} \, .$$

The <u>overflow principle</u> - a basic nonstandard principle - tells us that there is an <u>infinite</u> N and internal control U_N such that

$$|J(U_N) - J_o| < N^{-1} \approx 0$$

(Simply take $N = M - 1$ where M is the least element in ${}^*\mathbb{N}$ (if any) for which (5.2) fails).

Thus ${}^oJ(U_N) = J_o$; i.e. U_N is an <u>internal (nonstandard) optimal control</u>.

Now we ask, does U_N correspond to anything standard?

For a fixed observation path y and real r, the values ${}^{o}U_N(t,y)$ for $t \approx r$ are distributed throughout A , as in the following diagram.

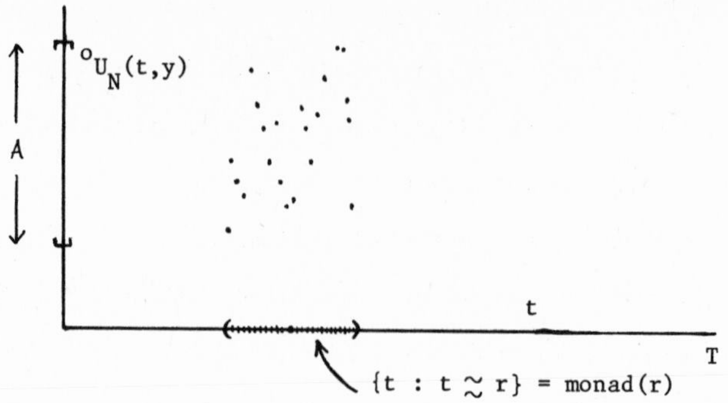

Using regular conditional distributions we can define a Borel probability $v_{r,y}$ on A, given roughly by

$$v_{r,y}(B) = \text{"}P_T({}^{o}U_N(t,y) \in B \mid t \approx r)\text{"}$$

where P_T is the Loeb counting measure on T. Thus $v(r,y) = v_{r,y}$ is a relaxed control.

Now let X^{U_N} be the process defined by (5.1) with $U = U_N$ and take v as just defined from U_N. Then we obtain

<u>5.2 Theorem</u> [4] (a) ${}^{o}X^{U_N}$ is a solution to (1.1) for the control v;

(b) $J(v) = {}^{o}J(U_N) = J_o$; i.e. v is an optimal relaxed control.

Using similar techniques we establish further:

<u>5.3 Theorem</u> [4] (a) Given <u>any</u> relaxed control $v' \in V$ there is an internal U' corresponding to it as in Theorem 5.2 (a), (b), and vice versa;

(b) $J(U)$ is dense in $J(V)$.

The result of Theorem 1.2 (c) is then easily derived by standard methods.

<u>§6 ANOTHER CONTROL SYSTEM</u> The nonstandard methods described above can be used to obtain an optimal relaxed control for the following system, which is also considered in its Markovian version in [5].

<u>Dynamics and cost:</u> as in §1, with the drift f taking the form

$$f(t,x,a) = f_o(t,x) + f_1(t,x)h(t,a)$$

where f_o, f_1 are measurable, adapted; h is bounded, measurable and $h(t,\cdot)$ is continuous.

Observations $y_t = \Pi x_t$, a projection onto a q-dimensional subspace of the state space.

Admissible controls Functions $u : [0,1] \times C([0,1], \mathbb{R}^q) \to A$ which are measurable, adapted, and satisfy the following Lipschitz condition: there is a function $\tilde{h}(t,y)$ such that

(a) $\tilde{h}(t,y) = \int_o^t h(s,u(s,y))ds \qquad (a.a.y)$

(b) $\|\tilde{h}(t,y) - \tilde{h}(t,y')\| \leq k\|y - y'\|_t$.

(k is uniform; $\tilde{h}$ depends on u.)

Again, if $h(t,A)$ is convex we obtain an ordinary optimal control.

REFERENCES

[1] Anderson, R. M., A non-standard representation for Brownian motion and Itô integration, Israel J. Math 25 (1976), 15-46.

[2] Cutland, N. J., Infinitesimal methods in measure theory, probability theory and stochastic analysis, Bulletin of the Institute of Maths. and its Appn., 18 (1982), 52-57.

[3] Cutland, N. J., On the existence of solutions to stochastic differential equations on Loeb spaces, Z. f. Wahrscheinlichketstheorie verw. Geb., to appear.

[4] Cutland, N. J., Optimal controls for partially observed stochastic systems: an infinitesimal approach, Stochastics, to appear.

[5] Elliott, R. J. and Kohlmann, M., On the existence of optimal partially, observed controls, J. Applied Math.& Opt., to appear.

[6] Keisler, H. J., Elementary Calculus, Prindle, Weber & Scmidt, Boston,1976.

[7] Keisler, H. J., Foundations of Infinitesimal Calculus, Prindle, Weber & Schmidt, Boston 1976.

[8] Keisler, H. J., An infinitesimal approach to stochastic analysis, Univ. of Wisconsin preprint; to appear as AMS memoir.

[9] Loeb, P. A., Conversion from nonstandard to standard measure spaces and

applications in probability theory, Tran. Amer. Math. Soc. 211 (1975), 113-122.

[10] Loeb, P. A., An introduction to nonstandard analysis and hyperfinite probability theory, in Probabilistic Analysis and Related Topics, Vol. 2 (Ed. A. T. Bharucha-Reid), Academic Press, 1979, 105-142.

[11] Stroyan, K. D., and Luxemburg, W. A. J., Introduction to the Theory of Infinitesimals, Academic Press, New York, 1976.

STOCHASTIC CONTROL WITH TRACKING OF EXOGENOUS PARAMETERS

M.H.A. DAVIS

Department of Electrical Engineering
Imperial College
London SW7 2BT

This paper concerns a control problem in which a system represented by a stochastic differential equation is to be steered so as to follow a similar system the whereabouts of which is known only through noisy observations. It is shown that an optimal control exists and that this control depends in a Markovian fashion on the state of the controlled system and on the conditional distribution of the position of the target system.

I. INTRODUCTION

General stochastic control problems with noisy observations are "hard" because both the estimation process and the system dynamics are affected by the control action, so that optimal control is a trade-off between attempts to acquire information and to steer the system along a low-cost trajectory. The purpose of this paper is to draw attention to a class of problems, of practical as well as theoretical interest, where no such interaction occurs and a fairly complete theory of optimal control can be worked out. To set the scene, let us consider the following two classes of "tractable" stochastic control problems:

(a) *Beneš' "small investor" problem* [1]. Here one wishes to minimize

$$E\int_0^T c(x_s,u_s)ds \tag{1.1}$$

over control processes (u_t) such that u_t is H_t-measurable for each t, where (H_t) is a given, fixed family of σ-fields (not necessarily increasing). The process (x_t) - the vector of stock prices in the small investor interpretation - evolves autonomously, i.e. is unaffected by (u_t). In this case the cost (1.1) is equal to $E\int_0^T \hat{c}(s,\omega,u_s)ds$ where $\hat{c}(s,\omega,u) \doteq E[c(x_s,u)|H_s]$ and Beneš shows that under appropriate conditions the "obvious" control $u_s^* = \arg\min_u \hat{c}(s,\omega,u)$ is optimal.

(b) *Stochastic control with complete observations.* Suppose the state process (x_t) satisfies the stochastic differential equation

$$dx_t = b(x_t,u_t)dt + dw_t^u \tag{1.2}$$

where (w_t^u) is Brownian motion and (u_t) is adapted to $X_t = \sigma\{x_s, 0 \le s \le t\}$; solutions are defined in the weak sense. This problem is treated in [3] [6]. One finds that the "value process" (W_t) is a semimartingale whose martingale part has a representation in the form $\int_0^t g_s dw_s^u$ where the integrand (g_s) does not depend on the control (u_t) being used. The optimal control is then

$$u_t^* = \arg_u \min (g_s b(x_s,u) + c(x_s,u))$$

What is proposed here is a kind of combination of the above cases : the system will consists of two parts, a noisily-observed part which is unaffected by control action, and a completely observed controlled part. This situation arises whenever, for example, exogenous parameters are being estimated on-line. A case in point would be the capacity expansion model considered in the paper of Dempster, Sethi and Vermes in this volume. They model the demand for some public utility, for example water supply, as a Poisson process. One might wish to estimate the rate of this process from the observed data, and this estimation proceeds autonomously under the arguably reasonable assumption that demand is unaffected by the Water Department's current construction programme.

In this paper a tracking problem related to (a) and (b) above will be analysed. (z_t) will be a Markov process which is measured via scalar observations (y_t) given by

$$dy_t = h(z_t)dt + dw_t \tag{1.3}$$

where h is a bounded function and w_t is Brownian motion independent of (z_t). A controlled process (x_t) is given by (1.2) above in which (u_t) is now to be adapted to $G_t = \sigma\{x_s, y_s, s \leq t\}$. The objective is to minimize the cost

$$J(u) = E^u\left(\int_0^T c(z_s,x_s,u_s)ds + \Phi(z_T,x_T)\right) \tag{1.4}$$

where c, Φ are bounded non-negative functions and T is a fixed final time. Various interpretations are possible. (z_t) could be the state of another system that "our" system is trying to track; or, somewhat more generally, (z_t) could be a message transmitted to us by our ground station to tell us what our control objective is. It will be shown that under weak conditions an optimal control (u_t^*) exists; further, this control takes the Markovian form $u_t^* = u^*(t,\sigma_t,x_t)$, where σ_t is an unnormalized conditional distribution of z_t given G_t, and it is constructed by minimizing a certain Hamiltonian function, (3.4) below. Strictly speaking, these results are not new, since the problem is a special case of the general control model studied by El Karoui, Lepeltier and Marchal [8], [9]. However, some of the argument differs from that of [8], [9] and in any case it is perhaps of interest to note that a significant class of partial-observations problems falls within the scope of their general model. Due to exigencies of space the argument below is given only in outline form; a full account will appear elsewhere.

II. PROBLEM FORMULATION

Let (z_t) be a diffusion process in R^p satisfying the stochastic differential equation

$$dz_t = \alpha(z_t)dt + \beta(z_t)dv_t$$

where (v_t) is a vector Brownian motion and α,β satisfy the standard Ito or Stroock-Varadhan conditions. The initial random variable x is independent of (v_t) and has a given distribution π. We denote by P_π the measure on the canonical space $C[0,T;R^p]$ induced by this process; from now on (z_t) will be the coordinate process on this space. The basic probability space for the problem will be $\Omega = C[0,T;R^{p+1+d}]$ where T is a finite positive number. The coordinate functions are $(z_t,y_t,x_t)_{0\le t\le T}$. We denote $F_t^0 = \sigma\{z_s,y_s,x_s,0\le s\le t\}$ and $G_t^0 = \sigma\{y_s,x_s,x_s,0\le s\le t\}$. Let P_0 be the measure on (Ω,F_T^0) given by $P_0 = P_\pi * \mu_w^1 * \mu_w^d$ where $*$ denotes product measure and μ_w^m is m-dimensional Wiener measure. (F_t) and (G_t) are the completions of $(F_t^0)(G_t^0)$ with all P_0-null sets of F_T.

Under measure P_0 the processes (z_t) $(y_t)(x_t)$ are independent ; (z_t) is the Markov process described above and (y_t), (x_t) are independent 1- and d-dimensional Brownian motions respectively. Let U be a compact space and $b:R^d \times U \to R^d$ be a measurable function such that $b(x,\cdot)$ is continuous for each x and $|b(x,u)| \le K(1+|x|)$ for some constant K. *Admissible controls* U are all U-valued G_t-predictable processes. For $u \in U$, define measure P_0^u by the Radon-Nikodym derivative

$$\frac{dP_0^u}{dP_0} = \exp(\int_0^T b'(x_s,u_s)dx_s - \tfrac{1}{2}\int |b(x_s,u_s)|^2)$$

Under P_0^u, (x_t) satisfies equation (1.2), where (w_t^u) is a Brownian motion independent of (z_t, y_t), whose distributions are the same as under P_0. Finally, define measure P^u by

$$\frac{dP^u}{dP_0^u} = \exp(\int_0^T h(z_s)dy_s - \tfrac{1}{2}\int_0^T h^2(z_s)ds) =: \Lambda_T$$

where $h \in C_b^2(R^p)$. With this definition of P^u, (y_t) satisfies (1.3), with (w_t) a Brownian motion independent of (w_t^u). Thus the processes (z_t,y_t,x_t) conform under measure P^u to the problem description of §I, and the control problem is to choose $u \in U$ so as to minimize $J(u)$ given by (1.4) in which E^u denotes integration with respect to P^u.

Let us now reformulate this problem entirely in terms of observable - i.e. G_T-measurable - quantities. Denote $\sigma_t(f) = E_0^u[\Lambda_t f(z_t)|G_t]$ for arbitrary bounded measurable functions $f : R^p \to R$. Then $E^u[f(z_t)|G_t] = \sigma_t(f)/\sigma_t(1)$ and σ_t satisfies the Zakai equation

$$d\sigma_t(f) = \sigma_t(Af)dt + \sigma_t(hf)dy_t$$

(2.1)

$$\sigma_0(f) = \pi(f)$$

for $f \in \mathcal{D}(A)$, where A is the generator of the process (z_t) :

$$Af(z) = \sum_{i=1}^{p} \alpha_i(z) \frac{\partial f}{\partial z_i} + \sum_{i,j=1}^{p} (\beta(z)\beta'(z))_{ij} \frac{\partial^2 f}{\partial z_i \partial z_j}$$

Equation (2.1) defines a positive-measure-valued process (σ_t) such that σ_t is an unnormalized conditional distribution of z_t given G_t. These facts are standard when $G_t = \sigma\{y_s, s\leq t\}$. Here $G_t = \sigma\{x_s, y_s\ s\leq t\}$ but the additional conditioning variables do not affect the estimation process since (W_t^u), (y_t) and (z_t) are independent under P_0^u. Similarly one can show, as in [5] that

$$E_0^u[\Lambda_t | G_t] = \sigma_t(1) \tag{2.2}$$

Define, for positive measures μ on R^p

$$\hat{c}(\mu,x,u) = \int_{R^p} c(z,x,u)\mu(dz)$$

$$\hat{\Phi}(\mu,x) = \int_{R^p} \Phi(z,x)\mu(dz)$$

Then

$$E^u[\Phi(z_T,x_T)] = E^u[\sigma_T(\Phi(\cdot,x_T))/\sigma_T(1)]$$

$$= E_0^u[E_0^u(\Lambda_T|G_T)\sigma_T(\Phi(\cdot,x_T))/\sigma_T(1)]$$

$$= E_0^u[\sigma_T(\Phi(\cdot,x_T))]$$

$$= E_0^u[\hat{\Phi}(\sigma_T,x_T)]$$

(The third equality above follows from (2.2).) A similar calculation applies to the integral cost term in $J(u)$. Thus an equivalent form of the optimization problem is : choose $u \in \mathcal{U}$ to minimize

$$J(u) = E_0^u[\int_0^T \hat{c}(\sigma_t,x_t,u_t)dt + \hat{\Phi}(\sigma_T,x_T)] \tag{2.3}$$

where (σ_t,x_t) satisfy under measure P_0^u the equations

$$d\sigma_t(f) = \sigma_t(Af)dt + \sigma_t(hf)dy \tag{2.4}$$

$$dx_t = b(x_t,u_t)dt + dw_t^u \tag{2.5}$$

We shall not need the measures P^u again. Formulating the problem in terms of P_0^u is convenient since (y_t) is then just a Brownian motion and the "observation equation" (1.3) disappears. The solution (σ_t) of the Zakai equation can be written as an explicit functional of (y_t) via the "pathwise" formula [4]

$$\sigma_t(f)(y) = \int_{C[0,t;R^p]} f(z_t)\exp(y_t h(z_t) - \int_0^t y_s dh(z_s) - \tfrac{1}{2}\int_0^t h^2(z_s)ds)P_\pi(dz) \quad \ldots \tag{2.6}$$

The formulation (2.3) - (2.5) shows that the "state" of the problem is the pair (σ_t, x_t), which splits conveniently into an uncontrolled infinite-dimensional part and a controlled, "uniformly elliptic" finite-dimensional part. It would, incidentally, be a simple matter to add a positive definite diffusion matrix in (2.5); we take this as the identity for notational convenience.

III EXISTENCE BY MARTINGALE METHODS

In view of the fact that σ_t is an explicit functional (2.6) of $(y_s, 0\leq s\leq t)$, the control problem falls within the class considered in the author's previous Bad Honnef contribution [3], for which existence of an optimal control is assured. The argument is as follows : introduce the *value process*

$$V_t = P_0\text{-ess-inf}_{u\in\mathcal{U}} E_0^u[\int_t^T \hat{c}(\sigma_s, x_s, u_s)ds + \hat{\Phi}(\sigma_T, x_T)|G_t], \tag{3.1}$$

and for each $u \in \mathcal{U}$ define

$$M_t^u = \int_0^t \hat{c}(\sigma_s, x_s, u_s)ds + V_t - V_0 \tag{3.2}$$

Then (M_t^u) is a G_t-submartingale for any u, and is a martingale if and only if u is optimal. For $u\in\mathcal{U}$, form the decomposition

$$M_t^u = A_t^u + N_t^u \tag{3.2}$$

where (N_t^u) is a martingale and (A_t^u) a predictable increasing process. (N_t^u) has the stochastic integral representation

$$N_t^u = \int_0^s \psi_s\, dy_s + \int_0^t g_s dw_s^u$$

The integrands (ψ_s, g_s) do not depend on u. The relation between A_t^u and A_t^v for different controls $u, v \in \mathcal{U}$ is

$$A_t^u = A_t^v + \int_0^t (H_s^u - H_s^v)ds \tag{3.3}$$

where

$$H_s^u = g_s b(x_s, u_s) + \hat{c}(\sigma_s, x_s, u_s) \tag{3.4}$$

This gives us a necessary condition for optimality : if v is optimal then

$$g_s b(x_s, v_s) + \hat{c}(\sigma_s, x_s, v_s) = \min_{q\in U} g_s\, b(x_s, q) + \hat{c}(\sigma_s, x_s, q) \qquad dP_0\times dt \text{ a.e.}$$

... (3.4)

This follows from (3.3) since $A_t^v = 0$ if v is optimal, so that (3.5) is just the condition that A_t^u be an increasing process. It is shown in [3] that the control $u^*(s,\omega)$ obtained by minimizing H_s^u pointwise is both admissible and optimal. Thus we have the following result

Theorem 3.1: *There exists an optimal control* u^* *in the class* $\mathcal{U}$ *of* G_t*-adapted processes.*

This result is less than satisfactory, since clearly the optimal control ought to be of the form $u^*(s,\sigma_s,x_s)$ whereas we have only shown that it is $\mathcal{G}_t$-adapted. Referring to (3.4), the required form is obtained if and only if $g_s = \tilde{g}(s,\sigma_s,x_s)$ for some function $\tilde{g}$. We show this in the next section. An alternative approach, using a "differentiation" formula of Airault and Föllmer, is given by Lepeltier and and Marchal in [10]

IV MARKOVIAN CONTROLS

Let B be the set of finite positive measures on R^p, with the weak* topology: $\mu_n \to \mu$ iff $\mu_n(f) \to \mu(f)$ for all $f \in C_b(R^p)$ such that $f(z) \to 0$ as $|z| \to \infty$. Let $\mathcal{B}$ denote the Borel σ-field. Then (σ_t) is a B-valued process.

Definition 4.1: The set of *Markovian controls* $\mathcal{U}_M$ consists of all measurable functions $u = [0,T] \times B \times R^d \to U$.

For $u \in \mathcal{U}_M$, define $\tilde{u}_t = u(t,\sigma_t,x_t)$; then $(\tilde{u}_t)$ is $\mathcal{G}_t$-adapted. This shows that $\mathcal{U}_M \subset \mathcal{U}$.

Proposition 4.2: *The process* $\xi_t = (t, \sigma_t,x_t)$ *is a Feller Markov process under measure* P_0^u, *for* $u \in \mathcal{U}_M$.

Proof: It is shown by Fleming [7, Lemma 3.6] that (σ_t) is a Feller process under the stated conditions. The result follows from this together with standard results about finite-dimensional diffusion processes. □

The Feller property means of course that the semigroup

$$T_t\phi(s,\sigma,x) = E^u_{s,\sigma,x}\,\phi(\tau_t,\sigma_t,x_t)$$

maps C(S) into C(S). Here $S = [0,T]\times B \times R^p$ is the state space, (τ_t) is the "time" process $\tau_t = \tau_0+t$, and $E^u_{s,\sigma,x}$ denotes expectation for the process starting at (s,σ,x) and using control $u(\tau_t,\sigma_t,x_t)$ for $u \in \mathcal{U}_M$.

We now wish to study the extended generator $(L\ \mathcal{D}(L))$ of the process (ξ_t) corresponding to some fixed $u^0 \in \mathcal{U}_M$. We denote $W_t^0 := W_t^{u^0}$. Following Fleming [7], It is useful to introduce a certain subset $\mathcal{D}$ of C(S). $\mathcal{D}$ consists of those functions ϕ which take the form

$$\phi(t,\sigma,x) = F(t,x,\sigma(f_1), \ldots ,\sigma(f_\ell)) \tag{4.1}$$

for some positive integer ℓ and functions $F \in C_b^\infty(R^{1+d+\ell})$, $f_i \in C_0^\infty(R^p)$, $i = 1,2,\ldots,\ell$. Referring to (2.4), (2.5) we can expand $\phi(t,\sigma_t,x_t)$ for $u^0 \in \mathcal{U}_M$ using the Ito formula as follows:

$$\phi(t,\sigma_t,x_t) - \phi(s,\sigma_s,x_s) = \int_0^t L\phi(r,\sigma_r,x_r)dr + \sum_{i=1}^{d}\int_s^t \frac{\partial F}{\partial x_i}\,dW^{0,i} + \sum_{k=1}^{\ell}\int\int \frac{\partial F}{\partial \zeta_k}\,\sigma_r(hf_k)dy_r$$

$$\ldots\ldots \tag{4.2}$$

In this expression

$$\frac{\partial F}{\partial \zeta_k} = \frac{\partial}{\partial \zeta_k} F(r,x,\zeta_1,\ldots,\zeta_\ell)\Big|_{(r,x_r,\sigma_r(f_1),\ldots,\sigma_r(f_\ell))}$$

and

$$L\phi(t,\sigma,x) = \frac{\partial F}{\partial t} + \sum_{i=1}^{d} \frac{\partial F}{\partial x_i} b(x,u^0(t,\sigma,x)) + \sum_{k=1}^{\ell} \frac{\partial F}{\partial \zeta_k} \sigma(Af_k)$$

$$+ \tfrac{1}{2} \sum_{i=1}^{d} \frac{\partial^2 F}{\partial x_i^2} + \tfrac{1}{2} \sum_{i,j=1}^{\ell} \frac{\partial^2 F}{\partial \zeta_i \partial \zeta_j} \sigma(hf_i)\sigma(hf_j) \qquad (4.3)$$

In (4.2), $\partial F/\partial t$ etc are evaluated at $(t,x,\sigma_t(f_1),\ldots,\sigma_t(f_\ell))$

Since the last two terms in (4.1) are martingales, this shows that $\mathcal{D} \subset \mathcal{D}(L)$ and that L is given by expression (4.3) for $\phi \in \mathcal{D}$, where ϕ is defined by (4.1)

Lemma 4.3: *$\mathcal{D}$ is dense in $C_b(S)$*

Proof; This is shown by Fleming [7, Lemma 2.4] for functions as in (4.1) but without the (t,x) dependence. The extension to include these is immediate.

Returning to the control problem, we have the following fundamental result of El Karoui, Lepeltier and Marchal [9, Theorem 14]

Theorem 4.4: *There exists a universally measurable function $\rho : S \to R$ such that for all stopping times τ.*

$$V_\tau = \rho(\tau,\sigma_\tau,x_\tau)$$

where V_t is the value process given by (3.1).

(In [8], [9] control is studied for process which are "right processes" for each fixed $u \in U$. This class includes Feller processes, and the hypotheses of [8] , [9] are satisfied).

In view of Theorem (4.4), the process $M_t^{u^0}$ of (3.2) becomes

$$M_t^{u^0} = \int_0^t \tilde{c}(\xi_s)ds + \rho(\xi_t) - \rho(\xi_0)$$

where $\xi_t = (t,\sigma_t,x_t)$ and $\tilde{c}(t,\sigma,x) = \hat{c}(\sigma,x,u^0(t,\sigma,x))$. Thus $(M_t^{u^0})$ is an *additive functional* (a.f) of (ξ_t). We would like to know that the a.f. property is inherited by the two components of the decomposition (3.2). Such a result was proved by Çinlar *et al.* [2].

Lemma 4.5 [2, Theorem 3.18]. *There exist additive functionals A_t^0, N_t^0* such that $A^{u^0} = A_t^0$, $N^{u^0} = N_t^0$, a.s. .

We are interested in the structure of the martingale additive functional (N_t^0). The basic result is the following, due essentially to Motoo and Watanabe [11].

Theorem 4.6: *There exist real-valued functions $g^i : S \to R$, $i = 0, 1,\ldots,d$ such that, with $g = (g^1,\ldots,g^d)$,*

$$N_t^0 = \int_0^t g^0(\xi_s)dy_s + \int_0^t g(\xi_s)\, dW_s^0$$

Proof: We already know from §III that (N_t^0) has such a representation for adapted integrands ψ, g. This shows in particular that (N_t^0) is continuous and hence locally square integrable. Motoo and Watanabe [11, Theorem 12.2] show that the space M of square integrable martingale additive functionals is spanned by $\{X[\gamma]: \gamma \in \mathcal{D}(L)\}$ where

$$X[\gamma] := \gamma(\xi_t) - \gamma(\xi_0) - \int_0^t L\gamma(\xi_s)ds$$

Let I be the subset of M

$$I = \{\int_0^t \beta^0(\xi_s)dy_s + \int_0^t \beta(\xi_s)dW_s^{u^0} \; ; \; E\int_0^\infty \beta^2(\xi_s)ds < \infty\}$$

Then I is a closed subspace of M. From the expansion (4.2) we see, after localization, $X[\phi] \in I$ if $\phi \in \mathcal{D}$. But $\mathcal{D}$ is dense in $C_b(S)$. If follows that $X(\gamma) \in I$ for all $\gamma \in \mathcal{D}(L)$ and hence that $I = M$.

Corollary 4.7: *There exists a Markovian control* $u^* \in U_M$ *which is optimal in* U.

Proof: As pointed out in §III, an optimal control can be constructed by selecting the pointwise minimum of H_s^u in (3.4). Since, however, $g_s = g(\xi_s)$ an application of Beneš implicit function lemma [1, Lemma 1] gives the existence of a measurable function $u^*(\xi_s)$ which achieves this minimum.

REFERENCES

[1] V.E. Beneš, Existence of optimal strategies, based on specified information, for a class of stochastic decision problems, SIAM J. Control 8(1970), 179-188.

[2] E. Çinlar, J. Jacod, P. Protter and M.J. Sharpe, Semimartingales and Markov processes , Z. Wahrscheinlichkeitstheorie ver. Geb. 54(1980) 161-219.

[3] M.H.A. Davis, Martingale methods in stochastic control, in *Stochastic Control and Stochastic Differential Systems*, Lecture Notes in Mathematics 16 Springer-Verlag, Berlin, 1980.

[4] M.H.A. Davis, Pathwise nonlinear filtering, in *Stochastic Systems*, ed. M. Hazewinkel and J.C. Willems, Reidel, Dordrecht, 1981.

[5] M.H.A. Davis and S.I. Marcus, An introduction to nonlinear filtering, in *Stochastic Systems*, ed. M. Hazewinkel and J.C. Willems, Reidel, Dordrecht 1981.

[6] R.J. Elliott, *Stochastic Calculus and Applications*, Springer-Verlag, New York, 1982.

[7] W.H. Fleming, Nonlinear semigroup for controlled partially observed diffusions, SIAM J. Control and Optimization 20(1982) 286-301.

[8] N. El Karoui, Methodes probabilistes en contrôle stochastique, in *École d'Eté de St. Flour*, Lecture Notes in Mathematics, Springer-Verlag, Berlin-Heidelberg-New York, 1980.

[9] N. El Karoui, J.P. Lepeltier and B. Marchal, Arret optimal dependant d'un paramètre et contrôle continu markovian,preprint, 1980.

[10] J.P. Lepeltier and B. Marchal, Sur l'existence de politiques optimales dans le contrôle integro-differentiel, Ann. Inst. H. Poincaré 13B(1979) 45-97

[11] M. Motoo and S. Watanabe, On a class of additive functionals of Markov processes, J. Math. Kyoto University 4(1965) 429-470.

NISIO SEMI-GROUP ASSOCIATED TO THE CONTROL OF MARKOV PROCESSES

N. El Karoui
Ecole Normale Supérieure
5, rue Boucicaut
92260 Fontenay aux roses

J.P. Lepeltier
Université du Mans
Département de Mathématiques
Route de Laval
72017 Le Mans cedex

B.Marchal
Université Paris Val de Marne
U.E.R. de Sciences économiques et de Gestion
58,avenue Didier
94210 La Varenne Saint-Hilaire

I-INTRODUCTION

We study the nonlinear semigroup introduced by M. Nisio [4] relatively to the control of Markov process by step-processes.

This semi-group is obtained as the value function of an optimization problem stopped at time t. The step controls along stopping times define impulse policies which allow to use impulse control technics to study this value function. We construct the generator of this semi-group.

The main interest of Nisio's semi-group is that in most cases where it is studied we prove that it is associated to a wider class of policies.

II-THE OPTIMIZATION PROBLEM

All processes are defined on the same probability space $(\Omega,\underline{\underline{F}},P)$ with a filtration $(\underline{\underline{F}}_t)$ satisfying the "usual conditions". A step control is a process valued in a compact space U :

(2-1) $$u = \Sigma_{i\geq 0}\, u_i\, 1\!\!1_{]\!]T^u_i,T^u_{i+1}]\!]}\,,$$

where (T^u_n) is an increasing sequence of $\underline{\underline{F}}_t$-stopping times such that $\sup_n T_n = +\infty$, $T_0 \equiv 0$, and the random functions u_n are $\underline{\underline{F}}_{T_n}$-mesurable, $n\geq 0$. Let $\mathfrak{D}_e$ be the set of these controls.

The set of controlled processes $\mathfrak{X}^u = (\Omega, \underline{\underline{F}}_t, X_t, \theta_t, P^u_x, x \in E)$ (Eis a Lusinian space) has the following properties :

(H_1) For any constant control $u \equiv a \in U$, $\mathfrak{X}^a$ is a right Markov process, i.e. particularly $P^a_x(X_0 = x) = 1$, and P^a_x-excessive functions have right continuous trajectories;

(H_2) $(a,x) \rightsquigarrow P^a_x$ is a transition function from UxE to Ω;

(H_3) For any initial law μ, any $u \in \mathfrak{D}_e$ and $n \geq 1$:

$$(2\text{-}2) \qquad P^u_\mu(\theta^{-1}_{T^u_n}(A)/_{\underline{\underline{F}}_{T^u_n}}) = P^{u(.)}_{X_{T^u_n}}(A) \quad \text{on} \quad (T^u_n < +\infty),$$

if $\theta^{-1}_{T^u_n}(A)$ belongs to $\underline{\underline{F}}_{T^u_{n+1}}$.

REMARKS

1- A right process is a Feller process with branching points for a good topology. This generalization is useful to handle many situations : Diffusions on $\mathbb{R}^n$, diffusions with jumps, diffusions reflected on a convex set of $\mathbb{R}^n$, diffusions killed and then time changed, ect.

2- Assumption (H_2) is to get a good measurability property in $a \in U$.

3- With assumption (H_3), between two jumping times of $u \in \mathfrak{D}_e$, the evolution of $\mathfrak{X}^u$ is Markovian and depends only of u in this interval.

4-If needed, we can define a wider family P^u_x of probabilities; for example the set of predictable processes.

5- In particular, we often control by absolutely continuous laws, and this assumptions are traduced by compatibility assumptions on the densities (see [1] , [2]).

We control the process until a terminal time D, Markov in the following sense :

$$(2\text{-}3) \qquad D \circ \theta_t = D - t \qquad \text{a.s.} \qquad \text{on} \quad (D < t)$$

(for example the exit time of a set). The objective function of our problem is :

$$(2\text{-}4) \qquad J^u_x(D) = E^u_x(C^u_D + e^{-H^u_D} g(X_D)),$$

where

$$(2\text{-}5) \qquad C^u_t = \int_0^t e^{-H^u_s} c(X_s, u_s)\, dK_s$$

and the actualization rate

(2-6) $$H_t^u = \int_0^t \alpha(X_s,u_s)\, dA_s .$$

The processes A and K are additive predictable functionnals:

(2-7) $$A(t+s) - A(s) = A_s \circ \theta_t \qquad \text{a.s. .}$$

We suppose also "good" measurability and boundness assumptions for α, c, g. In particular, the measurability property in x is intermediate between being borelian and universally measurable; this intermediate measurability, natural in right processes theory, is the $\mathcal{B}_e$-measurability ($\mathcal{B}_e$ is the σ-algebra generated by excessive functions).

In the next part, we construct the value function :

(2-8) $$W_D g(x) = \sup \left\{ J_x^u(D) ;\ u \in \mathcal{D}_e \right\}$$

by technics of impulse control; we also prove that the dynamic programmation principle is satisfied by W_D.

III CONSTRUCTION OF VALUE FUNCTION

In this part D is fixed (so we forget about it). Given some u in $\mathcal{D}_e$ the evolution of $\mathcal{X}^u$ is Markovian and its law is $P^{u_i(\cdot)}_{X_{T_i}(\cdot)}$ between T_i^u and T_{i+1}^u. Any control u can be considered as an impulse policy at times $(T_n^u)_{n \geqslant 1}$. We use impulse control technics ([3], [6]) without impulse cost. So we introduce controls with a fixed number of impulse times, let :

(3-1) $$\mathcal{D}_n = \left\{ u \in \mathcal{D}_e ;\ T_{n+1}^u = +\infty \right\}$$

and the value function relative to $\mathcal{D}_n$:

(3-2) $$W_n g(x) = \sup \left\{ J_x^u ;\ u \in \mathcal{D}_n \right\} .$$

For the study between two impulsions, we introduce the operator :

(3-3) $$Rh(x) = \sup \left\{ J_x^a(T,h) = E_x^a \left(\int_\theta^{T\wedge D} e^{-H_s^a} c(X_s,a)\, ds + e^{-H^a_{T\wedge D}} h(X_{T\wedge D}) \right) ;\ a \in U \text{ and } T\ \underline{\underline{F}}_t\text{-stopping time} \right\}$$

for any analytic function h.

We set :

$$W_n g = R^n \hat{g},$$

with :

(3-4) $$\hat{g}(x) = \sup \left\{ J_x^a(D) ;\ a \in U \right\} ,$$

and we get Wg as the increasing limit of $W_n g$, $n\to\infty$. This impulse control technic is the same as in [2]. We use it here in a more general setting.

The next lemma is essential for the result.

LEMMA 1

Let f be analytic on ExU;

i/ $\sup\{f(.,a);\ a\in U\}$ is analytic;

ii/ for any probability law μ on $(E,\underline{\underline{E}})$:

(3-5) $$\mu(\sup_a f(.,a)) = \sup\left\{\int_E \mu(dx)\ f(x,u(x));\ u \text{ Borel function}\right\}$$

if $(\Omega,\underline{\underline{G}},Q)$ is a complete probability space and Z is valued in E, $\underline{\underline{G}}$-measurable :

$$\sup_a f(Z,a) = Q\text{-ess}\sup\{f(Z,v);\ v\ \underline{\underline{G}}\text{-measurable valued in } U\}.$$

THEOREM 2

For any $n\geq 0$, $W_n g \equiv R^n\hat{g}$ and this sequence is increasing to Wg :

(3-7) $$W_n g(x) = R^n\hat{g}(x) \underset{n\to\infty}{\nearrow} Wg(x),$$

if

(3-8) $$\sup_a U_K^a 1(x) = \sup_a E_x^a\left(\int_0^D e^{-H_s^a}\,dK_s\right) \text{ is bounded.}$$

PROOF :

The identification of $W_n g$ and $R^n g$ is the same as in [1], [2]. The main problem is in the transposition of supremum and integral symbol in the operator R, and is solved by the Markov property and lemma 1.

Denote the limit by Wg(x); for this let $u\in\mathcal{D}_e$:

$$u = \Sigma_{0\leq i}\, u_i\, 1\!\!1_{]\!]T_i^u,T_{i+1}^u]\!]}$$

and for any $n\geq 0$:

(3-9) $$\hat{u}_n = \Sigma_{0\leq i\leq n-1}\, u_i\, 1\!\!1_{]\!]T_i^u,T_{i+1}^u]\!]} + u_n\, 1\!\!1_{]\!]T_n,+\infty]\!]}.$$

By the decomposition of J_x^u and the compatibility property (H_3) we get :

$$J_x^u = J_x^{\hat{u}_n} + E_x^u\left(\int_{T_n^u}^D e^{-H_s^u}\,c(X_s,u_s)\,dK_s\right) - E_x^{\hat{u}_n}\left(\int_{T_n^u}^D e^{-H_s^{\hat{u}_n}}\,c(X_s,u_s^n)\,dK_s\right)$$

$$+ E_x^u\left(e^{-H_D^u}\,g(X_D)\,1\!\!1_{(D>T_n^u)}\right) - E_x^{\hat{u}_n}\left(e^{-H_D^{\hat{u}_n}}\,g(X_D)\,1\!\!1_{(D>T_n^u)}\right)$$

$$= J_x^{\hat{u}_n} + A_1 - A_2 + A_3 - A_4.$$

Using boundness of c, g, compatibility (H_3), Markov's property and (3-8), we obtain $|A_i| \underset{n\to\infty}{\longrightarrow} 0$ $i=1,\dots,4$. Indeed :

1/ $|A_1|\leq K\, E_x^u\left(\int_{T_n^u}^D e^{-H_s^u}\,dK_s\right)$.

2/ $|A_2| \leqslant K\ E_x^{\hat{u}_n}\left(\int_{T_n^u}^{D} e^{-H_s^{\hat{u}_n}}\, dK_s\right) = K\ E_x^u\left(e^{-H^u_{T^u_n}} \int_{T_n^u}^{D} e^{-(H_s^{u_n} - H^{u_n}_{T_n^u})}\, dK_s\right)$

$\leqslant K\ E_x^u\left(e^{-H^u_{T^u_n}}\ U_K^{\hat{u}_n} 1(X_{T_n^u})\right) = K\ E_x^u\left(e^{-H^u_{T^u_n}} \sup_a U_K^a 1(X_{T_n^u})\right)$

$\leqslant K'\ E_x^u\left(e^{-H^u_{T^u_n}}\right);$

3/ $|A_3| \leqslant K\ P_x^u\ (D > T_n^u);$

4/ $|A_4| \leqslant K\ P_x^{\hat{u}_n}(D > T_n^u) = K\ P_x^u(D > T_n^u).$

So for $n \geqslant 0$, $u \in \mathcal{D}_e$, there is $\Sigma_n^u \xrightarrow[n\to\infty]{} 0$ such that :

$$J_x^u \leqslant J_x^{\hat{u}_n} + \Sigma_n^u \leqslant W_n g(x) + \Sigma_n^u,$$

and letting $n \to \infty$:

$$J_x^u \leqslant \lim_n W_n g(x),$$

$$\sup\left\{J_x^u;\ u \in \mathcal{D}_e\right\} = Wg(x) \leqslant \lim_n W_n g(x).$$

The definition of $W_n g$ implies the converse inequality and the final result. ■

To complete this part, we give the dynamic programming principle ([2]), which is the most important to have the semi-group property.

PROPOSITION 3

Under (3-8) for any $\underline{\underline{F}}_t$-stopping time $T \leqslant D$:

(3-10) $$W_D g(x) = \sup\left\{ E_x^u\left(\int_0^T e^{-H_s^u} c(X_s, u_s)\, dK_s + e^{-H_T^u} W_D g(X_T)\right);\ u \in \mathcal{D}_e\right\}.$$

PROOF :

Equality (3-10) is easy by setting for the T-conditional value function [3]. With lemma 1 and the proof of theorem 2 we establish that this function is precisely $W_D g(X_T)$ P_μ^u a.s. . ■

IV-RECOVERY OF NISIO'S SEMI-GROUP

For any $t \geqslant 0$, any function g we define Nisio's semi-group by :

(4-1) $$Q_t g(x) = \sup\left\{ E_x^u\left(\int_0^t e^{-H_s^u} c(X_s, u_s)\, dK_s + e^{-H_t^u} g(X_t)\right);\ u \in \mathcal{D}_e\right\}$$

first introduced by M. Nisio [4].

To get the semi-group property we use the previous theory. Since t is not a "Markovian time", we have to change our problem. We solve this problem by introducing a space-time Markov process.

THEOREM 4

Q_t is a semi-group, i.e. for any t, r, $0<r<t$:

$$Q_t g(x) = Q_r(Q_{t-r}g)(x).$$

PROOF :

Let $\hat{\mathfrak{X}}^a$ be the space-time Markov process associated to $\mathfrak{X}^a$, i.e. :

$$\hat{\mathfrak{X}}^a = (\Omega \times \mathbb{R}^+, \underline{\underline{F}}_t \otimes B_{\mathbb{R}^+}, \theta_t, X_t, P_x \otimes \varepsilon_x)$$

with $\qquad \hat{X}_t(\omega,r) = (X_t(\omega), r+t), \quad \hat{\theta}_t(\omega,r) = (\theta_t(\omega), r+t).$

The semi-group associated to $\hat{\mathfrak{X}}^a$ is $\hat{P}^a_t f(x,s) = E^a_x(f(X_t, s+t))$ and $\hat{t}(\omega,r) = (t-r)^+$ is a Markov terminal time.

The value function for step controls defined on $\Omega \times \mathbb{R}^+$, $\hat{\mathfrak{D}}_e$, is defined by (2-4), (2-8) :

$$\text{(4-3)} \qquad \hat{Q}_t g(x,s) = \sup\left\{E^u_x\left(\int_0^{t-s} e^{-H^u_r} c(X_r,u_r)\, dK_r + e^{-H^u_{t-s}} g(X_{t-s},t)\right);\ u \in \hat{\mathfrak{D}}_e(s)\right\}$$

with $\hat{\mathfrak{D}}_e(s) = \{u = \hat{u}(.,s);\ \hat{u} \in \hat{\mathfrak{D}}_e\} \equiv \mathfrak{D}_e$.

Let $s \leq r \leq t$; by the dynamic programming principle (proposition 4) we have (3-10) :

$$\text{(4-4)} \qquad \hat{Q}_t g(x,s) = \sup\left\{E^u_x\left(\int_0^{r-s} e^{-H^u_v} c(X_v,u_v)\, dK_v + e^{-H^u_{r-s}} Q_t g(X_{r-s},r)\right);\ u \in \hat{\mathfrak{D}}_e(s)\right\}.$$

Finally, if g does not depend of t, we notice that $Q_{t-s}g(x) = \hat{Q}_t(x,s)$; by (4-4) we deduce :

$$Q_{t-s}g(x) = \sup\left\{E^u_x\left(\int_0^{r-s} e^{-H^u_v} c(X_v,u_v)\, dK_v + e^{-H^u_{r-s}} Q_{t-r}g(X_{r-s})\right);\ u \in \mathfrak{D}_e\right\}$$
$$= Q_{r-s}(Q_{t-r}g)(x)$$

which is the semi-group property (4-2) when s = 0. ∎

We now study the generator of this semi-group when $K_t = t$ and $H^u_t = \alpha t$, $\alpha > 0$.

Let A^a be the generator of $\mathfrak{X}^a$ and for all g in $\bigcap_a \mathfrak{D}(A^a)$:

$$F^* g(x) = \sup_a \left(c(x,a) + A^a g(x) - \alpha g(x)\right).$$

The main result is obtained by in next proposition which gives Q_t in terms of A^a, a U, and by continuity in t of Nisio's semi-group.

PROPOSITION 5

If $A^a g(x)$ is $\underline{\underline{U}}\otimes\underline{\underline{E}}$-measurable, for any $g \in \bigcap_a \mathcal{D}(A^a)$:

(4-5) $$Q_t g(x) = g(x) + \sup\left\{E_x^u\left(\int_0^t e^{-\alpha s}(c(X_s,u_s) + A^{u_s} g(X_s) - \alpha g(X_s))\, ds\right);\ u \in \mathcal{D}_e\right\}.$$

PROOF :

By definition of A^a, we have the following equality :

$$Rg(x) = g(x) + \sup\left\{E_x^a\left(\int_0^t e^{-\alpha s}(c(X_s,a) + A^a g(X_s) - \alpha g(X_s))\, ds\right);\ a \in U\right\}.$$

The method used in theorem 2, using sophisticated results about optimal stopping, yields for any $n \geq 0$:

$$W^n g(x) = g(x) + \sup\left\{E_x^u\left(\int_0^t e^{-\alpha s}(c(X_s,u_s) + A^{u_s} g(X_s) - \alpha g(X_s))\, ds\right);\ u \in \mathcal{D}_n\right\}$$

and the result follows by letting $n \to \infty$. ■

COROLLARY 6

If $C^2 \subset \bigcap_a \mathcal{D}(A^a)$, for any continuous function g :

(4-6) $$\lim_{t \searrow s} \|Q_t g - Q_s g\| = 0$$

PROOF :

By (4-5), this property is true for any $g \in C^2$. Since Q_t is contracting, the corollary follows by a density argument. ■

THEOREM 7

Under the assumptions of proposition 5 and corollary 6, and, if $c(x,a)$ is continuous in x uniformly in a, for any function g in $\bigcap_a \mathcal{D}(A^a)$ such that $\sup_a(A^a g - \alpha g)$ is continuous, the generator A of Q_t is :

(4-7) $$A\, g(x) = \sup_a (c(x,a) + A^a g(x) - \alpha g(x)).$$

PROOF :

By the inequality (4-5) we have that :

(4-8) $$Q_t g(x) - g(x) = \sup\left\{E_x^u\left(\int_0^t ds\ e^{-\alpha s}(X_s,u_s) + A^{u_s(\cdot)} g(X_s) - \alpha g(X_s)\right);\ u \in \mathcal{D}_e\right\}.$$

Hence :

$$\frac{Q_t g(x) - g(x)}{t} \geq \frac{1}{t} E_x^a\left(\int_0^t ds\ e^{-\alpha s}(c(X_s,u_s) + A^{u_s(\cdot)} g(x) - \alpha g(x)\right)\ \forall a \in U$$

and :

(4-9) $$\liminf_{t \to 0} \frac{Q_t g(x) - g(x)}{t} \geq \sup_a\ (c(x,a) + A^a g(x) - \alpha g(x)) = F^* g(x).$$

Conversely by (4-8) we have :

$$Q_t g(x) - g(x) \leqslant \sup \left\{ E_x^u \left(\int_0^t e^{-\alpha s} F^* g(X_s)\, ds \right);\ u \in \mathcal{D}_e \right\}.$$

On the right we have the semi-group associated to $c(x,a) \equiv 0$ and $g \equiv F^* g$; so corollary 6 implies that :

$$\text{(4-10)} \qquad \limsup_{t \to 0} \frac{Q_t(x) - g(x)}{t} \leqslant F^* g(x).$$

By (4-9), (4-10) we get (4-7) immediatly. ■

BIBLIOGRAPHY

[1] N. EL KAROUI Les aspects probabilistes du contrôle stochastique. Ecole d'été de probabilités de Saint-Flour-1979. Lect. Notes in Math. n° 876.Springer

[2] N. EL KAROUI, J.P.LEPELTIER, B.MARCHAL Optimal of controlled Markov processes. To appear in proceedings of the Conference of IFIP- Mexico-1982

[3] J.P.LEPELTIER, B.MARCHAL Théorie générale du contrôle impulsionnel.Submitted to publication.

[4] M. NISIO On Stochastic Optimal Controls and Envelope of Markovian Semi-groups. Proc. of Intern. Symp. SDE. Kyoto 1976-pp 297-325.

[5] M.NISIO On non linear semi-groups for Markov processes associated with optimal stopping. Applied Math. and Opt.- 4-pp143-169-1978.

[6] M.ROBIN Contrôle impulsionnel des processus de Markov. Thèse de doctorat-Paris IX-1978.

OPTIMAL CONTROL OF PARTIALLY OBSERVED DIFFUSIONS VIA THE SEPARATION PRINCIPLE *

U. G. Haussmann

Department of Mathematics
University of British Columbia
Vancouver, B.C., V6T 1Y4, Canada

1. Introduction

The question of existence of optimal controls for diffusion processes with partial observation has long been unresolved. Recently the case when the system is nonlinear in the state has received some attention, [1], [2], [3], but only the existence of optimal randomized controls was established. On the other hand for the case of linear system and quadratic cost (the linear regulator) an explicit solution was found long ago using the "separation principle", c.f. [4], [5]. The intermediate case of linear system and non-quadratic cost has tantalized researchers for some time [4], [6], [7], since much of the theory of the quadratic case holds true: the separation principle "reduces" the problem to a completely observable one for which we have existence theorems. The difficulty is that these optimal controls may not make sense for the original problem. A recent result of Veretennikov [8] allows us to resolve this point.

The system is described by

(1.1) $$dx = [A(t)x + b(t,u(t))]dt + \sigma(t)dw, \quad x(0) = x_o ,$$

(1.2) $$dy = H(t)x\, dt + \bar{\sigma}(t)d\bar{w} , \quad y(0) = 0 ,$$

and the cost functional is

(1.3) $$J[u] = E\left\{ \int_0^T L(t, x(t), u(t))dt + c(x(t))\right\} .$$

Here $x(\cdot)$, $y(\cdot)$, $w(\cdot)$, $\bar{w}(\cdot)$ are $\mathbb{R}_n$-valued processes, $(w,\bar{w})$ being Brownian motion. T is a fixed finite time $0 < T$. More detailed assumptions are made in the next section.

If U is the set of control points then the admissible controls $\mathcal{U}$ consist of all measurable, separable, U valued stochastic processes which are y-adapted, such that (1.1) (1.2) have a strong solution. Now the problem to solve is

$$\mathcal{P} : \inf \left\{J[u]: u \in \mathcal{U}, \ (x,y) \text{ satisfy } (1.1), (1.2)\right\} .$$

* This work was supported by NSERC under grant A8051.

We shall write inf (P) for this infimum and we shall be concerned with finding a control $u \in \mathcal{U}$ such that $J[u] = \inf (P)$.

The definition of $\mathcal{U}$ is somewhat cyclical: u must be y-adapted but y depends on u through x. However (under our assumptions, c.f. section 2) for any $(w,\bar{w})$ adapted process u, (1.1) (1.2) has a unique solution (Itô's conditions hold). Now $\mathcal{U}$ is that subset of processes which also happen to be y adapted. Hence $\mathcal{U}$ is well defined although in an implicit manner. We add that the result of [8] implies that $\mathcal{U}$ contains processes of the form $u(t,\omega) = v(t,y(t,\omega))$ with $v: [o,T] \times \mathbb{R}_n \to U$ Borel measurable.

In section 2 we derive the separated problem P_s along well known lines and show that $\inf (P) \geq \inf (P_s)$. In section 3 we then show that an optimal control exists for P, and we give two examples.

2. The Separated Problem

We will now show the problem P induces a completely observable problem P_s such that $\inf(P) \geq \inf(P_s)$. To begin with let us collect our hypotheses:

(H)(i) (Ω, F, P) is a complete probability space with right continuous filtration $\{F_t\}$ such that F_o contains the null sets of F, carrying two independent n-dimensional Brownian motions w, $\bar{w}$.

(ii) $A, H, \sigma, \bar{\sigma}$ are $n \times n$ matrix valued functions on $[o,T]$, A is bounded, Borel measurable, and $H, \sigma, \bar{\sigma}$ are continuous, invertible.

(iii) U is a separable metric space, and $b: [o,T] \times U \to \mathbb{R}_n$ is Borel measurable, bounded, and continuous in u for each t.

(iv) x_o is a non-degenerate gaussian random variable $N(m_o, P_o)$ which is F_o measurable. (The notation means that m_o is the mean and P_o is the covariance.)

(v) L, c are scalar valued, Borel measurable, continuous in (x,u), continuous in x uniformly in u for each t, and for some $K, p < \infty$

$$|L(t, x, u)| + |c(x)| \leq K (1 + |x|^p).$$

We will write F_t^y for the σ-algebra generated by $\left\{\{\omega : y(s,\omega) \in B\} : s \leq t \text{ , B Borel}\right\}$, M^* for the transpose of a matrix M.

Lemma 2.1. <u>Assume</u> (H). <u>For any</u> $u \in \mathcal{U}$ <u>the conditional distribution</u> $P\left\{x(t) \middle| F_t^y\right\}$ is gaussian $N(m(t), P(t))$, where $m(t) = E\left\{x(t) \mid F_t^y\right\}$ satisfies

(2.1) $$dm = [A(t)m + b(t,u(t))]dt + F(t)\,\bar{\sigma}(t)\,d\nu, \quad m(o) = m_o ,$$

and $P(t) = E\left\{(x(t) - m(t))\,(x(t) - m(t))^* \mid F_t^y\right\}$ satisfies

(2.2) $$\dot{P} = A(t)P + P\,A(t)^* - F(t)H(t)P + \sigma(t)\sigma(t)^*, \quad P(o) = P_o ,$$

$$F(t) = P(t)\,H(t)^*\,[\bar{\sigma}(t)\bar{\sigma}(t)^*]^{-1}, \quad P(t) > 0 ,$$

and where ν is a Brownian motion on the filtration $\left\{F_t^y\right\}$, and

(2.3) $$\bar{\sigma}(t)\,d\nu = dy - H(t)\,m(t)\,dt.$$

Proof: The result follows from [9], lemma 4.9 and theorem 12.7. Note that §12.1 holds even if (in that notation) $a_o(t,\cdot)$ is $\mathcal{B}_{t+}$ measurable rather than $\mathcal{B}_t$ measurable.

Now let $g(t,x)$ denote the density $N(0,P(t))$,

$$g(t,x) = (2\pi)^{-n/2}[\det P(t)]^{-\frac{1}{2}}\exp[-x^*P(t)^{-1}x(t)/2] ,$$

and write

$$\bar{L}(t,m,u) = \int_{\mathbb{R}_n} L(t,x,u)\,g(t,x-m)\,dx ,$$

$$\bar{c}(m) = \int_{\mathbb{R}_n} c(x)\,g(T,x-m)\,dx .$$

Lemma 2.2. Assume (H). For any $u \in \mathcal{U}$

$$J[u] = \bar{J}[u] \equiv E\left\{\int_o^T \bar{L}(t,m(t),u(t))dt + \bar{c}(m(T))\right\} .$$

Proof: Since $u(t)$ is F_t^y measurable, then by lemma 2.1

$$E\left\{L(t, x(t), u(t)) \mid F_t^y\right\} = \bar{L}\,(t, m(t), u(t)) ,$$

and similarly for c, so the result follows.

The Brownian motion ν of lemma 2.1 is called the innovation. We show that the same ν arises for all $u \in \mathcal{U}$. Let x^o, y^o be the solution of

$$dx^o = A(t)x^o dt + \sigma(t)dw, \quad x^o(o) = x_0 ,$$

$$dy^o = H(t)x^o dt + \bar{\sigma}(t)d\bar{w} , \quad y^o(o) = 0 ,$$

and let $m^o(t) = E\left\{x^o(t) \middle| F_t^{y^o}\right\}$. If we set

$$dv^o = \bar{\sigma}(t)^{-1} \left\{dy^o - H(t)m^o(t)dt\right\} , \tag{2.4}$$

then from filtering theory it follows that m^o is the unique strong solution of

$$dm^o = A(t)m^o \, dt + F(t) \, \bar{\sigma}(t) \, dv^o , \quad m^o(o) = m_o \tag{2.5}$$

and hence that

$$F_t^{\nu^o} = F_t^{m^o} = F_t^{y^o} .$$

Lemma 2.3. <u>For any</u> $u \in U$, $\nu = \nu^o$.

Proof: If we set $x^u = x - x^o$, $y^u = y - y^o$, $m^u = m - m^o$, then (2.1) - (2.5) give

$$dm^u = [(A - FH)m^u + b]dt + F \, dy^u .$$

Since $dx^u = (Ax^u + b)dt$, $dy^u = H \, x^u dt$, then

$$d(x^u - m^u) = (A - FH) \, (x^u - m^u)dt , \quad x^u(o) - m^u(o) = 0 ,$$

so that $x^u(t) \equiv m^u(t)$ and $x(t) - m(t) = x^o(t) - m^o(t)$.

To complete the proof, observe that (2.3), (2.4) give

$$\begin{aligned} \bar{\sigma} \, d\nu &= H(x-m)dt + \bar{\sigma} \, d\bar{w} \\ &= H(x^o - m^o)dt + \bar{\sigma} \, d\bar{w} \\ &= \bar{\sigma} \, d\nu^o . \end{aligned}$$

We can now imbed the problem P in a new problem P'.
Define

$$F_t' = \bigvee_{u \in U} F_t^y ,$$

and let U' be the class of all U valued, F'_t progressively measurable stochastic processes. Note that lemmas 2.1 and 2.3 establish that ν is a Brownian motion on $\{F'_t\}$. We define the problem

$$P' : \inf \left\{ \bar{J}[u] : u \in U', \text{ m as in } (2.1) \right\} .$$

The problem P' is still not agreeable enough because of the definition of F'_t, wherein the observations are hidden. Instead let us define U_s, the class of separated controls, to be the Borel measurable, $\{G_t\}$ progressively measurable functions u: $[0,T] \times C \to U$, such that (2.1) has a strong solution. Here C is the space of continuous functions: $[0,T] \to \mathbb{R}$, $G = G_T$ is the σ-algebra of Borel sets (under the sup norm) and G_t is the sub-σ-algebra generated by

$$\left\{ \{u \in C : u(s) \in B\} : s \le t, \text{ B Borel} \right\} .$$

Now define the separated problem

$$P_s : \inf \left\{ \bar{J}[u] : u \in U_s \right\} .$$

Theorem 2.1. Assume (H). Then

$$\inf (P) \ge \inf (P') = \inf (P_s) .$$

Proof: The inequality follows from lemma 2.2 and the definition of U'. The equality follows from [10], chapter 3, theorem 1.7. [observe that U_s is not quite $A_{(E)}(0,m_o)$ of [10], but the result still holds].

The direction of attack now is to show that there are optimal controls for P_s which can be interpreted to lie in U, hence are optimal for P by theorem 2.1.

3. The Main Result

Let U_M be the set of elements of U_s which have the form $u(t, m) = u(t, m(t))$. For such u (1.1), (1.2), (2.1) become

(3.1) $$dx = [A(t)x(t) + b(t,u(t,m(t)))]\,dt + \sigma(t)\,dw \;,\; x(o)=x,$$

(3.2) $$dy = H(t)x(t)dt + \bar{\sigma}(t)d\bar{w}(t) \;,\; y(o) = 0 ,$$

(3.3) $$dm = [A(t)m(t) + b(t, u(t,m(t)))]dt + F(t)\bar{\sigma}(t)d\nu, \; m(o) = m_o .$$

Recall that we can think of ν as defined by x^o, y^o. Let us write $\bar{F}_t$ for $F_t^{x_o,w,\bar{w}}$

completed with the null sets of F.

Theorem 3.1. <u>Assume</u> (H). <u>For</u> $u \in \mathcal{U}_M$ (3.1) - (3.3) <u>have a unique</u> $\{\bar{F}_t\}$ <u>adapted solution such that</u>

(3.4) $$m(t) = E\{x(t) \mid F_t^y\} \quad \text{w.p.1} ,$$

(3.5) $$dy = H(t)\, m(t)dt + \bar{\sigma}(t)d\nu .$$

Proof: Let (x,m) be the unique $\{\bar{F}_t\}$ adapted solution of

(3.6) $$dx = [A(t)x + b(t, u(t,m))]dt + \sigma(t)dw, \; x(o) = x_o ,$$

(3.7) $$dm = [F(t)H(t)x + (A(t) - F(t)H(t))m + b(t,u(t,m))]dt$$
$$+ F(t)\bar{\sigma}(t)d\bar{w}, \; m(o) = m_o .$$

This solution exists by [8], theorem 1 (the fact that the drift is linear rather than bounded does not matter). Now define

$$dy = H(t)\, x(t)dt + \bar{\sigma}(t)d\bar{w} , \; y(0) = 0 ,$$

so that

(3.8) $$dm = [A(t)m + b(t,u(t,m))]dt + F(t)[dy - H(t)m(t)dt] .$$

As in the proof of lemma 2.3 but with (3.8) replacing (2.1), (2.3), it follows that $x-m = x^o-m^o$ and

$$dy - Hmdt = H(x-m)dt + \bar{\sigma}\, d\bar{w}$$
$$= H(x^o-m^o)dt + \bar{\sigma}\, d\bar{w}$$
$$= \bar{\sigma}\, d\nu .$$

Hence by [8] again, m is the unique strong solution of (3.3), i.e. the theorem is established with the exception of (3.4). Note, since m_o is fixed, "strong" means m is $\{\bar{F}_t^\nu\}$ adapted, where $\bar{F}_t^\nu$ is F_t^ν completed with the null sets of F.

By (3.3) we have $F_t^\nu \subset F_t^m \subset \bar{F}_t^\nu$, and by (3.5) $F_t^y \subset F_t^m$.

Conversely, observe that if we use the Girsanov transformation, then $z(t) \equiv \int_0^t \bar{\sigma}^{-1}\, dy$

is a Brownian motion under a new measure, and by (3.8)

$$dm = [(A-FH)m + b]dt + F\,\bar{\sigma}dz \quad .$$

Again from [8] it follows that the unique solution m is $\bar{F}_t^z$ (= $\bar{F}_t^y$) adapted; hence $F_t^m \subset \bar{F}_t^y$. It follows that $\bar{F}_t^y = \bar{F}_t^m = \bar{F}_t^\nu$.

If we apply [9], theorem 12.7 to (3.6), (3.7) then we conclude that $\bar{m}(t) \equiv E\left\{x(t) \mid F_t^m\right\}$ satisfies

$$d\bar{m} = [A\bar{m} + b]dt + \gamma(FH)^*(F\bar{\sigma}\bar{\sigma}^*F^*)^{-1}[dm -$$

$$((A-FH)\ m + b + FH\bar{m})dt] \ , \ \bar{m}(0) = m_o \ ,$$

$$\dot{\gamma} = A\gamma + \gamma A^* + \sigma\sigma^* - \gamma(FH)^*(F\bar{\sigma}\bar{\sigma}^*F^*)^{-1}FH\gamma, \ \gamma(0) = P_o \ .$$

It follows that $\gamma \equiv P$ (substitute for F in the above equation) and

$$d(\bar{m}-m) = (A-FH)(\bar{m}-m)dt, \ \bar{m}(0)-m(0) = 0 \ ,$$

so that $m(t) = \bar{m}(t) = E\left\{x(t) \mid F_t^y\right\}$ w.p.1.

Corollary 1. Assume (H). If $\hat{u} \in U_M$ achieves the inf for P_s, then $\hat{u}(t, \hat{m}(t,\omega))$ achieves the inf for P, where $\hat{m}$ is the solution of (2.1) generated by $\hat{u}$.

Proof: Theorem (3.1) shows that $U_M \hookrightarrow U$ and lemma 2.2 and theorem 2.1 yield

$$\inf\ (P) \geq \inf\ (P_s) = \bar{J}[\hat{u}] = J[\hat{u}] \geq \inf\ (P) \ .$$

Since equality must hold, the corollary is established.

Corollary 2. Assume (H), U compact, and L, c bounded. Then there exists $\hat{u} \in U$ such that $\hat{u}$ is optimal for P.

Proof: It follows from [11], theorem IV-2, that $\hat{u}$ as in corollary 1 exists. Note that a standard transformation involving the fundamental matrix of A eliminates the unbounded term Am in (2.1).

Remark 1. The boundedness of L, c in the above corollary can be replaced by some smoothness hypotheses, c.f. [12], corollaries 3.8, 3.9.

Remark 2. The requirement that H(t) be non-singular is severe. It requires us to observe all of x, although the observations are noise corrupted. If the dimension of y is less than n, then we expect randomized controls. In [13] Davis gives one approach to this problem.

Remark 3. The linear regulator does not quite satisfy (H). However the invertibility of H and σ can be dispensed with because $\hat{u}(t,m)$ is linear in m so existence of solutions of (3.6), (3.7) follows. Moreover the boundedness in u of b and L is only required to show

$$\inf (P') = \inf (P_s) , \tag{3.9}$$

but in fact for the regulator we take as admissible controls $u \in \mathcal{U}$ such that $|u(t,\omega)| \leq K\,(1 + \sup_{s \leq t} |y(t,\omega|)$. Direct computation and [4], chapter VI, corollary 4.2 show that (3.9) holds.

Example 1. (Predicted Miss) The problem P is

$$\inf E\, k(v \cdot x(T)) ,$$

$$dx = [A(t)x + B(t)u]dt + \sigma(t)dw, \quad x(0) = x_o ,$$

$$dy = H(t)x\, dt + \bar{\sigma}(t)d\bar{w} , \; y(0) = 0 ,$$

$$U = \left\{u \in \mathbb{R}_m : |u_i| \leq 1,\ i=1, \ldots, m\right\} ,$$

where k is even, non-negative, convex, continuous, $k(\xi) \leq K(1 + |\xi|^p)$ for some $p < \infty$, $v \in \mathbb{R}_n$, A, B, H, σ, $\bar{\sigma}$ are continuous, H, σ, $\bar{\sigma}$ invertible. Now (H) holds and $\bar{J}[u] = E\,\bar{k}(v \cdot m(T))$ where

$$\bar{k}(\xi) = \int k(\xi + v \cdot x)g(T,x)dx$$

is even, non-negative, convex. It follows, c.f. [14] or [15] and corollary 1, that the optimal control is

$$\hat{u}(t,\omega) = -\,\mathrm{sgn}[B(t)^*\, s(t)\; s(t)^*\, m(t,\omega)]$$

where

$$\dot{s} = -\,A(t)^*s, \quad s(T) = v ,$$

$$dm = [Am - B\,\mathrm{sgn}(B^*ss^*m)]dt + F\,[dy - Hm\,dt],\; m(0) = m_o .$$

Recall F is given in lemma 2.1.

Example 2. Here the problem P is

$$\inf E\left\{k(|x(T)|) + \int_0^T \ell(t, |x(t)|)dt\right\} ,$$

$$dx = [(D(t) + d(t)I)x + \beta(t)u]dt + \alpha(t)dw, \; x(0) = x_o ,$$

$$dy = h(t)x \, dt + \delta(t)d\bar{w} , \; y(0) = 0$$

$$U = \left\{u \in \mathbb{R}_n : |u| \le 1\right\} ,$$

where $k(\cdot)$ and $\ell(t, \cdot)$ are non-negative, non-decreasing and continuous functions for each t with $|k(\xi)| + |\ell(t,\xi)| \le K(1 + |\xi|^p)$, and where $D(t) + D(t)^* = 0$, α, β, δ, d, h are scalar-valued with D, d bounded, measurable, α, β, δ, h continuous and $\alpha(t)\beta(t)\, h(t) \neq 0$ for any t. As usual x_o should satisfy (H)(iv) but with $P_o = p_o I$, $p_o > 0$ scalar. Then (H) holds and $P(t) = p(t)I$ where

$$\dot{p} = 2d(t)p + \alpha(t)^2 - h(t)^2 \delta(t)^{-2} p^2$$

$$dm = [(D + dI)m + \beta u]dt + ph\,\delta^{-1} d\nu \quad .$$

It can be shown that (c is a constant)

$$\bar{k}(m) = c\, p(t)^{-\frac{1}{2}} \int k(|\xi+m|) \exp[-|\xi|^2/2p(t)]d\xi = \tilde{k}(|m|)$$

where $\tilde{k}$ is non-negative and non-decreasing. Similarly for $\tilde{\ell}(t, |m|)$. If $\Phi(t)$ is the fundamental matrix of $D(t)$ then it is orthogonal and $\bar{\nu}(t) \equiv \int_0^t \Phi(s)^* d\nu$ is a Brownian motion. If we let $\bar{m}(t) = \exp[-\int_0^t d(s)ds]\Phi(t)^* m(t)$ and $\bar{u}(t) = \Phi(t)^* u(t)$, then $|\bar{u}(t)| \le 1$ and

$$d\bar{m} = \exp[-\int_0^t d(s)ds][\beta(t)\bar{u}\, dt + p(t)\,\delta(t)^{-1}d\bar{\nu}] ,$$

$$\tilde{k}(|m|) = \tilde{k}(\exp[\int_0^T d(s)ds]|\bar{m}|) \equiv k^o(|\bar{m}|) ,$$

where k^o is non-negative and non-decreasing. Similarly for $\ell^o(t, |\bar{m}|)$. It follows from [16] that $\bar{u}(t,\bar{m}) = -\text{sgn}[\beta(t)]\,\bar{m}/|\bar{m}|$ is optimal for P given by $\bar{m}$, k^o, ℓ^o. Hence (corollary 1)

$$\hat{u}(t,\omega) = -\text{sgn}[\beta(t)]\, m(t,\omega) / |m(t,\omega)|$$

is optimal for P.

References

[1] W. H. Fleming, E. Pardoux, Optimal control for partially observed diffusions, SIAM J. Control and Optimization, 12 (1982), pp. 261-285.

[2] J-M. Bismut, Partially observed diffusions and their control, SIAM J. Control and Optimization, 12 (1982), pp. 302-309.

[3] U. G. Haussmann, On the existence of optimal controls for partially observed diffusions, SIAM J. Control and Optimization, 12 (1982), pp. 385-407.

[4] W. H. Fleming, R. W. Rishel, Deterministic and Stochastic Control, Springer Verlag, New York, 1975.

[5] M. H. A. Davis, Linear Estimation and Stochastic Control, Chapman and Hall, London, 1977.

[6] H. J. Kushner, Probability Methods for Approximations in Stochastic Control and for Elliptic Equations, Academic Press, New York, 1977.

[7] J. Ruzicka, On the separation principle with bounded controls, J. App. Math and Optimization, 3 (1977), pp 243-261.

[8] A. Ju.Veretennikov, On strong solutions and explicit formulas for solutions of stochastic integral equations, Math. USSR Sbornik, 39 (1981), pp.387-403.

[9] R. S. Liptser, A.N. Shiryayev, Statistics of Random Processes, Springer Verlag, New York, 1978.

[10] N. V. Krylov, Controlled Diffusion Processes, Springer Verlag, New York, 1980.

[11] J-M. Bismut, Théorie probabiliste du contrôle des diffusions, Mem. AMS, Vol 4, No. 167, 1976.

[12] U. G. Haussmann, On the adjoint process for optimal control of diffusion processes, SIAM J. Control and Optimization, 19 (1981), pp. 221-243.

[13] M. H. A. Davis, The separation principle in stochastic control via Girsanov solutions, SIAM J. Control and Optimization, 14 (1976), pp 176-188.

[14] V. E. Beneš, Composition and invariance methods for solving some stochastic control problems, Adv. App. Prob, 7 (1975), pp. 299-329.

[15] U. G. Haussmann, Extremal controls for completely observable diffusions, Proceedings of IFIP Working Conference on Filtering and Optimization, Cocoyoc, Mexico, 1982.

[16] N. Ikeda, S. Watanabe, A comparison theorem for solutions of stochastic differential equations and its applications, Osaka J. Math., 14 (1977), pp. 619-633.

A CLASS OF SINGULAR STOCHASTIC CONTROL PROBLEMS†

IOANNIS KARATZAS
Department of Mathematical Statistics
Columbia University
New York, N.Y. 10027, USA

1. Introduction and Summary.

We consider stochastic control problems of the following kind: given a Brownian Motion $w=\{w_t;t\geq 0\}$ on a probability space $(\Omega,\mathcal{F},P;\mathcal{F}_t)$, one has to follow it by a process $\xi=\{\xi_t;t\geq 0\}$ of bounded variation and adapted to the Brownian past $\mathcal{F}_t=\sigma(w_s;0\leq s\leq t)$, so that the "state of affairs" at time $t\geq 0$ is $x_t=x+w_t+\xi_t$. If w is thought of as representing the random disturbance acting on a satellite which is designed to maintain a fixed position in space (or as a random demand which has to be met by the controller), and ξ is considered in its canonical form $\xi=\xi^+-\xi^-$ as the difference of two nondecreasing processes, then $\xi_t^-(\xi_t^+)$ represents the total "push" in the negative (positive) direction, or accordingly the cumulative investment (disinvestment) up to time t. The total variation process $\check{\xi}_t=\xi_t^++\xi_t^-$ measures the fuel spent to date.

We admit costs both of "using fuel" and "being away from the origin". The former is proportional to the fuel spent, while the latter is measured by the cost function $h(x)$, which is assumed to be even, strictly convex, with $h'(x)\uparrow\infty$ as $x\uparrow\infty$ and $h(0)=0$. Consequently, the total cost incurred up to time t is:
$C(t;x,\xi)=\check{\xi}_t+\int_0^t h(x+w_s+\xi_s)ds$. We discuss the

(i) discounted : $\quad v_\alpha(x)=\min_\xi E\int_0^\infty e^{-\alpha t}d_tC(t;x,\xi)$,

†Research supported in part by the National Foundation under NSF MCS-81-03435.

(ii) stationary : $\lambda = \min_{\xi} \lim_{T\uparrow\infty} \frac{1}{T} EC(T;x,\xi)$, and

(iii) finite-horizon : $V(T,x) = \min_{\xi} EC(T;x,\xi)$

variants of the problem of minimizing expected total cost.

The answer to all these questions takes the form of consuming fuel in a singular manner: "do not use fuel while in the interior of a certain region; when on the boundary of this region, use fuel only in order not to exit; if you happen to start outside the region, use as much fuel as necessary to get immediately on the boundary, and then continue as before". Explicit solutions are found for questions (i) and (ii), while the third is reduced to an optimal stopping problem. This reduction enables us to study the behaviour and the asymptotics of the associated moving boundary. Finally, we establish the Abelian and ergodic relationships

$$\lim_{\alpha\downarrow 0} \alpha v_\alpha(x) = \lambda \quad \text{and} \quad \lim_{T\uparrow\infty} \frac{V(T,x)}{T} = \lambda$$

respectively, for all $x \in \mathbb{R}$.

In what follows we present the basic results and we outline some of the methods used to establish them. Detailed proofs will appear elsewhere. Related questions have been studied in [1], [2] and [6].

2. The discounted problem.

In this section we assume that $h''(x)$ is decreasing on $\mathbb{R}^+$ and bounded away from zero. If the "cost of doing nothing" is: $p(x) \triangleq E\int_0^\infty e^{-\alpha t} h(x+w_t)dt = \frac{1}{\sqrt{2\alpha}} \int_{\mathbb{R}} \exp\{-|x-z|\sqrt{2\alpha}\} h(z)dz$, and $c = c_\alpha$ is the unique solution of the transcendental equation

$$\tanh(c\sqrt{2\alpha}) = \sqrt{2\alpha}\, \frac{p'(c)-1}{p''(c)}, \tag{1}$$

then it can be seen that the function

$$\nu(x) = -\frac{p''(c)}{2\alpha}\,\frac{\cosh(x\sqrt{2\alpha})}{\cosh(c\sqrt{2\alpha})} + p(x) \quad ; \quad 0\le x<c$$

$$(2) \qquad = \nu(c) + x - c \quad ; \quad x\ge c$$

$$= \nu(-x) \quad ; \quad x<0$$

is twice continuously differentiable on $\mathbb{R}$, strictly convex in $(-c,c)$, and satisfies the relations: $\frac{1}{2}\nu''(x) + h(x) - \alpha\nu(x) = 0$ (>0), according as $|x|\le c$ $(|x|>c)$. An application of the Doléans Dade-Meyer change of variable formula for semimartingales [4] to the process $\{e^{-\alpha t}\nu(x+w_t+\xi_t);\ t\ge 0\}$ then yields $E\int_0^\infty e^{-\alpha t}d_tC(t:x,\xi) \ge \nu(x)$; for all $x\in\mathbb{R}$ and and process ξ of bounded variation, with equality if ξ is the process $\theta_\alpha = \theta_\alpha^+ - \theta_\alpha^-$ given by

$$\theta^+_{\alpha,t} = \max[0, \max_{0\le u\le t}\{-x-w_u+\theta^-_{\alpha,u}-c\}]$$

$$(3)$$

$$\theta^-_{\alpha,t} = \max[0, \max_{0\le u\le t}\{x+w_u+\theta^+_{\alpha,u}-c\}] .$$

We conclude that $\nu_\alpha(x) \equiv \nu(x)$. It can be shown as in [2], Appendix 2, that (3) has a unique solution, and that $X = \{x+w_t+\theta_{\alpha,t};\ t\ge 0\}$ is Brownian Motion with two reflecting barriers at $\pm$ c; therefore, X admits an ergodic distribution which is uniform on $(-c,c)$. Besides, it follows from renewal theoretic considerations that $\lim_{t\uparrow\infty}\frac{1}{t}E\check{\theta}_{\alpha,t} = \frac{1}{2c}$.

3. The stationary problem.

We determine two positive constants λ and b uniquely from the pair of equations

$$(4) \qquad \lambda = h(b) = \frac{1}{2b} + \frac{1}{b}\int_0^b h(y)dy ,$$

as well as the "potential function" $v(\cdot)$ on $\mathbb{R}$, so that $v(0) = 0$ and

$$\nu'(x) = 2(\lambda x - \int_0^x h(y)dy) \quad ; \quad 0\le x<b$$

$$(5) \qquad = 1 \quad ; \quad x\ge b$$

$$= -v'(-x) \quad ; \quad x<0 .$$

From the remarks at the end of section 2, in conjunction with relation (4), it follows that the process $\theta = \theta^+ - \theta^-$ defined by

$$\theta_t^+ = \max[0, \max_{0\le u\le t}\{-x-w_u+\theta_u^- -b\}]$$

(6)

$$\theta_t^- = \max[0, \max_{0\le u\le t}\{x+w_u+\theta_u^+ -b\}]$$

satisfies: $\lim_{\tau\uparrow\infty} \frac{1}{\tau} E[\check{\theta}_\tau + \int_0^\tau h(x+w_t+\theta_t)dt] = \lambda$, for all $x\in\mathbf{R}$.

On the other hand, an application of the change of variable formula to $\{v(x_t); t\ge 0\}$ with $x_t = x+w_t+\xi_t$, along with the properties of the function $v(\cdot)$, gives: $E[\check{\xi}_\tau + \int_0^\tau h(x_t)dt] \ge \lambda\tau + v(x) - Ev(x_\tau)$, whence $\underline{\lim}_{\tau\uparrow\infty} \frac{1}{\tau} EC(\tau;x,\xi) \ge \lambda$, for all $x\in\mathbf{R}$ and at least those processes ξ of bounded variation that satisfy $Ev(x_t) = o(t)$, as $t\to\infty$. The process θ in (6) is then optimal in this class.

4. The finite-horizon problem.

We proceed by introducing the following optimal stopping problem: for any $x>0$, we let $S(x) = \inf\{t\ge 0;\ x+w_t = 0\}$ be the first hitting time of the origin by the Brownian path started at x. If σ denotes an a.s. finite Brownian stopping time, we seek to characterize the optimal risk

(7) $$u(\tau,x) = \min_\sigma E[\int_0^{S(x)\wedge\tau\wedge\sigma} h'(x+w_t)dt + 1_{\{\sigma<S(x)\wedge\tau\}}]$$

for a stopping problem with mandatory termination upon reaching the origin, running cost $h'(\cdot)$ while continuing, and fixed fee equal to unity for premature termination (stopping before hitting the origin or running out of time). According to standard theory [3], [5], [7], $u(\tau,x)$ is the solution of a certain variational inequality or, equivalently, the solution of a free boundary problem, along with the optimal stopping boundary $b(\tau) = \min\{x>0;\ u(\tau,x) = 1\}$. The stopping time

$$\sigma^*(\tau,x) = \inf\{0\le t\le\tau;\ x+w_t\ge b(\tau-t)\}$$
$$= \tau;\ \text{if}\ \ x+w_t<b(\tau-t),\ \text{all}\ \ 0\le t\le\tau$$

is then optimal for the problem.

Similarly, it can be shown that the function $v'(x);\ x>0$ introduced in (5) is the optimal risk for the stopping problem

(8) $$v'(x) = \min_{\sigma} E\left[\int_0^{S(x)\wedge\sigma} h'(x+w_t)dt + 1_{\{\sigma<S(x)\}}\right]$$

with optimal stopping time $\sigma^* = \inf\{t\geq 0;\ x+w_t\geq b\}$. From (7), (8) one can deduce that $b(\tau)$ is a decreasing function of $\tau>0$, with $\lim_{\tau\uparrow\infty} b(\tau)=b$ and $\lim_{\tau\downarrow 0} b(\tau) = +\infty$.

The connection of the stopping problem (7) with the finite-horizon control problem is that $V(\tau,x)$, the value function for the latter, can be obtained as

$$\begin{aligned} V(\tau,x) &= \frac{1}{2}\int_0^{\tau} u_x(s,0)ds + \int_0^x u(\tau,y)dy &&;\quad \tau>0,\ 0\leq x<b(\tau) \\ &= V(\tau,b(\tau)) + x-b(\tau) &&;\quad \tau>0,\ x\geq b(\tau) \\ &= V(\tau,-x) &&;\quad \tau>0,\ x<0 \\ &= 0 &&;\quad \tau=0,\ x\in\mathbb{R}\ , \end{aligned}$$

and that the process $\mu = \mu^+ - \mu^-$ given on $[0,\tau]$ by

(9) $$\begin{aligned} \mu_t^+ &= \max[0, \max_{0\leq u\leq t}\{-x-w_u+\mu_u^- -b(\tau-u)\}] \\ \mu_t^- &= \max[0, \max_{0\leq u\leq t}\{x+w_u+\mu_u^+ -b(\tau-u)\}] \end{aligned}$$

is optimal. In other words, after a possible initial jump on the boundary of the region $\{(t,x);\ 0\leq t\leq\tau,\ |x|<b(\tau-t)\}$, the processes $\mu^{\pm}$ act in such a way as to keep the state process $\{X_t = x+w_t+\mu_t; 0\leq t\leq\tau\}$ on the closure of this region. They can also be shown to achieve this goal "without waste of fuel", i.e. to be the minimal ones among non-decreasing processes with this property.

For particular choices of $h(\cdot)$, one can obtain <u>global</u> upper and lower bounds for the moving boundary function $b(\tau)$ which, if tight enough, might provide information about the asymptotic behaviour of the boundary for small and large times. In order to fix ideas, let us consider the case $h(x) = \frac{1}{2}x^2$. To obtain a <u>lower</u> bound, we construct a "less favourable" stopping problem by considering the function $\underline{u}(\tau,x) = \tau x$, and noticing that, by virtue of (7):

$$\underline{u}(\tau,x) = E[\int_0^{S(x)\wedge\underline{\sigma}} (x+w_t)dt + 1_{\{\underline{\sigma}<S(x)\}}] \geq u(\tau,x) ,$$

where $\underline{\sigma} = \inf\{0\le t\le\tau\,;\ x+w_t \geq \frac{1}{\tau-t}\}$
$= \tau$, if $\{\dots\} = \emptyset$.

Consequently, $b(\tau)\geq\tau^{-1}$ for $\tau>0$, or even

(10) $$\underline{b}(\tau) \triangleq \min[\tau^{-1}, (1.5)^{1/3}] \leq b(\tau) \quad ; \quad \tau>0$$

since $b=(1.5)^{1/3}$ in this case. In order to construct an upper bound, we have to concoct a "more favourable" stopping problem (e.g. with the same running cost, and termination fee less than unity) that can be solved relatively easily. Such a construction is feasible but very involved; it leads to the upper bound $\overline{b}(\tau)$ given implicitly via

(11) $$\tanh\left[\frac{2\overline{b}^3(\tau)}{\tau\overline{b}(\tau)-1}\right]^{1/2} = \left[2\,\frac{\overline{b}(\tau)}{\tau}\left(\overline{b}(\tau) - \frac{1}{\tau}\right)\right]^{1/2} .$$

Simple expansions now yield the asymptotics:

$b(\tau) = \tau^{-1} + O(\tau^2)$, as $\tau\downarrow 0$ and $b(\tau) = (1.5)^{1/3} + O(\tau^{-1})$, as $\tau\uparrow\infty$.

5. Connections among the various problems.

The nature of the three problems studied so far suggests that they should be related to one another in a limiting sense, with the stationary problem as the focal point of the discussion. In particular, one expects the ergodic relation

(12) $$\lim_{\tau\uparrow\infty} \frac{1}{\tau} V(\tau,x) = \lambda$$

and the Abelian relation

(13) $$\lim_{\alpha\downarrow 0} \alpha v_\alpha(x) = \lambda$$

to hold for each $x\in R$.

To establish (12), we first note that the optimal process θ for the stationary problem is suboptimal for the problem on $[0,\tau]$; therefore, by the change of variable formula and with $X_t = x+w_t+\theta_t$:

$$V(\tau,x) \leq E[\check{\theta}_\tau + \int_0^\tau h(X_t)dt] = v(x) + \lambda\tau - Ev(X_\tau) \leq v(x) + \lambda\tau,$$

whence: $\overline{\lim_{\tau\uparrow\infty}} \frac{1}{\tau} V(\tau,x) \le \lambda$. For the opposite inequality, let us note that if $\zeta^{\pm}$, $\eta^{\pm}$ are defined by

$$\zeta_t^+ = \max[0, \max_{0\le u\le t} \{-x-w_u+\zeta_u^-\}]$$

(14)

$$\zeta_t^- = \max[0, \max_{0\le u\le t} \{x+w_u+\zeta_u^+-b\}]$$

for $t\ge 0$, and

$$\eta_t^+ = \max[0, \max_{0\le u\le t} \{-x-w_u+\eta_u^-\}]$$

$$\eta_t^- = \max[0, \max_{0\le u\le t} \{x+w_u+\eta_u^+-b(\tau-u)\}]$$

for $0\le t\le\tau$ (i.e. in analogy to (6) and (9) but with reflection at the origin and at the positive branch of the boundary), then for any $x>0$,

$$\lambda = \lim_{\tau\uparrow\infty} \frac{1}{\tau} E[\zeta_\tau^- + \int_0^\tau h(x+w_t+\zeta_t)dt] \quad \text{and} \quad V(\tau,x) = E[\eta_\tau^- + \int_0^\tau h(x+w_t+\eta_t)dt].$$

Now, given $\epsilon>0$, a $T=T(\epsilon)>0$ can be so selected that $b<b(\tau)<b+\epsilon$, for all $\tau\ge T$. Constructing the processes $\zeta_\epsilon^{\pm}$ as in (14) with $b+\epsilon$ instead of b, one can establish the a.s. comparisons

$$\eta_t^{\pm} \le \zeta_t^{\pm}\ ;\quad 0\le t\le\tau \qquad\qquad \zeta_{\epsilon,t}^{\pm} \le \eta_t \le \zeta_t^{\pm}\ ;\quad 0\le t\le \tau-T$$

$$\zeta_{\epsilon,t}^{\pm} \le \zeta_t^{\pm}\ ;\quad t\ge 0 \qquad\qquad x+w_t+\eta_t \ge x+w_t+\zeta_t \ge 0;\ 0\le t\le\tau$$

and obtain as a corollary the bound

$$V(\tau,x) \ge E[\zeta_\tau^- + \int_0^\tau h(x+w_t+\zeta_t)dt] - E[\zeta_{\tau-T}^- - \zeta_{\epsilon,\tau-T}^-] - E[\zeta_\tau^- - \zeta_{\tau-T}^-]$$

$$\ge \lambda\tau + \text{const.}\epsilon\tau + o(\tau).$$

It follows that $\underline{\lim}_{\tau\uparrow\infty} \frac{1}{\tau} V(\tau,x) \ge \lambda + \text{const.}\epsilon$, for every $\epsilon>0$, and so $\underline{\lim}_{\tau\uparrow\infty} \frac{1}{\tau}V(\tau,x) \ge \lambda$, which establishes (12).

An important consequence of the ergodic relation is that by means of it the process θ in (6) can be shown to be optimal for the stationary problem against <u>any</u> $\mathfrak{F}_t$-adapted process ξ of bounded variation.

The proof of the Abelian relation is similar in spirit to that of (12) although slightly more delicate. It rests on the fact that the derivative of the value function for the discounted problem is the optimal risk for the stopping problem

$$\nu'_\alpha(x) = \min_\sigma E[\int_0^{S(x)\wedge\sigma} e^{-\alpha t} h'(x+w_t)dt + e^{-\alpha\sigma} 1_{\{\sigma<S(x)\}}]$$

with optimal stopping time $\sigma^*_\alpha(x) = \inf\{t\geq 0;\ x+w_t \geq c_\alpha\}$, as well as on the corollary: $c_\alpha \downarrow b$, as $\alpha\downarrow 0$.

REFERENCES

[1] J.A. BATHER and H. CHERNOFF. Sequential decisions in the control of a spaceship, Proc. 5th Berkeley Symposium on Mathematical Statistics and Probability III, p.181-207. University of California Press, Berkeley, 1966.

[2] V.E. BENEŠ, L.A. SHEPP and H.S. WITSENHAUSEN: Some solvable stocastic control problems, Stochastics 4 (1980), p.39-83.

[3] A. BENSOUSSAN and J.L. LIONS: Applications des Inéquations Varitionelles en Contrôle Stochastique. Dunod, Paris, 1978.

[4] C. DOLÉANS-DADE and P.A. MEYER: Intégrales stochastiques par rapport aux martingales locales, Séminaire de Probabilités IV, Lecture Notes in Mathematics 124, p.77-107. Springer Verlag, Berlin, 1970.

[5] A. FRIEDMAN: Stochastic Differential Equations and Applications (Vol. 2). Academic Press, New York, 1976.

[6] I. KARATZAS: The monotone follower problem in stochastic decision theory, Appl. Math. Optim. 7 (1981), p.175-189.

[7] P. VAN MOERBEKE: Optimal stopping and free boundary problems, Arch. Rat. Mech. Analysis 60 (1976), p.101-148.

SUR L'ARRET OPTIMAL DE

PROCESSUS A DEUX INDICES REELS

G. Mazziotto

PAA / ATR / MTI
Centre National d'Etudes des Télécommunications
38-40, rue du Général Leclerc
92 131 - ISSY LES MOULINEAUX
FRANCE

1 - INTRODUCTION

Le problème d'arrêt optimal d'une somme, sur $\mathbb{N}^2$, de variables aléatoires indépendantes équidistribuées est bien connu; une résolution particulièrement originale est donnée dans ([7]). Le cas de processus indexés sur un ensemble discret partiellement ordonné est étudié dans ([8]), et l'existence de solutions est établie dans des conditions particulières. Dans ([10]), on a posé le problème d'arrêt de processus indexés sur $\mathbb{N}^2$, dans un cadre général, et on l'a résolu sur le compactifié de $\mathbb{N}^2$. Le cas d'un ensemble d'indices partiellement ordonné continu est plus délicat. Cet article constitue une contribution à cette étude, mais il n'a pas la prétention de résoudre ce problème. Néanmoins, je crois qu'il met en relief certaines des difficultés séparant le cas discret du cas continu. Un résultat important dans cette direction, est la généralisation au cas de $\mathbb{R}_+^2$ du Théorème de Mertens, obtenue dans ([4]). L'approche du problème adoptée ici, s'inspire beaucoup de celles de la théorie classique (i.e. l'arrêt de processus à indices réels), notamment de ([12]), ([2]), ([3]), et surtout ici, de ([6]).

Après quelques rappels, on présente dans un premier paragraphe, le problème d'arrêt optimal sur $\mathbb{R}_+^2$, et le système des gains conditionnels associé. Une construction de l'enveloppe de Snell, par approximations successives, est proposée dans le second paragraphe. On rappelle dans le troisième paragraphe le critère d'optimalité, et une méthode de résolution est présentée. Dans le quatrième paragraphe, on étudie le comportement de l'enveloppe de Snell, en dehors de l'ensemble où elle coincide avec le processus de gain. On montre en quoi la méthode proposée aboutit dans le cas de $\mathbb{N}^2$, et ce qui pose problème dans $\mathbb{R}_+^2$.

2 - PRELIMINAIRES

Les processus considérés ici sont indexés sur l'ensemble $\mathbb{R}_+^2$, et prolongés à son compactifié, $\overline{\mathbb{R}}_+^2 = \mathbb{R}_+^2 \cup \{\infty\}$, par la valeur nulle à l'infini. L'ordre partiel est défini par:

$\forall\ s=(s_1,s_2),\ t=(t_1,t_2)\ :\ s \leq t \iff s_1 \leq t_1$ et $s_2 \leq t_2$,

et la relation stricte, par : $s < t \iff s_1 < t_1$ et $s_2 < t_2$.

Sur un espace de probabilité complet, $(\Omega,\underline{\underline{A}},\mathbb{P})$, une filtration est une famille, indexée sur $\overline{\mathbb{R}}_+^2$, de sous-tribus de $\underline{\underline{A}}$, $\underline{\underline{F}} = (\underline{\underline{F}}_t ; t \in \overline{\mathbb{R}}_+^2)$, satisfaisant aux conditions habituelles, (5), (14): $\underline{\underline{F}}_0$ contient tous les $\mathbb{P}$-négligeables (axiome F1), $\underline{\underline{F}}$ est croissante (axiome F2) et $\underline{\underline{F}}$ est continue à droite (axiome F3). On suppose aussi que la filtration $\underline{\underline{F}}$ possède la propriété d'indépendance conditionnelle (axiome F4) suivante:

| $\forall\ t=(t_1,t_2)$, les tribus $\underline{\underline{F}}^1_{t_1} = \bigvee_u \underline{\underline{F}}_{(t_1,u)}$ et $\underline{\underline{F}}^2_{t_2} = \bigvee_v \underline{\underline{F}}_{(v,t_2)}$
| sont indépendantes, conditionnellement à $\underline{\underline{F}}_t$.

La tribu optionnelle sur $\Omega \times \mathbb{R}_+^2$, ainsi que la projection optionnelle d'un processus borné, X, notée ${}^oX$, sont définies dans (1).

Un point d'arrêt (p.a.) est une variable aléatoire (v.a.), T, à valeurs dans $\overline{\mathbb{R}}_+^2$, telle que : $\forall\ t : \{T \leq t\} \in \underline{\underline{F}}_t$. L'ensemble des p.a. est noté $\underline{\underline{T}}$. A un p.a. T, on associe la tribu $\underline{\underline{F}}_T$, des événements, A, tels que $\forall\ t : A \cap \{T \leq t\} \in \underline{\underline{F}}_t$. Toutes les propriétés classiques des temps d'arrêt ne se retrouvent pas sur les p.a. (13). Le graphe d'un p.a. T, est l'ensemble optionnel $[\![T]\!] = \{(\omega,t) : T(\omega) = t < \infty\}$. Etant donné un ensemble aléatoire, H, on note $[\![H,\infty[\![$ l'ensemble $\{(\omega,t) : \exists s \leq t$ tel que $(\omega,s) \in H\}$. Le Début de H, L_H, est le bord inférieur de l'ensemble $[\![H,\infty[\![$, en convenant que $L_H = \infty$ si la coupe est vide. Une ligne d'arrêt (l.a.) est le Début d'un ensemble progressif (11). On note $\underline{\underline{L}}$ l'ensemble des l.a.; on peut y plonger $\underline{\underline{T}}$, en associant à tout p.a. T, la l.a. L_T, Début de l'ensemble $[\![T]\!]$. La relation d'ordre partiel se prolonge de façon évidente à $\underline{\underline{T}}$, et aussi à $\underline{\underline{L}}$, en posant:

$\forall\ L, L' \in \underline{\underline{L}}\ :\ L \leq L' \iff [\![L',\infty[\![\subset [\![L,\infty[\![$.

Une surmartingale forte optionnelle est un processus optionnel, X, tel que : $\forall\ T\ \ \underline{\underline{T}}$, la v.a. $\underline{\underline{F}}_T$-mesurable, X_T, est intégrable et $\forall\ S \in \underline{\underline{T}}$ tel que que $S \leq T$, on a $E(X_T/\underline{\underline{F}}_S) \leq X_S$ p.s.. Deux processus, X et Y, sont dits pseudo-indistinguables si pour tout p.a. T, on a $X_T = Y_T$ p.s..

Sous l'hypothèse F4, et grâce aux résultats de (7), (8) et (13), on peut vérifier que, pour toute l.a. L, et tout p.a. T tel que $T \geq L$, il existe un p.a. S, porté p.s. par L (i.e. $\mathbb{P}(\{S \in L\}) = 1$), tel que $S \leq T$. Dans la suite, on écrira plus simplement "$S \in L$" pour exprimer cela. En particulier, pour une l.a. L, on désigne par $\underline{\underline{F}}_L$ la tribu construite comme intersection des tribus $\underline{\underline{F}}_T$, quand T parcourt l'ensemble des p.a. portés p.s. par L (i.e. $\underline{\underline{F}}_L = \bigcap_{T \in L} \underline{\underline{F}}_T$) .

---

3 - SYSTEME DES GAINS CONDITIONNELS

On considère un processus indexé sur $\mathbb{R}_+^2$, Y, optionnel, non-négatif et de la classe (D) (i.e. la famille $(Y_T;\ T \in \underline{\underline{T}})$ est uniformément intégrable, avec $Y_\infty = 0$). Le problème d'arrêt optimal, associé au processus de gain Y, consiste à chercher un p.a. T^* dit optimal, tel que $E(Y_{T^*}) = \sup\{E(Y_T)\ ;\ T \in \underline{\underline{T}}\}$.
Comme dans la théorie classique, ([2]), ([3]), ([6]), ([12]), la résolution de ce problème s'appuie sur le système des gains conditionnels. On le définit ici comme une famille de v.a. indexée sur $\underline{\underline{L}}$, en posant:

$$\forall\, L \in \underline{\underline{L}} : J(L) = \operatorname{esssup}\{E(Y_S / \underline{\underline{F}}_L);\ S \geq L,\ S \in \underline{\underline{T}}\} .$$

On montre, par les mêmes arguments que pour les temps d'arrêt ([6]), ou les p.a. ([10]), que, pour toute l.a. fixée, L, l'ensemble des v.a. $\{E(Y_T/\underline{\underline{F}}_L);\ T \geq L\}$ est filtrant croissant, et qu'il existe une suite de p.a. $(T_n; n \in \mathbb{N})$ telle que la suite $(E(Y_{T_n}/\underline{\underline{F}}_L); n \in \mathbb{N})$ soit croissante, de limite J(L). En particulier

$$\forall\, L, L' \in \underline{\underline{L}} : L \leq L' \Rightarrow E(J(L')/\underline{\underline{F}}_L) \leq J(L) \quad \text{p.s.} .$$

PROPOSITION 1 : *Pour toute l.a., L, on a*

$$J(L) = \operatorname{esssup}\{E(J(T)/\underline{\underline{F}}_L)\ ;\ T \in L\} \quad p.s. .$$

Démonstration: Il est évident que: $\forall\, T \in L : J(L) \geq E(J(T)/\underline{\underline{F}}_L)$ p.s. . Mais, d'autre part $\forall\, S \geq L, \exists\, T \in L$ tel que $T \leq S$ et $E(Y_S/\underline{\underline{F}}_L) \leq E(J(T)/\underline{\underline{F}}_L)$
La conclusion en découle.

Ce résultat permet de se ramener au système des gains conditionnels indexé sur $\underline{\underline{T}}$, comme dans ([9]) et ([10]). En particulier, on rappelle que

$$\forall\, S, T \in \underline{\underline{T}} : J(S) = J(T) \text{ p.s. sur l'ensemble } \{S = T\}.$$

Le Théorème de Mertens ([6]) se généralise au cas de $\mathbb{R}_+^2$: c'est le résultat de ([4]). Il existe au moins un processus optionnel, J, tel que

$$J \geq Y \quad \text{et} \quad \forall\, T \in \underline{\underline{T}} : J(T) = J_T \quad \text{p.s.} .$$

Le processus J est une surmartingale forte optionnelle, unique à une pseudo-indistinguabilité près, appelée enveloppe de Snell de Y, et notée aussi J = SN(Y).

4 - CONSTRUCTION DE L'ENVELOPPE DE SNELL

Dans ce paragraphe, on propose une construction de l'enveloppe de Snell par un procédé d'approximations successives. La méthode est inspirée de celle de ([12]) pour des chaînes de Markov, et de celle de ([6]) pour le cas à un indice réel. Pour la mettre en oeuvre, on fera appel aux notions de tactique dans $\mathbb{N}^2$, ([7]), ([8]), et de chemin croissant optionnel dans $\mathbb{R}_+^2$, ([13]). Aussi, quelques rappels sont nécessaires.

Soit $\mathbb{D}$ (resp. $\mathbb{D}^n; n \in \mathbb{N}$) l'ensemble de tous les dyadiques (resp. les dyadiques d'ordre n) et le point à l'infini; soient $\varepsilon_n^1 = (2^{-n}, 0)$, $\varepsilon_n^2 = (0, 2^{-n})$.

Etant donné un p.a. T, on note $\underline{\underline{T}}_d(T)$ (resp. $\underline{\underline{T}}_d^n(T); n \in \mathbb{N}$), l'ensemble des p.a., S, tels que $S - T \in \mathbb{D}$ (resp. $\mathbb{D}^n; n \in \mathbb{N}$) p.s. . On rappelle que, si S est un p.a. tel que $S \geq T$, il existe une suite décroissante de p.a., $(S_n; n \in \mathbb{N})$, telle que : $S_n \in \underline{\underline{T}}_d^n(T)$, $\forall n$ et $\lim_n S_n = S$ p.s. .
D'autre part, pour n fixé, $T \in \underline{\underline{T}}$ et $S \in \underline{\underline{T}}_d^n(T)$ donnés, il existe, d'après ([7]), ([8]), ([13]), une suite croissante de p.a. de $\underline{\underline{T}}_d^n(T)$, $(T_k; k \in \mathbb{N})$, telle que

i) $T_0 = T$, $\forall k \in \mathbb{N} : T_{k+1} = T_k$ ou $T_k + \varepsilon_n^1$ ou $T_k + \varepsilon_n^2$, $\lim_k T_k = S$

ii) $\forall k$: T_{k+1} est une v.a. $\underline{\underline{F}}_{T_k}$-mesurable.

La suite $(T_k; k \in \mathbb{N})$ est appelée une tactique.

Etant donné un processus, Y, optionnel, non-négatif et borné, on lui associe le processus R(Y), défini par

$$\forall t \in \mathbb{R}_+^2 : R(Y)_t = \sup \{ {}^o(Y_{r+.})_t \; ; \; r \in \mathbb{D} \} \quad \text{et} \quad R(Y)_\infty = 0 .$$

Le processus R(Y) est optionnel, non-négatif et borné; il majore Y et

$$\forall T \in \underline{\underline{T}} \; : \; R(Y)_T = \sup \{ E(Y_{r+T}/\underline{\underline{F}}_T) \; ; \; r \in \mathbb{D} \} \quad \text{p.s. .}$$

L'opérateur R (dit "de Réduite") est positif, et il laisse les surmartingales fortes optionnelles invariantes, à une pseudo-indistinguabilité près. Par récurrence, on définit la suite, $(I^n; n \in \mathbb{N})$, de processus,

$$I^0 = Y \; , \quad \forall n : I^{n+1} = R(I^n) \; .$$

Cette suite est croissante, uniformément bornée: soit I le processus limite, qui est encore optionnel, non-négatif et borné.

Soit Y un processus de gain, optionnel, non-négatif et borné; et soit J son enveloppe de Snell. D'après ce qui précède, le processus I minore J sur tous les p.a. . On va montrer que si les trajectoires de Y sont suffisamment régulières, alors les processus I et J sont pseudo-indistinguables.

<u>PROPOSITION 2</u> : Si le processus Y est continu à l'infini et si ses trajectoires sont p.s. des fonctions semi-continues inférieurement (sci) pour la topologie droite, alors

$$\forall T \in \underline{\underline{T}} \; : \; I_T = J_T = \text{esssup} \{ E(Y_S / F_T) ; \; S \geq T \} \; .$$

<u>Démonstration</u>: Il suffit de vérifier que $I_T \geq J_T$ p.s. , $\forall T \in \underline{\underline{T}}$.
On établit d'abord que : $\forall T \in \underline{\underline{T}}$, $\forall n$, $\forall S \in \underline{\underline{T}}_d^n(T)$: $E(Y_S / \underline{\underline{F}}_T) \leq I_T$ p.s. .
Pour cela, on revient à la suite $(I^n; n \in \mathbb{N})$ qui définit I, et on utilise les propriétés i) et ii) des tactiques, rappelées au début de ce paragraphe, avec les mêmes notations. Pour $m, k \in \mathbb{N}$ fixés, on a

$$E(I^m_{T_k} / \underline{\underline{F}}_{T_{k-1}}) = I^m_{T_k} \, 1\!\!1_{\{T_k = T_{k-1}\}} + \sum_{i=1,2} E(I^m_{T_{k-1}+\varepsilon_n^i} / \underline{\underline{F}}_{T_{k-1}}) \, 1\!\!1_{\{T_k = T_{k-1}+\varepsilon_n^i\}}$$

or, par définition, $\quad \forall r \in \mathbb{D} : I^{m+1}_{T_{k-1}} \geq E(I^m_{T_{k-1}+r} / \underline{\underline{F}}_{T_{k-1}})$,

d'où : $\quad E(I^m_{T_k} / \underline{\underline{F}}_{T_{k-1}}) \leq I^{m+1}_{T_{k-1}}$.

En particulier, en partant de Y, et en itérant, on obtient, ∀ k fixé:

$$E(Y_{T_k}/\underline{\underline{F}}_T) = E(Y_{T_k}/\underline{\underline{F}}_{T_{k-1}}/\ldots/\underline{\underline{F}}_{T_0}) \leq E(I^1_{T_{k-1}}/\ldots/\underline{\underline{F}}_{T_0}) \leq E(I^k_{T_0}/\underline{\underline{F}}_T) \leq I_T.$$

Sur l'ensemble $\{S \neq \infty\}$, la suite $(T_k; k \in \mathbb{N})$ converge p.s. stationnairement et donc la suite $(\mathbb{1}_{\{S\neq\infty\}} Y_{T_k}; k \in \mathbb{N})$ tend vers $\mathbb{1}_{\{S\neq\infty\}} Y_S$. Comme Y converge vers 0 à l'infini, la suite $(\mathbb{1}_{\{S=\infty\}} Y_{T_k}; k \in \mathbb{N})$ a pour limite 0. Donc,

$$E(Y_S/\underline{\underline{F}}_T) = \lim E(Y_{T_k}/\underline{\underline{F}}_T) \leq I_T \quad .$$

Il reste à généraliser ce résultat à un p.a. quelconque: $S \geq T$. On considère pour cela, la suite $(S_n; n \in \mathbb{N})$ associée à l'approximation de S par la droite. Comme Y est à trajectoires sci à droite, et qu'il est borné,

$$E(Y_S/\underline{\underline{F}}_T) \leq E(\liminf Y_{S_n}/\underline{\underline{F}}_T) \leq \liminf E(Y_{S_n}/\underline{\underline{F}}_T) \leq I_T \quad \text{p.s.}\ .$$

On en déduit alors le résultat annoncé en prenant l'enveloppe supérieure sur tous les p.a. $S \geq T$.

5 - METHODES DE RESOLUTION

Le critère d'optimalité, portant sur le système des gains conditionnels, exprime le principe d'optimalité de Bellman. Il est analogue à celui de ([6]), et est établi dans ([10]).

PROPOSITION 3 : Un p.a. ,T, est optimal si et seulement si

i) $J(T) = Y_T$ *p.s.*

ii) $E(J(T)) = E(J(0))$.

Dans la théorie classique, en général, ce critère est employé de la manière suivante ([3]), ([6]). Sous des conditions de régularité appropriées, on montre que le plus petit temps d'arrêt tel que la condition i) soit remplie (i.e. le Début de {Y = J}), satisfait aussi à ii), et est donc optimal. Dans ([10]), on a suggéré que, pour des processus à plusieurs indices, il semblait plus pratique de chercher un p.a. qui vérifie i) parmi les p.a. maximaux de l'ensemble où ii) est vraie. Une telle approche est justifiée par le fait qu'ici, le Début de l'ensemble {Y = J} est en général une ligne d'arrêt faible ([11]), qui peut aussi bien contenir aucun p.a., ou plusieurs, pas tous optimaux. Le cas, non aléatoire suivant en constitue un exemple.

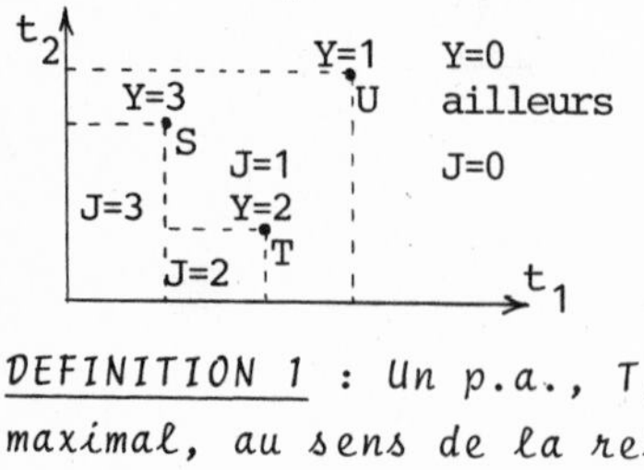

Début {Y = J} = S et T

S est optimal mais pas T

S est maximal.

DEFINITION 1 : Un p.a., T, est dit maximal si c'est un élément maximal, au sens de la relation d'ordre partiel sur $\underline{\underline{T}}$*, de l'ensemble* $\{S \in \underline{\underline{T}}$ *tel que* $E(J_S) = E(J_0)\}$*; i.e.*

$E(J_T) = E(J_0)$ *et* $E(J_S/\underline{\underline{F}}_T) < J_T \quad \forall\, S \geq T$ *avec* $\mathbb{P}(S = T) < 1$.

De tels p.a. maximaux existent, sous certaines conditions de régularités portant sur l'enveloppe de Snell, J, ou sur le processus de gain Y, indirectement.

PROPOSITION 4 : *Si l'enveloppe de Snell, J, est continue à gauche en espérance sur les p.a. (i.e. pour toute suite croissante de p.a. $(T_n; n\in\mathbb{N})$ de limite un p.a. T, on a $E(J_T) = \lim E(J_{T_n})$), il existe au moins un p.a. maximal.*

La démonstration est une application directe du Lemme de Zorn.

PROPOSITION 5 : *Si le processus de gain, Y, est à trajectoires p.s. continues sur le compact $\overline{\mathbb{R}}_+^2$, alors l'enveloppe de Snell, J, est continue à gauche en espérance sur les p.a..*

Démonstration: Soit $(T_n; n\in\mathbb{N})$ une suite croissante de p.a., de limite T. La suite $(E(J_{T_n}); n\in\mathbb{N})$ associée, est décroissante, soit j(T-) sa limite . On va montrer $E(J_T) = j(T-)$ p.s..
Pour chaque n, il existe un p.a. S_n, par définition du système des gains conditionnels, tel que : $S_n \geq T$ et $E(J_{T_n}) - E(Y_{S_n}) < 2^{-n}$. On en déduit que la suite $(E(Y_{S_n}); n\in\mathbb{N})$ a une limite, qui vaut évidemment j(T-).
Quand n tend vers l'infini, la distance entre le point S_n et le point $S_n \vee T = (\sup(S_{n1}, T_1), \sup(S_{n2}, T_2))$ tend p.s. vers 0; comme le processus Y est à trajectoires p.s. uniformément continues, on en déduit que la suite de v.a. $(Y_{S_n} - Y_{S_n \vee T}; n\in\mathbb{N})$ converge p.s. vers 0. Par l'intégrabilité uniforme, cela entraîne que $\lim(E(Y_{S_n}) - E(Y_{S_n\vee T})) = 0$. On conclut alors

$$j(T-) = \lim E(Y_{S_n}) = \lim E(Y_{S_n\vee T}) \leq E(J_T) \quad \text{et donc} \quad j(T-) = E(J_T) .$$

Pour résoudre le problème d'arrêt optimal, il faudrait maintenant montrer que les p.a. maximaux sont effectivement optimaux. Dans le cas des processus indexés sur $\mathbb{N}^2$, ce résultat est établi directement dans ([10]). Dans le cas général, les difficultés rencontrées proviennent, essentiellement, de ce que l'on connait mal le comportement de la surmartingale enveloppe de Snell.

6 - ENVELOPPE DE SNELL ET REDUITE DE SNELL

Dans ce paragraphe, on caractérise l'évolution de l'enveloppe de Snell, J, associée au processus de gain Y, en dehors de l'ensemble où ils coincident. Cette approche est inspirée de ([6]), elle est aussi apparentée à la méthode des pénalisations de la théorie classique ([2]). On en déduira une condition, satisfaite dans le cas $\mathbb{N}^2$, pour qu'un p.a. maximal soit optimal.

Pour $\lambda \in]0,1[$ et T un p.a., fixés, on pose

$$H_T^\lambda = \{Y \geq \lambda J\} \cap [\![T,\infty]\!] \quad \text{et} \quad SN(J\, 1\!\!1_{H_T^\lambda}) = J^{H_T^\lambda} .$$

On notera simplement H pour H_T^λ quand il n'y a pas de confusion possible; on appelle la surmartingale forte J^H, la Réduite de Snell de J sur H.

LEMME 6 : $\forall\, S \geq T$: $J_S^{H_T^\lambda} = J_S$ *p.s.* .

Démonstration: Elle est analogue à celle de ([6]). Tout d'abord, comme J est une surmartingale qui majore $J\, 1\!\!1_H$, elle majore aussi son enveloppe de Snell J^H. En particulier, pour tout p.a. S, on a

$$J_S^H \geq 1\!\!1_{\{S\in H\}} J_S \geq 1\!\!1_{\{S\in H\}} J_S^H \quad \text{i.e.} \quad J_S^H = J_S \ \text{ sur } \{S\in H\} \text{ p.s..}$$

Soit I la surmartingale forte optionnelle $\lambda J + (1-\lambda)J^H$; on a $J_S \geq I_S\ \forall\, S$. Inversement, il faut montrer que I majore Y, donc J, sur tout p.a. S tel que $S \geq T$. Or, sur $\{S \in H\}$: $J_S = J_S^H$ donc $I_S \geq Y_S$ sur cet ensemble. Mais, d'autre part, sur $\{S \notin H;\ S \geq T\}$: $Y_S < \lambda J_S$, et a fortiori $Y_S \leq I_S$. Ceci achève la démonstration.

PROPOSITION 7 : *Pour* $\lambda \in\,]0,1[$ *et* $T \in \underline{T}$ *fixés, soit* L_T^λ *la ligne d'arrêt Début de* H_T^λ. *On a*

$$J_T = E(J(L_T^\lambda)\, /\, \underline{\underline{F}}_T) \quad p.s. \,.$$

Démonstration: On va établir cette égalité sur le processus J^H et le Lemme 6 permettra alors de conclure. Par définition, on a

$$J_T^H = \text{esssup}\, \{E(J_S\, 1\!\!1_{\{S\,\in\, H\}}/\underline{\underline{F}}_T);\ S \geq T\} \quad \text{et} \quad T \leq L_T^\lambda$$

or, $\forall\, S \geq T$, $\exists\ \tilde{S} \geq L_T$ tel que $E(J_S\, 1\!\!1_{\{S\,\in\, H\}}/\underline{\underline{F}}_T) \leq E(J_{\tilde S}\, 1\!\!1_{\{\tilde S\,\in\, H\}}/\underline{\underline{F}}_T)$; il suffit, pour cela, de prendre $\tilde{S} = S$ sur $\{S \geq L_T^\lambda\}$ et ∞ ailleurs. D'où, nécessairement,

$$J_T^H = \text{esssup}\, \{E(J_{\tilde S}\, 1\!\!1_{\{\tilde S\,\in\, H\}}/\underline{\underline{F}}_T);\ \tilde{S} \geq L_T^\lambda\} = E(J^H(L_T^\lambda)/\underline{\underline{F}}_T)$$

Or, en utilisant la Proposition 1 et le Lemme 6, on montre que

$$J^H(L_T^\lambda) = J(L_T^\lambda) \quad \text{et} \quad J_T^H = J_T \quad \text{p.s.} \,,$$

ce qui achève la démonstration.

On considère maintenant l'hypothèse suivante sur les p.a. maximaux.

(H) | Pour tout p.a. maximal, T, on a $E(J_T) > E(J(L))$, pour toute l.a. $L \geq T$ telle que $\mathbb{P}(T = L) < 1$.

PROPOSITION 8 : *Sous l'hypothèse (H), et en supposant les processus J et Y à trajectoires continues à droite, tout p.a. maximal est optimal.*

Démonstration: Soit $\lambda \in\,]0,1[$ et T un p.a. maximal. D'après la Proposition 7, $E(J_T) = E(J(L_T^\lambda))$. Par hypothèse, ceci n'est possible que si $L_T^\lambda = T$ p.s. . En utilisant la continuité à droite des processus Y et J, on en déduit alors que $\lambda J_T \leq Y_T$ p.s. . En faisant varier λ, on obtient $J_T = Y_T$ p.s. . Le résultat annoncé en découle aisément.

Tout ce qui vient d'être dit s'applique directement à l'arrêt de processus indexés sur $\mathbb{N}^2$. Dans ce cas la continuité à droite des processus n'a plus d'objet; l'hypothèse (H) est toujours satisfaite: on s'en assure par le raisonnement suivant. Soit T un p.a. maximal, on désigne par T' et T" les p.a. (T_1+1,T_2) et (T_1,T_2+1), et par L* la l.a. Début de $[\![T']\!] \cup [\![T'']\!]$. Il est facile de voir que

$$J(L^*) = \text{esssup}\{E(J_S/\underline{\underline{F}}_{L^*});\ S \in L^*\} = \sup\{E(J_{T'}/\underline{\underline{F}}_{L^*}), E(J_{T''}/\underline{\underline{F}}_{L^*})\}$$

et donc, si T est maximal, $E(J(L^*)/\underline{\underline{F}}_T) < J_T$ p.s. .

Si maintenant L est une l.a. quelconque, telle que $L \geq T$, on a

$$E(J(L)/\underline{\underline{F}}_T) \leq J_T \, 1\!\!1_{\{T = L\}} + E(J(L^*)/\underline{\underline{F}}_T) \, 1\!\!1_{\{T \neq L\}} < J_T \quad \text{p.s.},$$

et on retrouve ainsi l'hypothèse (H).

Dans le cas général de $\mathbb{R}_+^2$, la justification de telles hypothèses ne peut s'appuyer que sur des connaissances encore plus précises des surmartingales à deux indices. En particulier remarquons que l'on est en train d'essayer de résoudre le problème d'arrêt optimal sans jamais faire appel à une décomposition des surmartingales, analogue par exemple à celle de Mertens ([6]) de la théorie classique.

7 - BIBLIOGRAPHIE

([1]) BAKRY, D.:"Théorèmes de section et de projection pour processus à deux indices". Z. Wahr. V. Geb. 55; 51-71; (1981).

([2]) BENSOUSSAN, A. - LIONS, J.L.:"Applications des inéquations variationnelles au controle stochastique". Dunod, Paris, (1978).

([3]) BISMUT, J.M. - SKALLI, B.:"Temps d'arrêt optimal, théorie générale des processus et processus de Markov. Z. Wahr. V. Geb. 39; 301-313; (1977).

([4]) CAIROLI, R.:"Enveloppe de Snell d'un processus à paramètre bidimensionnel". Ann. Inst. H. Poincaré 18,1; 47-54; (1982).

([5]) CAIROLI, R. - WALSH, J.B.:"Stochastic Integrals in the Plane" Acta Math. 134; 111-183; (1975).

([6]) ELKAROUI, N.:"Les aspects probabilistes du controle stochastique". Lect. N. Maths 876; 73-238; Springer Verlag (1981).

([7]) KRENGEL, U. - SUCHESTON, L.:"Stopping Rules and Tactics for Processes Indexed by a Directed Set". J. Mult. Anal. 11,2; 199-229; (1981).

([8]) MANDELBAUM, A. - VANDERBEI, R.J.:"Optimal Stopping and Supermartingales over Partially Ordered Sets". Z. Wahr. V. Geb. 57; 252 - 264; (1981).

([9]) MAZZIOTTO, G.:"Processus bimarkoviens et arrêt optimal sur $\mathbb{R}_+^2$" Preprint.

([10]) MAZZIOTTO, G. - SZPIRGLAS, J.:"Arrêt optimal sur le plan" Preprint.

([11]) MERZBACH, E.:"Stopping for Two-dimensional Stochastic Processes" Stoch. Proc. & Appl. 10; 49-63; (1980).

([12]) SHIRYAYEV, A.N.:"Optimal Stopping Rules" Springer verlag; Berlin; (1978).

([13]) WALSH, J.B.:"Optional Increasing Paths". Lect. N. Maths 863; 172-201; Springer Verlag, Berlin; (1981).

([14]) WONG, E. - ZAKAI, M.:"Martingales and Stochastic Integrals for Processes with a Multidimensional Parameter". Z. Wahr. V. Geb. 29; 109-122; (1974).

DUALITY THEORY FOR SOME STOCHASTIC CONTROL MODELS

Stanley R. Pliska
Northwestern University
Evanston, IL 60201/USA

1. Introduction

This paper is concerned with a duality theory for discrete time stochastic control problems. The principal result is that, for rather general models, the dual variables are martingales.

The methods and main ideas of this paper originate from several papers in the literature. Although his results are stated in the context of a continuous time stochastic control problem with the underlying process being Brownian motion, Bismut's [1973] approach has several similarities to the one here. In particular, the variables are in spaces of stochastic processes, namely, well measurable functions on $\Omega \times [0, \infty)$, and convex optimization theory is used to establish the duality between the primal and dual problems. Although conditions are stated which guarantee the equality of the primal and dual optimal values, Bismut [1973] does not derive the kind of characterization of the dual variables sought in this paper.

With regard to discrete time models, there are two relevant lines of research in the literature. To briefly describe the first, consider an adapted stochastic process $Z = \{Z_n;\ n = 0, 1, \ldots\}$ defined on a filtered probability space $(\Omega, \mathcal{F}, \mathbb{F}, P)$. Let $X = \{X_n;\ n = 1, 2, \ldots\}$ be a predictable stochastic process on the same space. The <u>predictable transform</u> of Z by the process X is defined to be the stochastic integral of X with respect to Z evaluated at some time N, that is,

$$X \cdot Z_N = \sum_{n=1}^{N} X_n \Delta Z_n .$$

The problem is to maximize $E[X \cdot Z_N]$ over a class of predictable processes X for which this expectation is defined.

Problems of this sort were studied by Millar [1968], who assumed $N < \infty$, and Alloin [1969], who allowed $N = \infty$. Alloin also observed this problem can be viewed as a generalized optimal stopping problem because any stopping time T, taking positive values, gives rise to a predictable process X defined by setting $X_n = I_{\{T \geq n\}}$ and for which the predictable transform is the process $Z - Z_0$ stopped at time T.

Alloin [1969] assumed the admissible processes X in the optimization problem are those bounded by a specified scalar. More recently, Kennedy [1981] studied the more general problem where the admissible processes are those in the unit ball of an L^p space, where $p > 0$. He related the solution of the problem to an optimal stopping problem and showed that its form depends on whether $p \leq 1$. In a second paper, Kennedy [1982] applied his results to the economic problem of optimally dividing (each period) a resource between consumption and investment, with the value next period of the invested portion being random. Using concave programming, he showed the Lagrange multipliers for his problem form a stochastic process that can be decomposed into the product of a martingale and a particular random discount factor (which corresponds to the rate of return for the investments). He also showed this dual process can be interpreted as a price system, for in an associated optimization problem where the decision maker can buy or sell unlimited quantitites of the resource at the prevailing price (i.e., current value of the dual process) the optimal consumption schedule turns out to be the same as before.

This last result by Kennedy, that the Lagrange multipliers are related to martingales, gets close to the focus of this paper. The other line of research on duality theory for discrete time stochastic models gets even closer and, indeed, was the original stimulus for this paper.

Rockafellar and Wets [1976] took a general version of a stochastic programming problem and derived a dual optimization problem in which the dual variables are stochastic processes satisfying the martingale type of conditional expectation relationship. To be more specific, and after transforming some of their stochastic programming terminology into probabilistic terms, their result is as follows.

A filtered probability space is specified, where the sample space Ω is a Borel subset of $\mathbb{R}^m$ and the filtration $\mathbb{F} = \{\mathcal{F}_0, \mathcal{F}_1, .., \mathcal{F}_{T-1}\}$, $T < \infty$. The problem is to choose a bounded, predictable process $X = \{X_t;\ t = 1, 2, .., T\}$ so as to minimize the expected value of $g(\omega, X_1, .., X_T)$, where $\omega \in \Omega$ and g is an inf-compact normal convex integrand. In their Theorem 2, Rockafellar and Wets [1976] show that with additional hypotheses there exists a natural dual optimization problem for which the dual variables are stochastic processes $Y = \{Y_t;\ t = 0, 1, .., T\}$ satisfying the martingale relationship

$$E[Y_{t+1} - Y_t | \mathcal{F}_t] = 0 \qquad t = 0, 1, .., T-1.$$

Although the general objective of the present paper is the same as that by Rockafellar and Wets [1976], there are several important differences. First, various assumptions are different. Their predictable processes are bounded, whereas those here are elements of an L^p space, $1 \leq p < \infty$. Also, the objective function here is more general, and there is no assumption, such as made by Rockafellar and Wets [1976], implying the existence of a solution to the primal problem.

More importantly, the methods are significantly different. Rockafellar and Wets [1976] viewed each of the variables $X_1, X_2, .., X_T$ as an element of the space of bounded random variables and proved their result by induction on the time horizon T. The approach here is to view the whole stochastic process X as an element of an L^p space of stochastic processes and then derive the duality results in one step.

The approach taken here has two important consequences. First, the dual variables are shown to be martingales, the key point being that the dual stochastic processes are shown here to be adapted to the original filtration, a result Rockafellar and Wets [1976] did not state in their Theorem 2. Secondly, by not being an induction proof, the approach taken here has potential for being applied to continuous time stochastic control problems.

After some preliminaries, which include the key result that the orthogonal complement of the subspace of predictable processes consists of the martingale difference processes, the basic duality results are presented. These results are rather general, so two examples are discussed in the succeeding section. The paper concludes with some remarks about the economic interpretation of the duality results as well as how things might go with continuous time models.

2. Preliminary Results

Let $(\Omega, \mathcal{F}, \mathbb{F}, P)$ be a filtered probability space, where the filtration $\mathbb{F} = \{\mathcal{F}_t;\ t = 0, 1, .., T\}$, $T < \infty$, $\mathcal{F}_0$ consists of Ω and all the null sets of P, and $\mathcal{F}_T = \mathcal{F}$.

Let $(S, \underline{\underline{O}}(\mathbb{F}), m)$ be the measure space with $S = \Omega \times \{0, 1, .., T\}$ and $\underline{\underline{O}}(\mathbb{F})$ the optional σ-field, that is, the σ-field generated by the adapted stochastic processes. Moreover, m is the bounded measure defined for $A \in \underline{\underline{O}}(\mathbb{F})$ by

$$m(A) = E[\sum_{t=0}^{T} 1_A(\omega, t)]$$

For $1 \le p < \infty$, let $L^p = L^p(S, \underline{\underline{O}}(\mathbb{F}), m)$ denote the L^p-space corresponding to $(S, \underline{\underline{O}}(\mathbb{F}), m)$. Thus L^p consists of all adapted stochastic processes $X = \{X_t;\ t = 0, 1, .., T\}$ such that $|X_t(\omega)|^p$ is m-integrable over S.

As usual, let q denote the conjugate exponent of p so that L^q is the dual space of L^p. Thus each bounded linear functional f(X) on L^p can be represented in the form $f(X) = E[\sum_{t=0}^{T} X_t Y_t]$ for some $Y \in L^q$. In particular, if p = 1, then L^∞, the space of bounded, adapted processes, is the dual of L^1.

Let $\underline{\underline{D}}$ denote the set of all predictable stochastic processes in L^p. In other words, $\underline{\underline{D}}$ consists of all the stochastic processes X in L^p satisfying $X_t \in \mathcal{F}_{t-1}$ for t = 1, 2, .., T.

Let $\underline{\underline{D}}^\perp$ denote the orthogonal complement of $\underline{\underline{D}}$, that is, all the elements in L^q orthogonal to every element of $\underline{\underline{D}}$. To be more specific,

$$\underline{\underline{D}}^\perp = \{Y \in L^q\colon\ E[\sum_{t=0}^{T} X_t Y_t] = 0 \quad \text{for all } X \in \underline{\underline{D}}\}.$$

If X is a stochastic process, then let ΔX denote the corresponding difference process, that is, $\Delta X_0 = 0$ and $\Delta X_t = X_t - X_{t-1}$ for t = 1, 2, .., T.

The key result of this paper is the following.

(1) Proposition. The stochastic process $Y \in \underline{\underline{D}}^{\perp}$ if and only if there exists a martingale $M \in L^q$ such that $Y = \Delta M$.

Remark Since each bounded linear functional $f(X)$ on L^p can also be written in the form $f(X) = E[X_0Y_0 + \sum_{t=1}^{T} X_t \Delta Y_t]$ for some $Y \in L^q$, Proposition (1) can be restated to say that $\underline{\underline{D}}^{\perp}$ consists of all the martingales in L^q that are null at zero.

Proof. The sufficiency is easy to demonstrate. For an arbitrary martingale $M \in L^q$, set $Y = \Delta M$. Under the convention $\Delta M_0 = 0$, the linear functional $f(X)$ corresponding to Y can be written as $f(X) = E[\sum_{t=1}^{T} X_t \Delta M_t]$. Now for arbitrary $X \in \underline{\underline{D}}$ the stochastic process $\sum_{s=1}^{t} X_s \Delta M_s$, being the stochastic integral of a predictable process with respect to a martingale, is itself a martingale. It is also null at zero, so $E[\sum_{t=1}^{T} X_t \Delta M_t] = f(X) = 0$ for all $X \in \underline{\underline{D}}$, in which case $Y = \Delta M \in D^{\perp}$.

Conversely, suppose $Y \in \underline{\underline{D}}^{\perp}$. Clearly $Y_0 = 0$, for $Y_0 \in \mathcal{F}_0$ implies Y_0 is constant, and any non-zero constant Y_0 would lead to $E[\sum_{t=0}^{T} X_tY_t] \neq 0$ for some $X \in \underline{\underline{D}}$.

Setting $M_0 = 0$ and $M_t = Y_t + M_{t-1}$ for $t = 1, 2, .., T$, it remains to show that M is a martingale. Since M is clearly adapted, it suffices to show

(2) $$E[\Delta M_t | \mathcal{F}_{t-1}] = E[Y_t | \mathcal{F}_{t-1}] = 0, \qquad t = 1, 2, .., T.$$

To do this, let $t \geq 1$ and $B \in \mathcal{F}_{t-1}$ be arbitrary, and set $X_t = 1_B$ and $X_s = 0$ for all $s \neq t$. Since $X \in \underline{\underline{D}}$, it follows that

$$E[\sum_{s=0}^{T} X_sY_s] = E[1_BY_t] = 0.$$

This verifies (2) because $B \in \mathcal{F}_{t-1}$ is arbitrary, so this proof is completed.

3. Basic Duality Results

Many discrete time stochastic control problems can be cast in the following (primal) form:

$$\text{(P)} \qquad \text{minimize} \quad f(X) \qquad \text{subject to} \quad X \in \underline{\underline{C}} \cap \underline{\underline{D}}$$

where $\underline{\underline{C}}$ is a convex subset of L^p, $f\colon \underline{\underline{C}} \to \mathbb{R}$ is a convex functional, and $\underline{\underline{D}}$, as above, is the subspace of predictable processes. This being the case, it is natural to apply classical optimization theory in order to establish the duality theory for the primal problem (P).

One line of approach is as follows. Let $\underline{\underline{C}}^*$ denote the conjugate set

$$\underline{\underline{C}}^* = \{Y \in L^q: \sup_{X \in \underline{\underline{C}}} [E[\sum_{t=0}^{T} X_t Y_t] - f(X)] < \infty\},$$

and let f^* denote the conjugate functional on $\underline{\underline{C}}^*$

$$f^*(Y) = \sup_{X \in \underline{\underline{C}}} \{E[\sum_{t=0}^{T} X_t Y_t] - f(X)\}.$$

These lead to the dual problem

$$\text{(D)} \qquad \text{maximize} \qquad -f^*(Y)$$
$$\qquad\qquad \text{subject to} \qquad Y \in \underline{\underline{C}}^* \cap \underline{\underline{D}}^{\perp}.$$

Observe, by Proposition (1), that the variables in dual problem (D) are martingale difference processes. Moreover, for $X \in \underline{\underline{C}}$ and $Y \in \underline{\underline{C}}^*$ the definition of f^* gives

$$-f^*(Y) \le -E[\sum_{t=0}^{T} X_t Y_t] + f(X),$$

so $-f^*(Y) \le f(X)$ for all $Y \in \underline{\underline{C}}^* \cap \underline{\underline{D}}^{\perp}$ and all $X \in \underline{\underline{C}} \cap \underline{\underline{D}}$. In other words, the optimal value of the dual problem (D), abbreviated sup D, is less than or equal to the optimal value of the primal problem (P), abbreviated inf P. An application of the Fenchel Duality Theorem (see, for example, Luenberger [1969, p. 201]) then gives a sufficient condition for equality to actually hold:

(3) Theorem. (Fenchel Duality) Suppose $\underline{\underline{C}} \cap \underline{\underline{D}}$ contains points in the relative interior of $\underline{\underline{C}}$ and $\underline{\underline{D}}$, the epigraph of f over $\underline{\underline{C}}$ has a nonempty interior, and inf P is finite. Then

$$\inf P = \sup D,$$

and there exists a solution Y_0 to the dual problem (D).

Alternative conditions can be given for inf(P) = sup(D). See, for example, Rockafellar [1974, pp. 56-57]. Moreover, conditions can be given that are sufficient for ensuring there exists a solution to primal problem (P). Rather than pursuing these matters any further, however, the discussion will turn to the application of these duality results to some particular control problems.

4. Two Examples

This first example is similar to the stochastic programming model studied by Rockafellar and Wets [1976].

For the primal problem (P) one has $\underline{\underline{C}} = L^p$ and

$$f(X) = E[\sum_{t=0}^{T} g(t, \omega, X_t(\omega))],$$

where g: $\{0, 1, .., T\} \times \Omega \times \mathbb{R} \to \mathbb{R}$ is a function such that $g(t, \omega, \cdot)$ is convex for each fixed (t, ω). Thus f is a convex integral functional, provided it is well-defined in the sense that $g(t, \omega, X_t(\omega))$ is a measurable function of (t, ω) for every $X \in L^p$.

The functional f will be well-defined, according to Rockafellar [1968], if g is what is called a normal convex integrand, a general condition that is implied by a variety of specific situations. Under this condition, Rockafellar [1968] went on to derive an explicit formula for the conjugate functional f^*. Let g^* be the integrand conjugate to g, that is,

$$g^*(t, \omega, y) = \sup_{x \in \mathbb{R}} \{xy - g(t, \omega, x)\}.$$

If $g(t, \omega, X)$ is majorized by an integrable function of (t, ω) for at least one choice of $X \in L^p$ and if $g^*(t, \omega, Y)$ is majorized by an integrable function of (t, ω) for at least one choice of $Y \in L^q$, then Rockafellar [1968] showed that $\underline{\underline{C}}^* = L^q$ and

$$f^*(Y) = E[\sum_{t=0}^{T} g^*(t, \omega, Y_t(\omega))].$$

Thus the dual problem (D) is to maximize $-f^*(Y)$ over the subspace of martingale difference processes in L^q. The first hypothesis in Theorem (3) is automatically satisfied, and the other two are easy to check for particular cases, so if they both hold then inf(P) = sup(D) and there exists a solution to (D).

The second example comes from Pliska's [1982] discrete time stochastic decision model. Let the sample space Ω be the 2^T T-dimensional vectors whose components are either 1 or -1. Let the probability measure P be arbitrary, subject only to the requirement that $P(\omega) > 0$ for all $\omega \in \Omega$. Define a stochastic process $Z = \{Z_t;\ t = 1, 2, .., T\}$ by setting $Z_t(\omega) = \omega_t$, the $t^{\underline{th}}$ component of ω. Let the filtration $\mathbb{F}$ be the one generated by Z. The problem is to minimize $E[u(\sum_{t=1}^{T} X_t Z_t)]$ subject to $X \in \underline{\underline{D}}$, where u is a specified convex function.

This problem fits in the framework of primal problem (P), for $\underline{\underline{C}} = L^p$ and $f(X) = E[u(\sum_{t=1}^{T} X_t Z_t)]$ is convex on $\underline{\underline{C}}$. With suitable assumptions about u the hypotheses in Theorem (3) will be easy to verify, in which case inf(P) = max(D). This will be the case, for example, if u is strictly convex and decreasing, has a continuous second derivative, and satisfies either $u'(x) \to 0$ as $x \to \infty$ or $u'(x) \to -\infty$ as $x \to -\infty$. In fact, as shown in Pliska [1982], under these particular conditions there will exist a solution to the primal problem (P).

5. Concluding Remarks

Besides being of general theoretical interest, the primary importance of the duality theory has to do with applications of stochastic control models where the dual variables can be interpreted as prices. Knowing by the duality theory presented above that the price processes are actually martingales may have significant economic implications. Of course, the interpretation of Lagrange multipliers and dual variables as prices is well-known for economic models. This interpretation is discussed in the context of linear programming in Dantzig [1963], for example. Some references involving more general settings are Arrow, Hurwicz and Uzawa [1958], Baumol [1977], and Gale [1960]. The paper by Rockafellar and Wets [1976] enumerates several additional papers dealing with price systems associated with constraints appearing in multistage stochastic programs.

One economic application where duality theory may be of interest is the area of consumption-investment problems. Such problems, as well as variations such as optimal capital accumulation under uncertainty and resource allocation under uncertainty, have been extensively studied in the economics literature; see, for example, Arrow and Hurwicz [1977], Brock and Mirman [1973], Mirman [1971], and Lucas and Prescott [1971]. The basic idea is that each period the decision maker must divide his wealth between current consumption, yielding immediate utility, and investment, which becomes worth a random amount next period. This leads to a trade-off between current and future consumption. Welch [1979], Zilcha [1976], and, as mentioned at the beginning of this paper, Kennedy [1982] studied particular versions of this problem and showed the Lagrange multipliers could be interpreted as prices.

Rockafellar and Wets [1976] derived their results in the context of the variables being bounded stochastic processes. The specific approach taken here would not work in that case, because the dual of L^∞ is not L^1 and so one could not proceed as in Proposition (1). However, it is possible that one could apply one aspect of their approach in order to overcome this difficulty without resorting to their induction proof methodology. This would be to add more structure to the convex functional f in primal problem (P) and then apply the results of Rockafellar [1971] to show that $\underline{\underline{C}}^* \cap D^\perp$ is actually in L^1, even though L^1 is not the dual of L^∞.

The principal advantage of the approach here over that by Rockafellar and Wets [1976] is that it does not rely on a proof by induction on the time variable. Thus this method has the potential of being suitable for continuous time stochastic control problems. Indeed, for $p < \infty$ one could proceed as above and compute $\underline{\underline{D}}^\perp$ but with linear functionals of the form $E\left[\int_0^T X_t Y_t dt\right]$ one would get the uninteresting result

$\underline{\underline{D}}^{\perp} = \emptyset$. On the other hand, with linear functionals of the form $E\left[\int_0^T X_t dY_t\right]$ one sees that $\underline{\underline{D}}^{\perp}$ contains all the martingales in L^q null at zero. These interesting issues deserve further study.

References

1. Alloin, C., "Processus prévisible optimaux associés a un processus stochastique," Cahiers Centre Etud. Rech. Opér. 11(1969), 92-103.

2. Arrow, K. J., L. Hurwicz, editors, Studies in Resource Allocation Processes, Cambridge University Press, Cambridge-New York, 1977.

3. Arrow, K. J., and L. Hurwicz and H. Uzawa, Studies in Linear and Non-Linear Programming, Stanford University Press, Palo Alto, California, 1958.

4. Baumol, W. J., Economic Theory and Operations Analysis, Prentice-Hall, Englewood Cliffs, N.J., 1977.

5. Bismut, J. M., "Conjugate Convex Functions in Optimal Stochastic Control," J. Math. Anal. Appl. 44(1973), 384-404.

6. Brock, W. A., and L. J. Mirman, "Optimal Economic Growth and Uncertainty: The Discounted Case," J. Econ. Theory 4(1972), 479-513.

7. Dantzig, G. B., Linear Programming and Extensions, Princeton University Press, Princeton, N.J., 1963.

8. Gale, D., The Theory of Linear Economic Models, McGraw-Hill, New York, 1960.

9. Kennedy, D. P., "Optimal Predictable Transforms," Stochastics 5(1981), 323-334.

10. Kennedy, D. P., "Stimulating Prices in a Stochastic Model of Resource Allocation," preprint (1982).

11. Lucas, Jr., R. E., and E. C. Prescott, "Investment Under Uncertainty," Econometrica 39(1971), 659-681.

12. Luenberger, D. G., Optimization by Vector Space Methods, John Wiley, New York, 1969.

13. Millar, P. W., "Transforms of Stochastic Processes," Ann. Math. Stat. 39(1968), 372-376.

14. Mirman, L. J., "Uncertainty and Optimal Consumption Decisions," Econometrica 39(1971), 179-185.

15. Pliska, S. R., "A Discrete Time Stochastic Decision Model," Proceedings of the IFIP Working Conference on Recent Advances in Filtering and Optimization, edited by W. H. Fleming and L. G. Gorostiza, to be published by Springer-Verlag, New York-Berlin, 1982.

16. Rockafellar, R. T., "Integrals which are Convex Functionals," Pacific J. Math. 24(1968), 525-539.

17. Rockafellar, R. T., "Integrals which are Convex Functionals, II," Pacific J. Math. 39(1971), 439-469.

18. Rockafellar, R. T., "Conjugate Duality and Optimization," Regional Conference Series in Applied Mathematics 16, Society for Industrial and Applied Mathematics, Philadelphia, 1974.

19. Rockafellar, R. T., and R. Wets, "Nonanticipativity and L^1-Martingales in Stochastic Optimization Problems," Mathematical Programming Study 6, Stochastic Systems, Modeling, Identification and Optimization, II, North Holland, 1976.

20. Welch, R. L., "The Representation of Shadow Values in Resource Allocation Teams," J. Econ. Theory 20(1979), 23-30.

21. Zilcha, I., "Characterization by Prices of Optimal Programs Under Uncertainty," J. Math. Econ. 3(1976), 173-184.

ON THE CONTROL OF JUMP PROCESSES

H.Pragarauskas

§ 1. Introduction

In this paper we present several results concerning the solvability of the Bellman's equation for controlled jump processes.

Let R^d be a d-dimensional Euclidean space, $T\in(0,\infty)$, $H_T=(0,T)\times R^d$, $\bar{H}_T=[0,T]\times R^d$, A a separable metric space, $L^p(R^d,\Pi)$ a space of functions $u: R^d\to R^d$ such that $\|u\|_{p,\Pi}=\left\{\int|u(z)|^p\Pi(dz)\right\}^{1/p}<\infty$, $p\in(0,\infty)$, $\Pi(dz)=dz/|z|^{d+1}$, $L^p_{loc}(H_T)$ a class of functions $u:[0,T]\times R^d\to R^1$, such that $\int_Q|u(t,x)|^p dt\,dx<\infty$ for every compact subset $Q\subset H_T$, $S_R=\{x\in R^d: |x|<R\}$, $C_{T,R}=(0,T)\times S_R$.

Let for all $\alpha\in A$, $(t,x)\in\bar{H}_T$ be defined: $b(\alpha,t,x)\in R^d$, $c(\alpha,t,x,\cdot)$ an element of $L^2(R^d,\Pi)$, $d(\alpha,t,x,\cdot)$ an element of $L^1(R^d,\Pi)$ and real $r(\alpha,t,x)\geq 0$, $f(\alpha,t,x)$, $g(x)$.

To simplify the presentation, we will assume throughout this paper:

b,c,d,r,f,g are Borel measurable; b,r,f,g are continuous in α and continuous in x uniformly with respect to α for every t; c is continuous in α in sense of the norm $\|\cdot\|_{2,\Pi}$; d is continuous in α in sense of the norm $\|\cdot\|_{1,\Pi}$; for some constants $m,K\geq 0$ and all $\alpha\in A$, $t\in[0,T]$, $x,y\in R^d$

$$|b(\alpha,t,x)|+\|c(\alpha,t,x,\cdot)\|_{2,\Pi}+\|c(\alpha,t,x,\cdot)\|_{2\vee m,\Pi}+$$

$$+\|d(\alpha,t,x,\cdot)\|_{1,\Pi}+\|d(\alpha,t,x,\cdot)\|_{1\vee m,\Pi}\leq K(1+|x|),$$

$$|b(\alpha,t,x)-b(\alpha,t,y)|+\|c(\alpha,t,x,\cdot)-c(\alpha,t,y,\cdot)\|_{2,\Pi}+$$

$$+\|d(\alpha,t,x,\cdot)-d(\alpha,t,y,\cdot)\|_{1,\Pi} \le K|x-y|,$$

$$r(\alpha,t,x)+|f(\alpha,t,x)|+|g(x)| \le K(1+|x|)^m.$$

The constants m, K are fixed throughout this paper.

Let $(\Omega,\mathcal{F},P)$ be a complete probability space with a family $(\mathcal{F}_t)$ of complete non-decreasing σ-algebras $\mathcal{F}_t\subset\mathcal{F}$, $(z_t,\mathcal{F}_t)$ a d-dimensional Cauchy process with a Levy measure Π, $p(dtdz)$ a Poisson measure on $[0,\infty)\times R^d$ constructed from the jumps of z_t and $q(dtdz)$ a corresponding Poisson martingale measure.

Denote by $\mathfrak{A}$ a class of all processes $\alpha:\Omega\times[0,T]\to A$ progressively measurable with respect to $(\mathcal{F}_t)$.

For every strategy $\alpha\in\mathfrak{A}$ and $(s,x)\in H_T$ we set into correspondence the solution $x_t^{\alpha,s,x}$ of the equation

$$x_t = x+\int_0^t b(\alpha_r,s+r,x_r)dr + \int_0^t\int c(\alpha_r,s+r,x_r,z)q(drdz)$$

and the solution $y_t^{\alpha,s,x}$ of the equation

$$y_t = x+\int_0^t b(\alpha_r,s+r,y_r)dr + \int_0^t\int d(\alpha_r,s+r,y_r,z)p(drdz).$$

For $\alpha\in\mathfrak{A}$, $(s,x)\in\overline{H}_T$ set

$$v^\alpha(s,x) = E\left\{\int_0^{T-s} e^{-\varphi_t^{\alpha,s,x}} f(\alpha_t,s+t,x_t^{\alpha,s,x})dt + e^{-\varphi_{T-s}^{\alpha,s,x}} g(x_{T-s}^{\alpha,s,x})\right\}, \tag{1}$$

where $\varphi_t^{\alpha,s,x}=\int_0^t r(\alpha_u,s+u,x_u^{\alpha,s,x})du$. Define the function $w^\alpha(s,x)$ by the formula (1), replacing $x_t^{\alpha,s,x}$ with $y_t^{\alpha,s,x}$. Set

$$v(s,x)=\sup_{\alpha\in\mathfrak{A}} v^\alpha(s,x), \quad w(s,x)=\sup_{\alpha\in\mathfrak{A}} w^\alpha(s,x).$$

In this paper we will give sufficient conditions under which the payoff functions v, w are unique solutions of the problems

$$u_t + \sup_{\alpha\in A} (\mathcal{L}_1^\alpha u + f(\alpha)) = 0 \text{ a.e. } H_T, \; u(T,\cdot) = g(\cdot) \tag{2}$$

$$u_t + \sup_{\alpha\in A} (\mathcal{L}_2^\alpha u + f(\alpha)) = 0 \text{ a.e. } H_T, \; u(T,\cdot) = g(\cdot) \tag{3}$$

respectively in suitable classes of functions, where

$$\mathcal{L}_1^\alpha u(t,x) = \sum_{i=1}^d b_i(\alpha,t,x) u_{x_i}(t,x) + \int \nabla^2_{c(\alpha,t,x,z)} u(t,x)\, \Pi(dz) - r(\alpha,t,x) u(t,x),$$

$$\mathcal{L}_2^\alpha u(t,x) = \sum_{i=1}^d b_i(\alpha,t,x) u_{x_i}(t,x) + \int \nabla^1_{d(\alpha,t,x,z)} u(t,x)\, \Pi(dz) - r(\alpha,t,x) u(t,x),$$

$$\nabla_y^1 u(t,x) = u(t,x+y) - u(t,x),$$

$$\nabla_y^2 u(t,x) = \nabla_y^1 u(t,x) - \sum_{i=1}^d u_{x_i}(t,x) y_i$$

and u_t, u_{x_i}, $i=1,\ldots,d$ are partial derivatives in Sobolev sense.

§ 2. The problem (2). Existence of a solution

Assumption 2.1. For some constant $\delta \in (1,2]$ and all $\alpha \in A$, $t \in [0,T]$, $R>0$, $x, x\pm y \in S_R$

$$\|c(\alpha,t,x,\cdot)\|_{4m \vee 2(m+2),\Pi} \le K(1+R),$$

$$\|c(\alpha,t,x+y,\cdot) - c(\alpha,t,x,\cdot)\|_{2(m+2),\Pi} \le K|y|,$$

$$|\gamma(\alpha,t,x+y) - \gamma(\alpha,t,x)| \le K|y|(1+R)^m, \quad \gamma = r, f, g,$$

$$\Delta_y^2 \gamma(\alpha,t,x) \ge -K|y|^\delta (1+R)^m, \quad \gamma = f, g,$$

$$|\Delta_y^2 r(\alpha,t,x)| + |\Delta_y^2 b(\alpha,t,x)| + \|\Delta_y^2 c(\alpha,t,x,\cdot)\|_{2,\Pi} \le K|y|^\delta (1+R)^m,$$

where $\Delta_y^2 \gamma(\alpha,t,x,z) = \gamma(\alpha,t,x+y,z) + \gamma(\alpha,t,x-y,z) - 2\gamma(\alpha,t,x,z)$.

Assumption 2.2. For all $\alpha \in A$, $t_1, t_2 \in [0,T]$, $x \in R^d$

$$|b(\alpha,t_1,x) - b(\alpha,t_2,x)| + \|c(\alpha,t_1,x,\cdot) - c(\alpha,t_2,x,\cdot)\|_{2,\Pi} + |r(\alpha,t_1,x) - r(\alpha,t_2,x)| + |f(\alpha,t_1,x) - f(\alpha,t_2,x)| \le K|t_1-t_2|(1+|x|)^m.$$

Let us define the measure $\pi(\alpha,t,x,dy)$ on Borel subsets of R^d by the formulae:

$$\pi(\alpha,t,x,\{0\})=0,\quad \pi(\alpha,t,x,dy)=\Pi(z: c(\alpha,t,x,z)\in dy\setminus\{0\}).$$

Assumption 2.3. For every $R>0$ there exist a number $\rho(R)>0$ and a measure $\bar{\pi}_R(dy)$ on Borel subsets of $S_{\rho(R)}$ such that for all $\alpha\in A$, $(t,x)\in C_{T,R}$

$\pi(\alpha,t,x,dy)\leq\bar{\pi}_R(dy)$ on Borel subsets of $S_{\rho(R)}$ and

$\int |y|^{\delta_1}\bar{\pi}_R(dy)<\infty$ for some $\delta_1\in(1,2]$ and all $R>0$.

Definition 2.4. Let $\delta\in(1,2]$. Denote by $\Lambda_m^{1,\delta}$ a class of all continuous functions $u(\cdot,\cdot)$ on $\bar{H}_T$ such that:

1) for some constant N and all $(t,x)\in\bar{H}_T$

$$|u(t,x)|\leq N(1+|x|)^m;$$

2) there exists partial derivatives $u_t, u_{x_i}, i=1,\dots,d$ in Sobolev sense and for every $R>0$

$$\operatorname*{ess\,sup}_{C_{T,R}}|\operatorname{grad}_x u(t,x)|<\infty,\quad \operatorname*{ess\,sup}_{C_{T,R}} u_t(t,x)<\infty,$$

$$\operatorname*{ess\,inf}_{C_{T,R}}\ \inf_{|y|\leq R}|y|^{-\delta}\nabla_y^2 u(t,x)>-\infty.$$

Theorem 2.5. Let the assumption 2.1 with $\delta=2$ and the assumption 2.2 (respectively the assumption 2.3 with $\delta_1=2$) hold. Then $v\in\Lambda_m^{1,2}$ and satisfies the problem (2). Moreover $v_t\in L^\infty_{loc}(H_T)$ (respectively $v_t\in L^2_{loc}(H_T)$).

Theorem 2.6. Let $\delta_0\in(1,2)$ and for all $\alpha\in A$, $(t,x)\in\bar{H}_T$

$$\|c(\alpha,t,x,\cdot)\|_{\delta_0,\Pi}\leq K(1+|x|).$$

Let the assumption 2.1 for some $\delta\in(\delta_0,2)$ and the assumption 2.2 (respectively assumption 2.3 with $\delta_1=\delta_0$, respectively assumption 2.1 with $\delta=2$) hold. Then $v\in\Lambda_m^{1,\delta}$ and satisfies the problem (2). Moreover

$v_t \in L^{\infty}_{loc}(H_T)$ (respectively $v_t \in L^2_{loc}(H_T)$, respectively $v_t \in L^{p}_{loc}(H_T)$ for every $p \in [1, \frac{1}{\delta_0 - 1})$).

These results are obtained on the basis of the papers [1]-[3].

§ 3. The problem (2). Uniqueness of a solution

Let $\rho > 0$. Define the measure $\pi^{\rho}(\alpha,t,x,dy)$ on Borel subsets of R^d by the formulae:

$$\pi^{\rho}(\alpha,t,x,\{0\}) = 0, \quad \pi^{\rho}(\alpha,t,x,dy) = \Pi(z: |z| \le \rho,\ c(\alpha,t,x,z) \in dy \setminus \{0\}).$$

Assumption 3.1. For every $R > 0$ there exist: a number $\rho_0(R) > 0$ and measures $\bar{\pi}^{\rho,R}(dy)$, $\rho \in (0, \rho_0(R)]$ on Borel subsets of R^d such that for all $\alpha \in A$, $(t,x) \in C_{T,R}$, $\rho \in (0, \rho_0(R)]$

$$\pi^{\rho}(\alpha,t,x,dy) \le \bar{\pi}^{\rho,R}(dy)$$

and for some $\delta \in (1,2]$ and all $R > 0$

$$\int |y|^{\delta}\, \bar{\pi}^{\rho_0(R),R}(dy) < \infty,$$

$$\lim_{\rho \to 0} \int |y|^{\delta}\, \bar{\pi}^{\rho,R}(dy) = 0.$$

Theorem 3.2. Let the assumption 3.1 with $\delta = 2$ holds. Then the problem (2) has at most one solution in the class $\Lambda^{1,2}_{m}$.

Theorem 3.3. Let $\delta_0 \in (1,2)$ and for all $\alpha \in A$, $(t,x) \in \bar{H}_T$

$$\| c(\alpha,t,x,\cdot) \|_{\delta_0,\Pi} \le K(1+|x|).$$

Let the assumption 3.1 with $\delta = \delta_0$ holds. Then the problem (2) has at most one solution in the class $\bigcup_{\delta > \delta_0} \Lambda^{1,\delta}_{m}$.

These results are obtained on the basis of the paper [4].

§ 4. The problem (3)

Assumption 4.1. For some constant $\delta \in (0,1]$ and all $t \in [0,T]$, $R > 0$,

$x, x+y \in S_R$

$$|r(\alpha,t,x+y)-r(\alpha,t,x)| + |f(\alpha,t,x+y)-f(\alpha,t,x)| + \\ + |g(x+y)-g(x)| \le K|y|^{\delta}(1+R)^m.$$

Definition 4.2. Let $\delta \in (0,1]$. Denote by $\Lambda_m^{\bar{1},\bar{\delta}}$ a class of all functions $u(\cdot,\cdot)$ on $\bar{H}_T$ such that:

1) for some constant N and all $(t,x) \in \bar{H}_T$

$$|u(t,x)| \le N(1+|x|)^m,$$

2) for every $R>0$ there exists a constant N_R such that for all $s,t \in [0,T]$, $x,y \in S_R$

$$|u(t,x)-u(s,y)| \le N_R\left(|t-s|+|x-y|^{\delta}\right).$$

Theorem 4.3. Let the assumption 4.1 with $\delta = 1$ holds. Then $w \in \Lambda_m^{\bar{1},\bar{1}}$ and satisfies the problem (3).

Theorem 4.4. Let $\delta_0 \in (0,1)$ and for all $\alpha \in A$, $(t,x) \in \bar{H}_T$

$$\| d(\alpha,t,x,\cdot)\|_{\delta_0,\Pi} \le K(1+|x|).$$

Let the assumption 4.1 holds for some $\delta \in (\delta_0, 1]$, $b \equiv 0$. Then $w \in \Lambda_m^{\bar{1},\bar{\delta}}$ and satisfies the problem (3).

Definition 4.5. Let $\delta \in (0,1]$, $\delta_1 \in (\delta, 2]$. Denote by $\Lambda_m^{\bar{1},\bar{\delta},\underline{\delta_1}}$ a class of all functions $u \in \Lambda_m^{\bar{1},\bar{\delta}}$ such that for every $R>0$

$$\operatorname*{ess\,inf}_{C_{T,R}} \inf_{|y|\le R} |y|^{-\delta_1} \Delta_y^2 u(t,x) > -\infty .$$

Theorem 4.6. The problem (3) has at most one solution in the class $\bigcup_{\delta_1>1} \Lambda_m^{\bar{1},\bar{1},\underline{\delta_1}}$.

Theorem 4.7. Let $\delta_0 \in (0,1)$ and for all $\alpha \in A$, $(t,x) \in \bar{H}_T$

$$\| d(\alpha,t,x,\cdot)\|_{\delta_0,\Pi} \le K(1+|x|).$$

Let $b \equiv 0$. Then the problem (3) has at most one solution in the class $\bigcup_{\delta_1>\delta_0} \Lambda_m^{\bar{1},\bar{\delta_0},\underline{\delta_1}}$.

These results are obtained on the basis of the paper [5].

References

1. H.Pragarauskas, On Bellman equation for controlled degenerated general stochastic processes, Lecture Notes in Control and Information Sciences, vol.25, p.69-79, Springer-Verlag, 1980.
2. N.V.Krylov, H.Pragarauskas, On traditional derivation of Bellman equation for general controlled stochastic processes, "Lietuvos matem.rink.", XXI (1981), No 2, p.101-110 (in Russian).
3. H.Pragarauskas, On the Bellman equation in a lattice of measures for general controlled stochastic processes, I, II, "Lietuvos matem.rink.", XXI (1981), No 4, p.169-184, XXII (1982), No 1, p.138-145 (in Russian).
4. H.Pragarauskas, On the uniqueness of a solution of the Bellman equation associated with general controlled stochastic processes, "Lietuvos matem.rink.", XXII (1982), No 2, p. 137-149 (in Russian).
5. H.Pragarauskas, On first order singular Bellman equation, "Lecture Notes inControl and Information Sciences", vol.36, p. 175-188, Springer-Verlag, 1981.

H.Pragarauskas

Institute of Mathematics and Cybernetics,
Academy of Sciences of the Lithuanian SSR,
Vilnius, K.Poželos 54, USSR.

A PARTIALLY OBSERVED INVENTORY PROBLEM

Raymond Rishel
Department of Mathematics
University of Kentucky
Lexington, KY 40506

I. INTRODUCTION

There has been a great deal of research on impulse control of processes with complete observations of their past, for instance as in [1][2][3][5][6][9][10]. Impulse control of processes based on partial information about their past has been much less widely considered. The objective of this paper is to approach this area through consideration of a particular example which can be treated by relatively elementary means. The example is a particular partially observed inventory problem. The problem has special structure in that the measured process has a finite number of states, and that by restricting the class of controls a renewal structure can be exploited. However, it is felt that a number of problems will have similar structure and could be treated in an analogous manner. The method used is to proceed analogously to discrete time dynamic programming by conditioning with respect to the past measurements up to the renewal times. Through this conditioning a deterministic problem which characterizes the stochastic problem is obtained.

II. DESCRIPTION OF THE INVENTORY PROBLEM

Consider an inventory process in which it can only be observed whether the inventory is strictly greater than or less than or equal to a fixed level K. Decisions to order new stock are to be based on the past of these observations.

We shall also assume that the inventory process satisfies the following additional properties:

I. Each customer purchases one item. The arrivals of customers are given by a Poisson process $D(t)$ which has parameter μ and right continuous sample paths.

II. Orders for new stock must be of fixed size N and the new stock arrives instantaneously.

III. There is a holding cost h per unit time for each item in the inventory and a shortage cost s per unit time when the inventory is empty. Placing an order for new stock incurs a cost c. Future costs are discounted at the rate $e^{-\alpha t}$.

IV. Demands for goods made when the inventory is empty are not filled and are considered to be lost from the system.

V. The initial inventory I_o is a known constant greater than K.

VI. All constants which have been defined are positive.

Let τ_n denote the time that the n-th order for new stock is placed. The τ_n's must satisfy

$$\tau_1 < \tau_2 < \cdots < \tau_n < \cdots \tag{1}$$

We shall allow the τ_n's to assume the value $+\infty$. The interpretation of this is that when $\tau_n = +\infty$ fewer than n orders are placed. If $\tau_n = +\infty$ then (1) requires $\tau_{n+k} = +\infty$ for each $k > 0$. Define the corresponding inventory process I(t) by

$$\begin{aligned} I(t) &= [I_0 - D(t)]^+ && \text{if } 0 \le t \le \tau_1 \\ I(t) &= [I(\tau_{n-1}) + N - (D(t) - D(\tau_{n-1}))]^+ && \text{if } \tau_{n-1} < t \le \tau_n \end{aligned} \tag{2}$$

For I(t) to be defined on $[0,\infty)$ we must have

$$\lim_{n\to\infty} \tau_n = +\infty . \tag{3}$$

Define the measurement process m(t) by

$$m(t) = \begin{cases} 1 & \text{if } I(t) > K \\ 0 & \text{if } I(t) \le K \end{cases} \tag{4}$$

Let

$$\gamma_1, \gamma_2, \ldots \tag{5}$$

denote the successive times at which m(t) changes value from one to zero or from zero to one. The same convention about infinite values of τ_n applies to γ_n.

REMARK 1: The values of m(t) must alternate between 0 and 1. Assumption V. implies m(0)=1. Thus knowing

$$\gamma_1, \gamma_2, \ldots$$

completely determines m(t) through the formula

$$m(t) = \begin{cases} 1 & \text{if } 0 \le t < \gamma_1 \text{ or } \gamma_{2j-2} < t < \gamma_{2j-1} \\ 0 & \text{if } \gamma_{2j-1} \le t \le \gamma_{2j} \end{cases} \tag{6}$$

REMARK 2: Due to the assumption that orders for new stock are filled instantaneously, there are some problems in defining the inventory if orders are allowed to be placed exactly at a time at which the inventory has a downward jump. The definition (2) has been chosen to avoid these difficulties.

Consider the case in which the initial inventory is one unit and the ordering policy for new stock is *replace each unit as soon as a sale is made*. If K=0 the measured process m(t) is just the inventory process I(t) in this case. Definition (2) implies the inventory is one unit just before a sale, zero at the time of the sale and one unit just after the sale. If the inventory had been defined to be right continuous, the inventory would be the constant one and it would not be possible to tell when the sales were made from the inventory.

Next we shall describe the requirement that the order times be based on the past of the measurement process m(t). Notice that an order should never be placed while m(t)=1. Since the orders arrive instantaneously, the holding cost on the order over the interval until m(t) becomes zero can be saved by waiting and placing the order when m(t) becomes zero. It will simplify our formulation to require for all our policies that no orders be placed while m(t)=1. To meet this and the requirement that the order times be based on the past of the measured process, require that each τ_n have the following form.

For each $j \geq 1$ there are extended real valued Borel measurable functions T_n defined on E^{2j-1} such that[1]

$$\tau_n = \sum_{j=1}^{\infty} 1_{\{\gamma_{2j-1} \leq \tau_n \leq \gamma_{2j}\}} T_n(\gamma_1,\ldots,\gamma_{2j-1}) \tag{7}$$

and

$$\{\gamma_{2j-1} \leq \tau_n \leq \gamma_{2j}\} = \{\gamma_{2j-1} \leq T_n(\gamma_1,\ldots,\gamma_{2j-1}) \leq \gamma_{2j}\} . \tag{8}$$

Condition (7) asserts that τ_n only occurs on the sets where m(t) = 0 and that when it does occur on one of these sets it is expressible as a function of the past times of change $\gamma_1,\ldots,\gamma_{2j-1}$ of m(t). Remark 1 implies if $\gamma_{2j-1} \leq t \leq \gamma_{2j}$ that m(s) on $0 \leq s \leq t$ is completely determined by $\gamma_1,\ldots,\gamma_{2j-1}$ through (6). Thus (7) gives a general expression for an order time which depends on the past of m. The condition (8) asserts that τ_n occurs between γ_{2j-1} and γ_{2j} if and only if $T_n(\gamma_1,\ldots,\gamma_{2j-1})$ does. In particular if at time $T_n(\gamma_1,\ldots,\gamma_{2j-1})$, γ_{2j} has not yet occurred, then τ_n does occur at time $T_n(\gamma_1,\ldots,\gamma_{2j-1})$.

Define a sequence

$$\tau_1,\tau_2,\ldots,\tau_n,\ldots$$

of order times to be an admissible policy if (1) and (3) are satisfied and if each order time has the form (7)(8).

From assumption III. in our description of the inventory problem, we see that given an admissible policy

$$(\tau_1,\tau_2,\ldots,\tau_n,\ldots)$$

the corresponding long-term discounted expected cost over the infinite horizon is given by

$$E\left[\sum_{n=1}^{\infty} c\, e^{-\alpha\tau_n} + \int_0^{\infty} e^{-\alpha t}\,[hI(t)+s1_{\{I(t)=0\}}]dt\right] . \tag{9}$$

Consider the problem of choosing a policy in the class of admissible ordering policies so that the long-term discounted expected cost is minimized.

III. OBTAINING A DETERMINISTIC PROBLEM THROUGH CONDITIONING

Since customers buy one item at a time the inventory decreases by one unit each time a demand is made. Thus at each of the times γ_{2j-1} at which m(t) changes from

[1]The notation 1_A denotes the characteristic function of the set A.

one to zero the inventory I(t) must change from K+1 to K. Thus at each of the times γ_{2j-1} the value of the inventory is known to be the same value K.

Let $J(\gamma_1,\ldots,\gamma_{2j-1})$ denote the conditional remaining cost from the time γ_{2j-1} onward given the random variables $\gamma_1,\gamma_2,\ldots,\gamma_{2j-1}$. From III. and (9) we see that this is given by

$$J(\gamma_1,\ldots,\gamma_{2j-1}) = E\left[\sum_{\tau_n\geq\gamma_{2j-1}} c\, e^{-\alpha\tau_n} + \int_{\gamma_{2j-1}}^{\infty} e^{-\alpha t}(hI(t)+s1_{\{I(t)=0\}})dt \,\Big|\, \gamma_1,\ldots,\gamma_{2j-1}\right] . \tag{10}$$

In the remainder of the paper, to avoid lengthy expressions we shall abbreviate by letting

$$\Gamma_j = (\gamma_1,\ldots,\gamma_{2j-1}) . \tag{11}$$

Breaking up (10) into the cost from γ_{2j-1} to γ_{2j+1} and the cost from γ_{2j+1} onward, and using the law of iterated conditional expectations, (10) can be written as

$$\begin{aligned} J(\Gamma_j) &= E\left[\sum_n 1_{\{\gamma_{2j-1}\leq\tau_n\leq\gamma_{2j}\}} c\, e^{-\alpha\tau_n} \,\Big|\, \Gamma_j\right] \\ &\quad + E\left[\int_{\gamma_{2j-1}}^{\infty} 1_{\{\gamma_{2j-1}\leq t<\gamma_{2j+1}\}} e^{-\alpha t}(hI(t)+s1_{\{I(t)=0\}})dt \,\Big|\, \Gamma_j\right] + E[J(\Gamma_{j+1})\,|\,\Gamma_j] . \end{aligned} \tag{12}$$

Define $q_i(t,\Gamma_j)$ by

$$q_i(t,\Gamma_j) = E[1_{\{\gamma_{2j-1}\leq t<\gamma_{2j+1}\}} 1_{\{I(t)=i\}}\,|\,\Gamma_j] . \tag{13}$$

Consider the case in which m is the last integer for which $\tau_m < \tau_{2j-1}$, that is

$$\tau_m < \gamma_{2j-1} \leq \tau_{m+1} . \tag{14}$$

Then our assumption (7) implies that as long as $\gamma_{2j-1} \leq \tau_{m+k} \leq \gamma_{2j}$ there are functions $T_{m+k}(\Gamma_j)$ such that

$$\tau_{m+k} = T_{m+k}(\Gamma_j) .$$

From the definitions of γ_{2j} and γ_{2j+1} it follows that

$$1_{\{\gamma_{2j-1}\leq t\leq\gamma_{2j}\}} = \sum_{i=0}^{K} 1_{\{\gamma_{2j-1}\leq t<\gamma_{2j+1}\}} 1_{\{I(t)=i\}} . \tag{15}$$

Thus in the case above

$$E\left[\sum_n 1_{\{\gamma_{2j-1}\leq\tau_n\leq\gamma_{2j+1}\}} c\, e^{-\alpha\tau_n} \,\Big|\, \Gamma_j\right] = \sum_{k=1}^{\infty} c\, e^{-\alpha T_{m+k}(\Gamma_j)} \left[\sum_{n=0}^{K} q_i(T_{m+k}(\Gamma_j),\Gamma_j)\right] . \tag{16}$$

Now using (12), (16), (13) and interchanging integration and conditional expectation gives under the condition (14) that

$$\begin{aligned} J(\Gamma_j) &= \sum_{k=1}^{\infty} c\, e^{-\alpha T_{m+k}(\Gamma_j)} \left[\sum_{i=0}^{K} q_i(T_{m+k}(\Gamma_j)\right] \\ &\quad + \int_{\gamma_{2j-1}}^{\infty} e^{-\alpha t} \left(\sum_{i\geq1} i\, h\, q_i(t,\Gamma_j) + s\, q_0(t,\Gamma_j)\right) dt + E[J(\Gamma_{j+1})\,|\,\Gamma_j] . \end{aligned} \tag{17}$$

Explicit expressions can be given for $q_i(t,\Gamma_j)$ as follows. Consider the case when (14) holds. Since at time γ_{2j-1} the inventory must be at K and the arrivals of customers are Poisson processes, we have, on the interval $\gamma_{2j-1} \le t \le T_{m+1}(\Gamma_j)$ that

$$q_i(t,\Gamma_j) = \begin{cases} 0 & \text{if } i > K \\[2ex] \dfrac{[\mu(t-\gamma_{2j-1})]^{K-i}}{(K-i)!} e^{-\mu(t-\gamma_{2j-1})} & \text{if } K \ge i \ge 1 \\[2ex] 1 - \displaystyle\sum_{k=1}^{K} \dfrac{[\mu(t-\gamma_{2j-1})]^{(K-k)}}{(K-k)!} e^{-\mu(t-\gamma_{2j-1})} & \text{if } i = 0 \end{cases} \tag{18}$$

At each of the times $T_{m+k}(\Gamma_j)$ there are two possibilities; one that at some previous order time $T_{m+\ell}(\Gamma_j)$ with $1 \le \ell < k$ the inventory went above K, thus

$$\gamma_{2j} < T_{m+k}(\Gamma_j)$$

and (8) would imply no order for stock would be placed at time $T_{m+k}(\Gamma_j)$; the other that this had not happened, that

$$\gamma_{2j} \ge T_{m+k}(\Gamma_j) \quad ,$$

and (8) would imply an order for stock would be placed at time $T_{m+k}(\Gamma_j)$. Thus at time $t = T_{m+k}(\Gamma_j)$, $q(t,\Gamma_j)$ has a jump to $q^+(t,\Gamma_j)$ given by

$$q_i^+(t,\Gamma_j) = \begin{cases} q_i(t,\Gamma_j) + q_{i-N}(t,\Gamma_j) & \text{if } i > K \\ q_{i-N}(t,\Gamma_j) & \text{if } K \ge i \ge N \\ 0 & \text{if } i < N \end{cases} \tag{19}$$

Consider interval $T_{m+k}(\Gamma_j) < t \le T_{m+k+1}(\Gamma_j)$. In order to have $I(t)=i$ and $t < \gamma_{2j+1}$ if $i > K$, we must have $I(t) \ge i$ just after $T_{m+k}(\Gamma_j)$ and $T_{m+k}(\Gamma_j) < \gamma_{2j+1}$. In order to have $I(t)=i$ and $t < \gamma_{2j+1}$ if $0 \le i \le K$, we must have $I(t)$ between K and i just after $T_{m+k}(\Gamma_j)$ and $T_{m+k}(\Gamma_j) < \gamma_{2j+1}$. Thus on the interval $T_{m+k}(\Gamma_j) < t \le T_{m+k+1}(\Gamma_j)$

$$q_i(t,\Gamma_j) = \begin{cases} \displaystyle\sum_{\ell>i} q_\ell^+(T_{m+k}(\Gamma_j),\Gamma_j) \dfrac{[\mu(t-T_{m+k}(\Gamma_j)]^{\ell-i}}{(\ell-i)!} e^{-\mu(t-T_{m+k}(\Gamma_j))} & \text{if } i > K \\[2ex] \displaystyle\sum_{K\ge\ell\ge N} q_\ell^+(T_{m+k}(\Gamma_j),\Gamma_j) \dfrac{[\mu(t-T_{m+k}(\Gamma_j))]^{\ell-i}}{(\ell-i)!} e^{-\mu(t-T_{m+k}(\Gamma_j))} & \text{if } 1 \le i \le K \\[2ex] \displaystyle\sum_{K\ge\ell\ge N} q_\ell^+(T_{m+k}(\Gamma_j),\Gamma_j) \left[1 - \sum_{r=1}^{\ell} \dfrac{[\mu(t-T_{m+k}(\Gamma_j))]^{\ell-r}}{(\ell-r)!} e^{-\mu(t-T_{m+k}(\Gamma_j))}\right] & \text{if } i=0 \end{cases} \tag{20}$$

Since at each time γ_{2j-1} the inventory is known to be exactly K, it seems intuitive that the decision times should have the same functional form after each of these times. In line with this, consider the class of policies which have the following form. If

$$\tau_m < \gamma_{2j-1} \le \tau_{m+1} ,$$

and $\tau_{m+1}, \tau_{m+2}, \ldots$ are to be chosen after γ_{2j-1}, then there is a fixed sequence of constants, $\bar{T}_1, \ldots, \bar{T}_k, \ldots$, independent of m, j and $\gamma_1, \ldots, \gamma_{2j-1}$ such that as long as the measured process m(t) remains zero

$$\tau_{m+k} = \gamma_{2j-1} + \bar{T}_k . \tag{21}$$

In the notation of (7) the defining property of these controls is that if

$$\tau_m < \gamma_{2j-1} \le \tau_{m+1} \quad \text{and} \quad \gamma_{2j-1} \le \tau_{m+k} \le \gamma_{2j} ,$$

then

$$T_{m+k}(\Gamma_j) = \gamma_{2j-1} + \bar{T}_k . \tag{22}$$

For policies of this type, the times between two successive returns of the inventory to level K are independent identically distributed random variables and the inventory between each of these pairs of times has the same statistical properties.

If we multiply (10) by $e^{\alpha\gamma_{2j-1}}$, this has the effect of discounting from time γ_{2j-1}. Since, at each of these times, the inventory is at K and for policies of this type it has the same statistical properties as a function of the time after these times, it must follow that

$$e^{\alpha\gamma_{2j-1}} J(\Gamma_j) = e^{\alpha\gamma_{2j+1}} J(\Gamma_{j+1}) \tag{23}$$

or that these quantities are constant. Let J denote the value of this constant. Notice that J will depend on the policy used but not on j or Γ_j. Formula (23) implies

$$J(\Gamma_j) = e^{-\alpha\gamma_{2j-1}} J . \tag{24}$$

For policies of this type, if we let $r = t-\gamma_{2j-1}$, we see from (18), (19), (20) that

$$q_i(t,\Gamma_j) = q_i(r) \tag{25}$$

and $q_i(r)$ is given by:

On the interval $0 \le r \le \bar{T}_1$,

$$q_i(r) = \begin{cases} 0 & \text{if } i > K \\ \dfrac{[\mu r]^{K-i}}{(K-i)!} e^{-\mu r} & \text{if } K \ge i \ge 1 \\ 1 - \displaystyle\sum_{\ell=1}^{K} \dfrac{[\mu r]^{(K-\ell)}}{(K-\ell)!} e^{-\mu r} & \text{if } i = 0 \end{cases} \tag{26}$$

For each $\bar{T}_k$

$$q_i^+(\bar{T}_k) = \begin{cases} q_i(\bar{T}_k) + q_{i-N}(\bar{T}_k) & \text{if } i > K \\ q_{i-N}(\bar{T}_k) & \text{if } K \geq i \geq N \\ 0 & \text{if } i < N \end{cases} \tag{27}$$

On the interval $\bar{T}_k < r \leq \bar{T}_{k+1}$

$$q_i(r) = \begin{cases} \sum_{\ell > i} q^+(\bar{T}_k) \dfrac{[\mu(r-\bar{T}_k)]^{\ell-i}}{(\ell-i)!} e^{-\mu(r-\bar{T}_k)} & \text{if } i > K \\ \sum_{K\geq\ell\geq N} q_\ell^+(\bar{T}_k) \dfrac{[\mu(r-\bar{T}_k)]^{\ell-i}}{(\ell-i)!} e^{-\mu(r-\bar{T}_k)} & \text{if } 1 \leq i \leq K \\ \sum_{K\geq\ell\geq N} q_\ell^+(\bar{T}_k) \left[1 - \sum_{j=1}^{\ell} \dfrac{[\mu(r-\bar{T}_k)]^{\ell-j}}{(\ell-j)!} e^{-\mu(r-\bar{T}_k)}\right] & \text{if } i = 0 \end{cases} \tag{28}$$

In addition, it is not difficult to show that the conditional density of γ_{2j+1} given Γ_j is given as a function of $r = \gamma_{2j+1} - \gamma_{2j-1}$ by

$$\mu\, q_{k+1}(r) \tag{29}$$

on $[0,\infty)$.

Substituting (24), (25), (29) in (17) and multiplying by $e^{\alpha\gamma_{2j-1}}$ we see that

$$\begin{aligned} J = \sum_k c\, e^{-\alpha\bar{T}_k} &\left[\sum_{i=0}^{K} q_i(\bar{T}_k)\right] \\ &+ \int_0^\infty e^{-\alpha r} \left(\sum_{i=1} i\, h\, q_i(r) + s\, q_0(r)\right) dr \\ &+ J \int_0^\infty e^{-\alpha r} \mu\, q_{K+1}(r)\, dr\,. \end{aligned} \tag{30}$$

The principle of optimality asserts that to minimize the expected cost, we must minimize the conditional remaining cost from a given time onward. Thus if we wish to find the optimum in the class of controls of the form (22), we must find $\bar{T}_1, \bar{T}_2, \ldots, \bar{T}_k, \ldots$ to minimize J where J is a solution of (30) and $q_i(r)$ satisfies (26), (27), (28). Call this problem *the deterministic problem.*

On each of the intervals $0 \leq r < \bar{T}_1$ or $\bar{T}_k < r \leq \bar{T}_{k+1}$, the function $q_i(r)$ in (26) and (29) is a polynomial in r times an exponential function of r. Thus each of the integrals in (30) could be explicitly evaluated over these intervals, and (30) could be used to express J explicitly as a function $\bar{T}_1, \bar{T}_2, \bar{T}_3, \ldots$ Solving the deterministic problem then is the problem of minimizing this function.

Since in any computation it is only possible to exhibit the values of finitely many variables, some approximation may be necessary to reduce the problem to a finite one. One way would be to restrict the number of orders which could be placed on each interval $\gamma_{2j-1} \le t \le \gamma_{2j}$ to some fixed finite number.

A case in which this is not necessary and for which the deterministic problem becomes quite simple is that in which N > K. In this case at each order time the measured process changes from 0 to 1, so that there is exactly one order in in each interval $\gamma_{2j-1} \le t \le \gamma_{2j}$. In this case the deterministic problem reduces to: Find the value of $\bar{T}$ which minimizes J where

$$J = c\, e^{-\alpha\bar{T}} + J \int_0^\infty e^{-\alpha r} \mu\, q_{K+1}(r)\, dr + \int_0^\infty e^{-\alpha r} \left(\sum_{i\ge 1} i\, h\, q_i(r) + s\, q_0(r) \right) dr \tag{31}$$

and $q_i(r)$ satisfies, if $0 \le t \le \bar{T}$

$$q_i(r) = \begin{cases} 0 & \text{if } i > K \\ \dfrac{[\mu r]^{K-i}}{(K-i)!} e^{-\mu r} & \text{if } K \ge i \ge 1 \\ 1 - \sum\limits_{j=1}^{K} \dfrac{[\mu r]^{K-j}}{(K-j)!} e^{-\mu r} & \text{if } i = 0 \end{cases} \tag{32}$$

$$q_i^+(\bar{T}) = \begin{cases} q_{i-N}(\bar{T}) & \text{if } i \ge N \\ 0 & \text{if } i < N \end{cases} \tag{33}$$

if $\bar{T} < r < \infty$,

$$q_i(r) = \begin{cases} \sum\limits_{j>i} q_j^+(\bar{T}) \dfrac{[\mu(r-\bar{T})]^{j-i}}{(j-i)!} e^{-\mu(r-\bar{T})} & \text{if } i > K \\ 0 & \text{if } i \le K\ . \end{cases} \tag{34}$$

The solution of the deterministic problem determines the solution of the stochastic problem in a strong sense as indicated in the following theorem.

THEOREM 1: Let $T_1^*, T_2^*, \ldots, T_k^*, \ldots$ be values of $\bar{T}_1, \bar{T}_2, \ldots, \bar{T}_k, \ldots$ which give a solution of the deterministic problem. Then the control of the form (7)(8) defined by

$$T_{m+k}(\Gamma_j) = \gamma_{2j-1} + T_k^*$$

when $\tau_m < \gamma_{2j-1} \le \tau_{m+1}$ and $\gamma_{2j-1} \le \tau_{m+k} \le \gamma_{2j}$ is optimal for the stochastic control problem in the class of all controls satisfying (1),(3),(7),(8).

We shall not give the proof of Theorem 1, but a proof can be given entirely analogously to standard proofs in discrete time discounted cost dynamic programming as in, for instance [3][4][8].

The operator T which is involved in these proofs in this case is the operator which maps functions $f(\Gamma_{j+1})$ into functions of Γ_j defined by

$$T[f](\Gamma_j) = \sum_{k=1}^{\infty} c\, e^{-\alpha T_{m+K}(\Gamma_j)} \left[\sum_{i=0}^{K} q_i(T_{m+k}(\Gamma_j))\right]$$

$$+ \int_{\gamma_{2j-1}}^{\infty} e^{-\alpha t} \left(\sum_{i=1} i\, h\, q_i(t,\Gamma_j) + s\, q_o(t,\Gamma_j)\right) dt + E[f(\Gamma_{j+1})|\Gamma_j] \,. \tag{35}$$

REFERENCES

[1] A. Bensoussan and J.L. Lions, "Control Impulsionnel et Temps d'Arret: Inequations Variationnelles et Quasi-Variationnelles d'Evolution," Cahlur de Mathematiques de la Decision #7523, University de Dauphine - Paris IX, (1975).

[2] A. Bensoussan and J.L. Lions, "Applications des Inequations Variationnelles en Controle Stochastique," Dunod, Paris, (1978).

[3] D. Bertsekis, "Dynamic Programming and Stochastic Control," Academic Press, (1976).

[4] H. Kushner, "Introduction to Stochastic Control," Holt, Rinehart, Winston, (1971).

[5] J. Lepeltier and B. Marchal, "Techniques Probabilistes dans le Controle Impulsionnel," Stochastics, 2 (1979), pp. 243-286.

[6] A. Makowski, "Dynamic Programming for Problems of Impulse Control," University of Kentucky, Ph.D. Thesis, (1981).

[7] M. Robin, "Controle Impulsionnel des Processus de Markov," These de Doctorat d'Etat, (1978), University de Dauphine - Paris IX.

[8] S. Ross, "Applied Probability Models with Optimization Applications," Holden-Day, 1970.

[9] A.N. Shiryayev, "Optimal Stopping Rules," Springer-Verlag, 1978.

[10] J. Zabczyk, "Introduction to the Theory of Optimal Stopping," in Stochastic Control Theory and Stochastic Differential Systems, Lecture Notes in Control and Information Sciences, Vol.16, pp. 227-250, Springer-Verlag, (1979).

ON IMPULSIVE CONTROL WITH LONG RUN AVERAGE COST CRITERION

Łukasz Stettner

Institute of Mathematics
Polish Academy of Sciences
Sniadeckich 8, 00-950 Warsaw
Poland

1. SUMMARY

The impulsive control with long run average cost criterion was considered first by M. Robin in [2], for Markov processes having nice ergodic properties. The aim of this paper is to complete and extend the results of paper [2]. In particular we show that for Fellerien Markov processes the optimal value is constant and find optimal or ε-optimal strategies. We also prove, that the use of general stopping times, instead of those of the form $\tau_i = \tau_{i-1} + \sigma_i \circ \theta_{\tau_{i-1}}$ as in the paper [2] does not change the optimal value of the functional. Results are only reported here and the detailed proofs will appear elsewhere, see [3].

2. INTRODUCTION

Let $\Omega = D(R^+, E)$ the space of right continuous, left limited functions from R^+ into E, a locally compact with countable base, state space.

Let $x_t(\omega) = \omega(t)$ for any $\omega \in \Omega$, $F_t^o = \sigma\{x_s, s \le t\}$, and F_t be universally completed σ-field of F_t^o.

Suppose $X = (\Omega, F_t, \theta_t, x_t, P_x)$ is a homogeneous Markov process with the semigroup $(\Phi(t))_{t \ge 0}$.

We assume

(1) $\Phi(t)C \subset C$ for $t \ge 0$

(2) $\Phi(t)C_o \subset C_o$ for $t \ge 0$

where $C(C_o)$ denotes the Banach space of continuous, bounded (vanishing at infinity in addition) functions on E.

Impulsive control consists in shifting current states of the Markov process x_t to new random states ξ_i at moment τ_i respectively. To describe the evolution of the controlled process we have to recall the construction of the new probability space due to M. Robin [1]. Let $\tilde{\Omega} = \Omega^N$.

The impulsive strategy $V = (\tau_i , \xi_i)_{i \in N}$ consists of pairs of Markov times τ_i and random variables ξ_i such that

τ_1 is $F_t \otimes \{\emptyset,\Omega\}\otimes\ldots\otimes\{\emptyset,\Omega\}\otimes\ldots$ Markov time

ξ_1 is $F_{\tau_1} \otimes \{\emptyset,\Omega\}\otimes\ldots\otimes\{\emptyset,\Omega\}\otimes\ldots$ measurable random variable

....

τ_n is $F^n_t \otimes \{\emptyset,\Omega\}\otimes\ldots\otimes\{\emptyset,\Omega\}\otimes\ldots$ Markov time

ξ_n is $F^n_{\tau_n} \otimes \{\emptyset,\Omega\}\otimes\ldots\otimes\{\emptyset,\Omega\}\otimes\ldots$ measurable random variable

where $F^n_t = F^{n-1}_t \otimes F_t$ and $F^1_t = F_t$.

So for $\omega \in \tilde{\Omega}$, $\omega = (\omega_1,\omega_2,\ldots,\omega_n,\ldots)$ we have

$$\tau_1(\omega) = \tau_1(\omega_1) \qquad \xi_1(\omega) = \xi_1(\omega_1)$$

$$\tau_2(\omega) = \tau_2(\omega_1,\omega_2) \qquad \xi_2(\omega) = \xi_2(\omega_1,\omega_2)$$

....

and the trajectory of the controlled process X is of the form

$$y_t(\omega) = x^{n-1}_t(\omega_n) \quad \text{for} \quad t\ [\tau_{n-1} , \tau_n[\ , \ \tau_o = 0$$

(3)

$$y_{\tau_n}(\omega) = \xi_n(\omega_1,\ldots,\omega_n) \quad .$$

The impulsive control V generates a probability measure P^V on the space $\tilde{\Omega}$, see [1].

With each strategy V is associated the following long run average cost functional

$$(4) \qquad J_x(V) = \liminf_{t \uparrow \infty} (1/t)\, E^V_x\, g_t(V)$$

where

$$(5) \quad 0 \leqq g_t(V) = \int_0^t f(y_s)ds + \sum_{i=1}^{\infty} \chi_{\tau_i \leqq t}\left[c\left(x^{i-1}_{\tau_i}\right) + d\left(y_{\tau_i}\right)\right]$$

and we assume $f,c,d \in C$, $f \geq 0$, $d \geq 0$, c is strictly positive, $c(x) \geq a > 0$. Our aim is minimization of $J_x(V)$, and characterization of $u(x) = \inf_V J_x(V)$.

3. THE CASE OF MARKOV IMPULSIVE STRATEGIES.

In this section we will consider impulsive control consisting of the special form stopping times. Namely, similarly as in [2], the times τ_i will be of the form

(6) $$\tau_i(\omega) = \tau_{i-1}(\omega_1,\dots,\omega_{i-1}) + \sigma_i(\omega_i)\circ\theta_{\tau_{i-1}(\omega_1,\dots,\omega_{i-1})}\quad,\quad i \ge 2$$

where σ_i is an arbitrary F_t Markov time. Such strategy we call Markov impulsive strategy.

Let

(7) $$\lambda = \inf_x \inf_\tau \frac{E_x \int_0^\tau f(x_s)ds + c(x_\tau) + d(x)}{E_x\{\tau\}}$$

(8) $$w(x) = \inf_\tau E_x\left\{\int_0^\tau (f(x_s)-\lambda)ds + c(x_\tau)\right\}$$

and

(9) $$\tau_\varepsilon = \inf\left\{s \ge 0 : w(x_s) \ge c(x_s) - \varepsilon\right\} \quad \text{for} \quad \varepsilon > 0 \quad .$$

We will assume henceforth, that for τ, such that $E_x\tau = \infty$ the notation $E_x\left\{\int_0^\tau f(x_s)ds + c(x_\tau)\right\}$ or

$$\frac{E_x \int_0^\tau f(x_s)ds + c(x_\tau)}{E_x\{\tau\}} \quad \text{means}$$

$$\liminf_{t\uparrow\infty} E_x\left\{\int_0^{\tau\wedge t} f(x_s)ds + c(x_{\tau\wedge t})\right\}$$

or

$$\liminf_{t\uparrow\infty} \frac{E_x \int_0^{\tau\wedge t} f(x_s)ds + c(x_{\tau\wedge t})}{E_x\{\tau\wedge t\}} \quad \text{respectively.}$$

Suppose $V(\tau,y)$ denotes impulsive strategy consisting of the times τ_i of the form (6), where $\tau_1 = \tau = \sigma_i$ for $i \in N$, and deterministic fix point y. Then the following theorem holds.

<u>Theorem 1.</u> The optimal value $u(x)$ with respect to Markov impulsive strategies is constant and equals λ.

Moreover

(a) if for some $y \in E$ $E_y\{\tau_\varepsilon\} \uparrow \infty$ as $\varepsilon \to 0$, then the strategy $V(\tau^*,y)$ with the stopping time $\tau^* = \lim_{\varepsilon\downarrow 0} \tau_\varepsilon$ is optimal

(b) if for each $y \in E$, $E_y\tau^* < \infty$, and

(10) $$\inf_y [w(y) + d(y)] = 0$$

then the strategy $V(\tau_\varepsilon, x^\varepsilon)$, where x^ε is such that $w(x^\varepsilon) + d(x^\varepsilon) \leq \varepsilon$ is $\frac{2\varepsilon}{E_{x^\varepsilon}\{\tau_\varepsilon\}}$ optimal

(c) if

$$(11) \qquad \sup_x \liminf_{t \uparrow \infty} E_x \left\{ \int_0^t (f(x_s) - \lambda) ds + c(x_t) \right\} < \infty$$

then the strategy "do not interfere in the run of process" $V(\infty, y)$ is optimal.

Proof (Sketch):

1. The trajectory y_t for $t < \tau_i$, together with the random variable x_{τ_i} is independent of the trajectory $y_{\tau_i + t}$ for $t \geq 0$, with respect to probability measure P^V . Similarly, for each $i \in N$, the stopping times τ_i and $\sigma_i \circ \theta_{\tau_i}$ are P^V independent. Thus using Blackwell renewal theorem we obtain $u(x) \leq \lambda$.

2. We will prove the reverse inequality $u(x) \geq \lambda$. Let us introduce the following function

$$(12) \qquad S_t(V)(x) = E_x^V \left\{ \int_0^t f(y_s) ds + \sum_{i=1}^\infty \chi_{\tau_i \leq t} \left[c\left(x_{\tau_i}^{i-1}\right) + d(\xi_i) \right] - t\lambda + \right.$$

$$\left. + w(y_t) \right\} = E_x^V \left\{ \left\{ \int_0^t (f(y_s) - \lambda) ds + \sum_{i=1}^\infty \chi_{\tau_i \leq t} \left[w\left(x_{\tau_i}^{i-1}\right) - w(\xi_i) \right] + \right.\right.$$

$$\left. + w(y_t) - w(x) \right\} + \left\{ \sum_{i=1}^\infty \chi_{\tau_i \leq t} \left[w(\xi_i) - w\left(x_{\tau_i}^{i-1}\right) + c\left(x_{\tau_i}^{i-1}\right) + d(\xi_i) \right] + \right.$$

$$\left.\left. + w(x) \right\} \right\} .$$

Since

<u>Lemma 1.</u> The function w satisfies the inequalities $w(x) \leq c(x)$ and $w(x) + d(x) \geq 0$.

We have

$$\liminf_{t \uparrow \infty} (1/t)\, S_t(V)(x) = \liminf_{t \uparrow \infty} (1/t)\, E_x^V g_t(V) - \lambda$$

and it remains to show

$$\liminf_{t \uparrow \infty} (1/t)\, S_t(V)(x) \geq 0 .$$

The last inequality follows from Lemma 1 and the following proposition.

<u>Proposition 1.</u> The function w is C_0 continuous, that is for each $x \in E$

$$(13) \qquad P_x \left\{ \lim_{t \downarrow 0} w(x_t) = w(x) \right\} = 1 .$$

Moreover

$$(14) \qquad \Psi(t) = \int_0^t (f(x_s) - \lambda)ds + w(x_t)$$

is the right continuous submartingale.

3. To find optimal or ε-optimal strategies we use the suitable form of the function $S_t(V)$ and the following lemma.

Lemma 2. If τ_ε defined in (9) is such that $E_x\tau_\varepsilon < \infty$, then τ_ε is ε-optimal stopping time in the definition of the function w .

Moreover

$$(15) \qquad w(x) = E_x \left\{ \int_0^{\tau_\varepsilon} (f(x_s) - \lambda)ds + w\left(x_{\tau_\varepsilon}\right) \right\} .$$

The next theorem completes Robin's result.

Theorem 2. Suppose E is compact and there exists invariant probability measure m , constants K , $\gamma > 0$ such that

$$(16) \qquad |P_x(x_t \in \Gamma) - m(\Gamma)| \le K e^{-\gamma t}$$

for any Borel set Γ . Let $\bar{f} = \int_E f(x)\, m(dx)$.

Then one of the conditions (10) or (11) is satisfied.

Moreover, if $\lambda < \bar{f}$ then (10) and $E_y\tau^* < \infty$, $y \in E$, if $\lambda = \bar{f}$, then (11) holds.

4. THE CASE OF GENERAL IMPULSIVE STRATEGIES.

So far we considered impulsive control with the Markov times of the form (6). Now we will interest the general case, the times τ_i will be arbitrary Markov times.

Theorem 3. The use of the general impulsive control does not change the optimal value $u(x)$ of the functional (4), $u(x) = \lambda$.

Proof:

Let us define first, the finite time functional

$$(17) \qquad J_x(V,t) = (1/t)\, E_x^V \left\{ \int_0^t f(y_s)ds + \sum_{i=1}^{\infty} \chi_{\tau_i \le t} \left[c\left(x_{\tau_i}^{i-1}\right) + d\left(y_{\tau_i}\right) \right] \right\} .$$

Suppose $u(x)$ denotes now the optimal value of the functional (4) with respect to arbitrary impulsive controls. Then there exists ε-optimal strategy V_ε

$$(18) \qquad J_x(V_\varepsilon) \le u(x) + \varepsilon .$$

Moreover from the definition of lim inf , there exists a sequence t_n ,

$t_n \to \infty$ as $n \to \infty$, such that

(19) $J_x(V_\varepsilon, t_n) \to J_x(V_\varepsilon)$.

The following result will play an important role in the proof of this theorem.

Lemma 3. For each t there exists the impulsive strategy V_t consisting of the times of the form (6) such that

(20) $J_x(V_t, t) - J_x(V_\varepsilon, t) \le (1/3)\,\varepsilon$.

Taking into account these facts we will construct the strategy $\bar{V}_\varepsilon$, consisting of the stopping of the form (6), satisfying the inequality

(21) $J_x(\bar{V}_\varepsilon) \le \varepsilon + J_x(V_\varepsilon)$.

We choose the subsequence $\left(t_{n_k}\right)_{k \in N}$ from the sequence $(t_n)_{n \in N}$ successively in the following way

$t_{n_1} = t_1$

for given t_{n_i} we take $t_{n_{i+1}}$ such that

$t_{n_i} \le (1/3)\, t_{n_{i+1}}$

$2\,\|f\|\, t_{n_i} + [\|c\| + \|d\|] N_{V_\varepsilon}\left(0, t_{n_i}\right) \le (1/3)\,\varepsilon\, t_{n_{i+1}}$

where $N_{V_\varepsilon}(0,t)$ denotes the number of shifts in the time interval $[0,t]$.

Let us denote by $V^{[t_1,t_2[}$ the impulsive strategy restricted to the time interval $[t_1, t_2[$.

Next we define the strategy $\bar{V}_\varepsilon$

(22) $\bar{V}_\varepsilon = \left(V_{t_{n_1}}^{[0,t_{n_1}[}, V_{t_{n_2}}^{[t_{n_1},t_{n_2}[}, \ldots, V_{t_{n_k}}^{[t_{n_{k-1}},t_{n_k}[}, \ldots\right)$.

This means, that in the time interval $[t_{n_i}, t_{n_{i+1}}[$ the impulsive strategy $V_{t_{n_{i+1}}}$ defined by Lemma 3 is adopted.

An easy calculation leads to the inequality

(23) $J_x\left(\bar{V}_\varepsilon, t_{n_{i+1}}\right) \le \varepsilon + J_x\left(V_\varepsilon, t_{n_{i+1}}\right)$.

Using (19) we see that (21) is satisfied. Since ε can be taken arbitrarily, the proof of theorem is complete.

REFERENCES

[1] Robin M., Contrôle impulsionnel des processus de Markov, thesis, University of Paris IX, 1978.

[2] Robin M., On some impulsive control problems with long run average cost, SIAM on Control 19, 1981, 333 - 358.

[3] Stettner L., On impulsive control with long run average cost criterion, submitted for publication to Studia Mathematica.

SEPARATION THEOREM FOR OPTIMAL IMPULSE CONTROL WITH DISCONTINUOUS OBSERVATIONS

J. SZPIRGLAS

PAA / ATR / MTI
Centre National d'Etudes des Télécommunications
38-40 rue du Général Leclerc
92131 - ISSY LES MOULINEAUX
FRANCE

A separation theorem between filtering and control is proved for a partially observed impulse control problem as in (6). But the observation process of an inventory is here a jump process, the intensity of which is a function of the controlled inventory. The difference with (6) arizes from the intensive use of the unnormalized filter associated to the "inventory-observation" system. This generalizes the cases with a finite dimensional filter of (1) and (7). Both uncontrolled and controlled models are constructed thanks to the reference probability method (12) first introduced in (13).

I- CONSTRUCTION OF THE UNCONTROLLED MODEL

I-a: The reference space . Let the inventory level be represented by a cadlag Feller process X with values in a compact E of $\mathbb{R}^d$, defined on its canonical space:

$$X = (\Omega^X, \underline{\underline{F}}^X_\infty, \underline{\underline{F}}^X_t, \theta^X_t, X_t, (\mathbb{P}^X_x; x \in E)) \text{ with } X_{t+s} = X_t \circ \theta^X_s \quad .$$

Let μ the initial law of X and $\mathbb{P}^X_\mu = \mu . \mathbb{P}^X_\cdot$.

Let Y be a Poisson process defined on its canonical space:

$$Y = (\Omega, \underline{\underline{G}}_\infty, \underline{\underline{G}}_t, \theta_t, Y_t, \mathbb{P}) \text{ with } Y_{t+s} - Y_s = Y_t \circ \theta_s \quad .$$

The reference probability space is the tensor product of the previous ones:

$$\Omega' = \Omega x \Omega^X \ , \ \underline{\underline{F}}_\infty = \underline{\underline{G}}_\infty \otimes \underline{\underline{F}}^X_\infty \ , \ \underline{\underline{F}}_t = \underline{\underline{G}}_t \otimes \underline{\underline{F}}^X_t \ , \ \mathbb{P}_\mu = \mathbb{P} \otimes \mathbb{P}^X_\mu$$

$$X(\omega') = X(\omega^X) \quad \text{and} \quad Y(\omega') = Y(\omega)$$

I-b: The uncontrolled model . Notice that X and Y are $\mathbb{P}_\mu$-independent. Let H be a continuous function on E such that:

$1 + H \geq \delta > 0$ and $\int_0^\infty (1+H^2(X_s))\,ds < \infty$ $\mathbb{P}_\mu^X$-a.s. $\forall\mu$.

Then define process L :

$$L_t = \exp\left(\int_0^t (\mathrm{Log}(1+H(X_{s^-}))\,dY_s - H(X_s)ds) \right)$$

Process L is a $(\underline{\underline{F}},\mathbb{P}_\mu)$uniformly integrable martingale.

Let $Q_\mu = L_\infty \cdot \mathbb{P}_\mu$. Then it is classical (see(2)),that X has the same law for $\mathbb{P}_\mu$ and Q_μ and Y is under Q_μ a jump process with intensity $(1 + H(X_t))\,dt$. It can be noticed that Y is no more a process with independent increments under probability Q_μ.

I-c:<u>The filtering processes</u>: Let K_μ denote:

$K_\mu(\omega;d\omega^Y,d\omega^X) = \varepsilon_\omega(d\omega^Y) \otimes \mathbb{P}_\mu^X(d\omega^X)$ where ε_ω is the Dirac measure in ω on Ω. Then the filter $\Pi^\mu = (\Pi_t^\mu;t\geq 0)$ of X given Y is such that:

$\forall f\in b(E)$, $\Pi_t^\mu(f) = K_\mu(L_t f(X_t)) \,/\, K_\mu(L_t)$ where b(E) denotes the set of bounded Borel functions on E. That is the so-called Kallianpur-Striebel formula, the numerator of which defines the unnormalized filter such that:

$\forall f\in b(E)$, $\tilde{\Pi}_t^\mu(f) = K_\mu(L_t f(X_t))$

The main result of this paragraph is (see (11)) :

<u>*Theorem 1*</u>*: Under the previous assumptions, the family $(\tilde{\Pi}^\mu;\ \mu\in\tilde{M}(E))$ where $\tilde{M}(E)$ is the set of positive bounded measures on E, is a Feller Markov family with respect to probability $\mathbb{P}$, filtration $\underline{\underline{G}}$, and translation operator θ_t.*

This can be proved by solving as (4), the stochastic differential equations which have $\tilde{\Pi}$ for solution.

II- CONSTRUCTION OF THE CONTROLLED MODEL

This is done by regenerative technics (9), and using once more the reference probability method. But, first define the admissible controls

II-a: <u>Admissible controls</u>: An admissible control is a double sequence $v = (T_n,\xi_n;n>0)$ of ordering times and ordering amounts at those times.

$(T_n;n>0)$ is a sequence of $\underline{\underline{G}}$-stopping times such that:

$T_n < T_{n+1}$ on $(T_n < \infty)$ and $\lim T_n = \infty$ $\mathbb{P}$-a.s..

$(\xi_n;n>0)$ is a sequence of $\underline{\underline{G}}_{T_n}$-measurable random variables on E.

As $\underline{\underline{G}}$ is a natural uncompleted filtration (10), there exists a sequence $(S_n;n>0)$ of $\underline{\underline{G}}_{T_n} \otimes \underline{\underline{G}}_\infty$-measurable random variables on Ω^2 such that:

$\forall\omega\in\Omega$, $S_n(\omega,.)$ is a $\underline{\underline{G}}$ stopping time

$\forall\omega\in\Omega$, $T_{n+1}(\omega) = T_n(\omega) + S_n(\omega,\theta_{T_n}\omega)$

For that reason, an admissible control will be written either $v = (T_n,\xi_n;n>0)$ or $v = (T_n,\xi_n,S_n;n>0)$. To each admissible control can be associated a family of truncated controls $(v/n;n>0)$

$v/n = ((T_1,\xi_1),\ldots,(T_n,\xi_n),(\infty,0),\ldots.)$

II-b: The reference space: Let $\overline{\Omega} = \overline{\Omega}x(\Omega^X)^N$, with $\overline{\omega} = (\omega,\omega_0,.,\omega_n,..)$ for running point, and $\overline{\underline{\underline{F}}}_\infty = \underline{\underline{G}}_\infty \otimes (\underline{\underline{F}}^X_\infty)^{\otimes N}$.

Given an admissible control v, define the controlled inventory level by:

$$X^v_t(\overline{\omega}) = X_t(\omega_0)\ \mathbb{1}\{0\le t\le T_1(\omega)\} + \sum_{n=1}^{n=\infty} X_{t-T_n(\omega)}(\omega_n)\ \mathbb{1}\{T_n(\omega)<t\le T_{n+1}(\omega)\}$$

This process is only ladlag to distinguish between jumps resulting from ordering and others (5).

Define also the observation process: $Y(\overline{\omega}) = Y(\omega)$

$(\overline{\Omega},\overline{\underline{\underline{F}}})$ is endowed with the usual filtration $\underline{\underline{F}}^v$ (see(9)), to which X^v and Y are adapted.

Consider now probability $\mathbb{P}^v_\mu$ on $(\overline{\Omega},\overline{\underline{\underline{F}}},\underline{\underline{F}}^v)$:

$\mathbb{P}^v_\mu(\overline{\omega}) = \mathbb{P}(d\omega)\ K^v_\mu(\omega;d\omega_0,..,d\omega_n,..)$ where the Markovian kernel K^v_μ is defined thanks to Ionescu-Tulcea theorem by:

$$K^v_\mu(\omega;d\omega_0,\ldots,d\omega_n,\ldots) = \mathbb{P}^X_\mu(d\omega_0)\ldots\mathbb{P}^X_{\Phi(X^v_{T_n}(\omega,..,\omega_{n-1}),\xi_n)}(d\omega_n)\ \ldots\ldots$$

where Φ is a continuous function from E^2 to E such that:

$\forall\ x,\xi_1,\xi_2\in E,\quad \Phi(\Phi(x,\xi_1),\xi_2) = \Phi(x,\Phi(\xi_1,\xi_2))$

That means that two simultaneous orders of ξ_1 and ξ_2 have the same effect on the inventory level that a unique order of $\Phi(\xi_1,\xi_2)$.

Then we get: $\quad X^v_{T_n^+} = \Phi(X^v_{T_n},\xi_n) \qquad \mathbb{P}^v_\mu$-a.s..

II-c: The controlled model: Define on $(\overline{\Omega},\overline{\underline{\underline{F}}}_\infty)$, the following $(\underline{\underline{F}}^v,\mathbb{P}^v_\mu)$-local martingale:

$$L^v_t = \exp(\textstyle\int_0^t(\mathrm{Log}(1+H(X^v_{s^-}))dY_s - H(X^v_s)ds))$$

L^v is generally not a martingale, but for all v/n, $L^{v/n}$ is a uniformly integrable martingale. This allows us to define a family $\mathbb{Q}^v_\mu$ of probabilities $\mathbb{Q}^{v/n}_\mu$ equivalent to $\mathbb{P}^{v/n}_\mu$:

Theorem 2: (11) *The family* $Q_\mu^\nu = (Q_\mu^{\nu/n}, n>0)$ *is such that:*
(*i*) *For all n and all* $\underline{\underline{G}}$-*stopping time T such that* $T_n < T \leq T_{n+1}$ *and* $T(\omega) = T_n(\omega) + S(\omega, \theta_{T_n}\omega)$ $\mathbb{P}$-*a.s.*, *for all f in* $b(E)$:
$$E_{Q_\mu^{\nu/n}}(f(X_T^\nu)/\underline{\underline{F}}_{T_n}^\nu)(\omega,..,\omega_{n-1}) = E_{Q_{\Phi(X_{T_n}^\nu,\xi_n)}(\omega,...,\omega_{n-1})}(f(X_{S(\omega,.)}))$$
(*ii*) *Process Y is under probability* $Q_\mu^{\nu/n}$ *a jump process with* $(1+H(X_t^{\nu/n}))dt$ *for Levy measure.*

(*iii*) *The family* Q_μ^ν *is projective, i.e. the restrictions of* $Q_\mu^{\nu/n}$ *and* $Q_\mu^{\nu/m}$ *to* σ-*field* $\underline{\underline{F}}_{T_m}^\nu$ *are equal for* $m \leq n$.

Point (i) is the regenerative property which binds the controlled system to the uncontrolled one. It expresses, that between two successive ordering times the inventory X^ν moves as the uncontrolled one.

Point (ii) tells, we have constructed a classical filtering model with a jump process for observation (2).

Point (iii) expresses that the family Q_μ^ν defines uniquely the law of the system on each observation interval $[0,T_n]$. Each probability $Q_\mu^{\nu/n}$ can be seen as the law of the controlled system up to time T_n and for the uncontrolled system after T_n. It is the probability considered by the controller, when having observed the system up to time T_n, he has to decide the best next ordering time T_{n+1}.

II-d: <u>The filtering processes:</u> From point (iii) of theorem 2 and definition of admissible controls, we can say that there exists a unique filter of X^ν given Y, $\Pi^{\nu,\mu}$ such that:
$$\Pi_t^{\nu,\mu} = \Pi_t^{\nu/n,\mu} \quad \text{on} \quad [0,T_{n+1}[$$
It is still given by Kallianpur-Striebel formula.
$$\forall f \in b(E),\ \Pi_t^{\nu,\mu}(f) = K_\mu^\nu(L_t^\nu f(X_t^\nu))/K_t^\nu(L_t^\nu)$$
and the same for the unnormalized filter:
$$\forall f \in b(E),\ \tilde{\Pi}_t^{\nu,\mu}(f) = K_\mu^\nu(L_t^\nu f(X_t^\nu))$$

From thoses properties, we get the following theorem:

Theorem 3: *Given any admissible control* $\nu = (T_n, \xi_n, S_n; n>0)$, *the unnormalized filter of* X^ν *given Y satisfies the following:*
(*i*) $\tilde{\Pi}^{\nu,\mu}$ *is a* $\underline{\underline{G}}$ *adapted process ladlag for weak convergence topology such that:*
$$\tilde{\Pi}_0^{\nu,\mu} = \mu, \quad \tilde{\Pi}_{T_n^+}^{\nu,\mu} = \Phi'(\tilde{\Pi}_{T_n}^{\nu,\mu}, \xi_n) \quad \text{for all } n.$$
where Φ' *denotes the continuous function from* $\tilde{M}(E) \times E$ *in* $\tilde{M}(E)$ *defined by:* $\forall \mu \in \tilde{M}(E)$, $\forall \xi \in E$, $\forall f \in b(E)$, $\Phi'(\mu,\xi)(f) = \mu(f(\Phi(.,\xi)))$

(ii) $\tilde{\Pi}^{\nu,\mu}_{T_{n+1}}(\omega) = \Pi_{S_n(\omega,\theta_{T_n}\omega)}^{\tilde{\Pi}^{\nu,\mu}_{T_n^+}(\omega)}(\theta_{T_n}\omega)$ *for all n,*

where $\tilde{\Pi}^{\mu}$ *is the filter associated to the uncontrolled system.*

(iii) For all bounded Borel functions on $\tilde{M}(E)$, f' *and* g'

$$E_{\mathbb{P}}\left(\int_{T_n}^{T_{n+1}} e^{-\alpha s} f'(\tilde{\Pi}^{\nu,\mu}_s)\,ds + e^{-\alpha T_{n+1}} g'(\tilde{\Pi}^{\nu,\mu}_{T_{n+1}}) \;/\; \underline{\underline{G}}_{T_n}\right)$$

$$= e^{-\alpha T_n(\omega)} E_{\mathbb{P}}\left(\int_0^{S_n(\omega,.)} e^{-\alpha s} f'(\tilde{\Pi}^{\nu(\omega)}_s)\,ds + e^{-\alpha S_n(\omega,.)} g'(\tilde{\Pi}^{\nu(\omega)}_{S_n(\omega,.)})\right)$$

where $\nu(\omega) = \tilde{\Pi}^{\nu,\mu}_{T_n^+}(\omega)$

The third point is the regenerative property of the filter $\tilde{\Pi}$ which will make possible a solution of our problem, similar to that one with complete information (10).

III-SOLUTION OF THE OPTIMAL IMPULSE CONTROL PROBLEM

After defining a cost function, we use a result of (10) to solve the problem.

III-a:The cost function:Given any admissible control v and initial law μ of M(E), we define the mean cost function by:

$$J(v,\mu) = \lim_{n\to\infty} E_{Q_\mu^{v/n}}\left(\int_0^{T_n} e^{-\alpha s} f(X_s^v)\,ds + \sum_{m=1}^{m=n} e^{-\alpha T_m} C(X_{T_m}^v, \xi_m)\right)$$

where f is a positive continuous function on E, the storage cost by unit of time, α is a positive real, the actualization parameter, and C is a continuous function on E^2, the impulse ordering cost such that:

$C \geq K > 0$ and $\forall x, \xi_1, \xi_2 \in E$, $C(x,\xi_1) + C(\Phi(x,\xi_1),\xi_2) >> C(x,\Phi(\xi_1,\xi_2))$

This last assumption will forbid any simultaneous orders.
It is easy to express function $J(v,\mu)$ with respect to the unnormalized filter.

$$J(v,\mu) = E_{\mathbb{P}}\left(\int_0^\infty e^{-\alpha s} \tilde{\Pi}_s^{v,\mu}(f)\,ds + \sum_{n=1}^{n=\infty} e^{-\alpha T_n} \tilde{\Pi}^{v,\mu}_{T_n}(C(.,\xi_n))\right)$$

Notice that we get a formulation very similar the one of (10) with complete information, except the fact the processes are not defined on their canonical spaces. Both controlled and uncontrolled filters are defined on the same observation space. But we shall apply the following result of (10).

Theorem 4: Let $\tilde{\underline{H}}$ be the natural filtration of the unnormalized filter $\tilde{\Pi}$. Then, there exists a unique continuous function on $\tilde{M}(E)$ such that:

$$\forall \mu \in \tilde{M}(E),\ u(\mu) = \inf_{T\ \underline{H}s.t.} E_{\mathbb{P}}\left(\int_0^T e^{-\alpha s}\tilde{\Pi}_s^{\mu}(f)ds + e^{-\alpha T}Mu(\tilde{\Pi}_T^{\mu})\right)$$

where M is the operator which associates to each continuous function h on $\tilde{M}(E)$ the continuous function Mh on $\tilde{M}(E)$ defined by:

$$\forall \mu \in \tilde{M}(E),\ Mh(\mu) = \inf_{\xi \in E} (\mu(C(.,\xi)) + h(\Phi'(\mu,\xi)))$$

Furthermore, the following stopping point $T(\omega,\mu)$ achieves the minimum defining $u(\mu)$:

$$T(\omega,\mu) = \inf(t \geq 0 \ / \ u(\tilde{\Pi}_t^{\mu}(\omega)) = Mu(\tilde{\Pi}_t^{\mu}(\omega))\ )$$

and there exists a measurable selection $\mu \to \xi(\mu)$ such that:

$$Mu(\mu) = \mu(C(.,\xi(\mu))) + h(\Phi'(\mu,\xi(\mu)))$$

Notice, we have implicitely extended the definition space of $\tilde{\Pi}$ to Ωx $\tilde{M}(E)$, as to transform the Markov family into a classical Markov process.

Then applying results of (8), we get the following corollary:

Corollary 5: With notations of theorem 4, we have:

$$\forall \mu \in \tilde{M}(E)\ ,\quad u(\mu) = \inf_{T\ \underline{G}s.t.} E_{\mathbb{P}}\left(\int_0^T e^{-\alpha s}\tilde{\Pi}_s^{\mu}(f)ds + e^{-\alpha T}Mu(\tilde{\Pi}_T^{\mu})\ \right)$$

III-b: Construction of an optimal control : We shall construct an optimal control as a Markovian function of the controlled filters. The proofs are similar to (6). We shall only correct some mistakes of it.

Given the initial inventory law μ in $\tilde{M}(E)$, define sequentially control v* :

$$T_1(\omega) = T(\mu,\omega),\ \xi_1(\omega) = \xi(\tilde{\Pi}_{T_1}^{\mu}(\omega)),\ \eta_1(\omega) = \Phi'(\tilde{\Pi}_{T_1}^{\mu}(\omega),\xi_1(\omega))$$

$$T_{n+1}(\omega) = T_n(\omega) + T(\eta_n(\omega),\theta_{T_n}\omega)$$

$$\xi_{n+1}(\omega) = \xi(\tilde{\Pi}^{\eta_n(\omega)}_{T(\eta_n(\omega),\theta_{T_n}\omega)}(\theta_{T_n}\omega)\ )$$

$$\eta_{n+1}(\omega) = \Phi'(\tilde{\Pi}^{\eta_n(\omega)}_{T(\eta_n(\omega),\theta_{T_n}\omega)}\ (\theta_{T_n}\omega),\xi_{n+1}(\omega)\ \ )$$

Theorem 6: Control v is an admissible control.*

Proof: From their construction, the sequence $(T_n;n>0)$ is an increasing sequence of $\underline{G}$-stopping times, and for all n ξ_n is $\underline{\underline{G}}_{T_n}$-measurable. We have to check that $\lim_n T_n = \infty$.

From the formula of corollary 5, with $T = \infty$, we get:

$$u(\mu) \leq \sup_E |f|/\alpha$$

Then by iterating regenerative property(iii) of theorem 3, and noticing that $\eta_n = \tilde{\Pi}^{v,\mu}_{T_n^+}$, we get formula (*) for all n:

$$(*) \quad u(\mu) = E_{\mathbb{P}}(\int_0^{T_n} e^{-\alpha s}\tilde{\Pi}^{v,\mu}_s(f)ds + \sum_{m=1}^{m=n} e^{-\alpha T_m}\tilde{\Pi}^{v,\mu}_{T_m}(C(.,\xi_m)) + e^{-\alpha T_n}u(\tilde{\Pi}^{v,\mu}_{T_n^+}))$$

This implies:

$$\forall n>0 \ , \ u(\mu) \geq E_{\mathbb{P}}(\sum_{m=1}^{m=n} e^{-\alpha T_m}\tilde{\Pi}^{v,\mu}_{T_m}(C(.,\xi_m)))$$

$$u(\mu) \geq nK\, E_{\mathbf{P}}(e^{-\alpha T_n})$$

because T_n is increasing. As $u(\mu)$ is bounded, we get: $\lim_n T_n = \infty$.
We have now to show that: $T_n<T_{n+1}$ on $\{T_n< \infty\}$ .It is sufficient to prove

$$\mathbb{P}(T(\mu,.) > 0) = 1$$

This is a consequence of hypotheses on Φ and C which prevent μ and $\Phi(\mu,\xi(\mu))$ of being simultaneously in the set $\{\mu/ u(\mu) = Mu(\mu)\}$.

Theorem 7: Control v is optimal.*

Proof: This can be seen from formula (*) of theorem 6

It can be proved that functions u and Mu are homogeneous; then by definition of control v*, and associated Π^{v^*} and $\tilde{\Pi}^{v^*}$, we get our last theorem.

Theorem 8: Given any initial law μ, the stopping times $(T_n;n>0)$ of control v are the successive entry times of processes $\Pi^{v^*,\mu}$ or equivalently $\tilde{\Pi}^{v^*,\mu}$ into the set of measures:*
$\{ \mu \in \tilde{M}(E)/ u(\mu) = Mu(\mu)\}$.

REFERENCES

(1) R.F. ANDERSON and A. FRIEDMAN: "Multidimensional Quality Control Problems and Quasi-Variational Inequalities" "Quality Control for Markov Chains and Free Boundary Problems" Trans.Amer.Math.Soc. Vol 246 (1978) 31-94.

(2) P. BREMAUD:"La méthode des semi-martingales en filtrage quand l'observation est un processus ponctuel marqué" Sém. Prob. X, Lect. Notes in Math. N°511 (1976) 1-18.

(3) C. DELLACHERIE and P.A. MEYER: "Probabilités et potentiel" Tomes 1 et2, Hermann, Paris 1975 et 1980.

(4) H. KUNITA: "Asymptotic Behaviour of the Non Linear Filtering Errors of Markov Processes" J.Mult.Anal. 1 N°4 (1971) 365-393.

(5) J.P. LEPELTIER and B. MARCHAL: "Techniques Probabilistes dans le Contrôle Impulsionnel" Stochastics Vol 2 (1979) 243-286.

(6) G. MAZZIOTTO and J. SZPIRGLAS: "Separation Theorems for Optimal Stopping and Impulse Control"Third IMA Conference on Control Theory, Academic Press, 1981, 567-586.

(7) J.L. MENALDI: "The Separation Principle for Impulse Problems" Proc. A.M.S. To appear.

(8) J.F. MERTENS :"Théorie des Processus Stochastiques Généraux, Applications aux Surmartingales" Z.f.Wahr.V.Geb.Vol 26 (1973) 119-139.

(9) P.A. MEYER :"Renaissance, recollements, mélanges, ralentissements des processus de Markov" Ann.Inst.Fourier, t.XXV, fasc.3-4 (1975) 465-497.

(10) M. ROBIN :"Contrôle Impulsionnel des Processus de Markov" Thèse Univ. Paris IX, 1978.

(11) J. SZPIRGLAS : "Principe de séparation pour le contrôle impulsionnel avec observation partielle discontinue" Thèse Univ. Paris VI 1982).

(12) J. SZPIRGLAS and G. MAZZIOTTO : "Modèle général de filtrage non linéaire et équations differentielles stochastiques associées" Ann.Inst.H. Poincaré Vol XV N°2 (1979) 147-173.

(13) M. ZAKAI: "On the Optimal Filtering of Diffusion Processes" Z.f. Wahr.V. Geb.Vol 11 (1969) 230-249.

OPTIMAL CONTROL BASED ON OBSERVATIONS ON THE BOUNDARY

Domokos Vermes

Bolyai Institute, University of Szeged
6720 Szeged, Aradi vèrtanuk tere 1
Hungary

1. INTRODUCTION

We consider an object whose motion is described by the stochastic differential equation

$$dx_t = b(x_t)dt + \sigma(x_t)dw_t \tag{1}$$

in a C^2-domain $\Omega \subset R^{n+1}$ and which is reflected from the boundary $\partial\Omega$. The controller situating outside of the domain, can observe the position of the object only when it hits the boundary $\partial\Omega$, and he can influence its motion by prescribing the angle of the (non-normal) reflexion at each point of the boundary. The controller's objective is to minimize the expected cost (composed of a terminal and an integral part) arising until the first hitting of a target set, where the process is killed. We assume, that the target set is an attainable subset of the boundary, this means that the process terminates almost surely in finite time and that its death is observable for the controller.

We shall show that the problem can be reduced to a Markovian control problem of the type dealt in [4]. As for this types of problems the non-linear bang-bang principle holds [6] one is forced to take non-continuous control strategies into consideration. But for stochastic differential equations with non-continuous boundary-conditions not even existence and uniqueness of the solutions is assured. So in the first part of the present communication we show that there exists a uniquely determined Markov process, which is the weak solution of Eq.(1), with arbitrary, measurably changing angle of reflection at the boundary. The technique of the proofs necessites to impose a technical condition which forces the interior part of the process to remain in an intensive connection with the boundary. We require that the outward normal component b_N of the drift vector tends in a prescribed manner to infinity as one approaches to the boundary. More precisely $b_N(x) = b_0(x) + b_1/\text{dist}(x,\partial\Omega)$ in a neighbourhood of the boundary, where b_0 is a bounded measurable function, while $b_1 \in]0,1[$ is strictly separated from 0 and 1 . This condition can be omitted, if the set of all admissible angles of reflection is strictly convex. But as this result needs an entirely different technique of proof, we restrict us to announcing it only.

In the second part of the paper we show that the solvability of a Hamilton - Jacobi - Bellman type equation is a necessary and sufficient condition of optimality. We also show that the value function is the unique solution of the equation in an appropriate Sobolev - Bessel space. The regularity of the value function expressed by the properties of general elements of this space makes the methods of [6] applicable to the present problem. This way a number of extremity and regularity properties of the optimal strategy become available, but we do not state them here as they coincide entirely with the results of [6]. The actual form of the HJB - equation is an elliptic boundary value problem with Dirichlet and reflecting boundary conditions. Hence as well for its further analysis as for its solution a great number of traditional methods are available.

In the special case where the evolution of the process in the interior of the domain gives rise of no costs, our original problem can be reduced to a control problem for the boundary process. As this latter can be represented as the solution of a stochastic differential equation with stable forcing term, the results of Pragarauskas [2] become applicable. Beside that we want to call attention by our simple example to the significance of Pragarauskas' sofisticated theory, a goal of the present paper is to show a way of solution of singular integro - differential equations arising in jump - diffusion theory. We can namely solve an elliptic boundary value problem with reflecting boundary conditions in a domain Ω instead of solving a singular integro - differential equation on the boundary $\partial\Omega$. The possibility of this imbedding approach was pointed out in an early paper by Molchanov and Ostrovskii[1]. On the way of the precise realization of this idea the author was guided by a great number of very helpful informal discussions with J.M. Bismut.

2. EXISTENCE AND UNIQUENESS OF THE PROCESSES

Now we turn to the proof of the existence and uniqueness of a diffusion with non-continuous boundary conditions. In most of this paragraph we follow the lines of the method of Sato - Ueno [3]. But in order to be able to include non - continuous boundary operators as well we use more powerful L_p - methods instead of continuity arguments and Feller property in analysing the trace of our process on the boundary.

As any boundary problem in a domain with C^2-boundary can be reduced by standard methods to an equivalent problem in R^+_{n+1}, it is no restriction of generality if we focus our attention to $\Omega = R^+_{n+1} = \{(x_0,\xi) \in R_1 \times R_n : x_0 > 0\}$ and $\partial\Omega = \{(x_0,\xi) \in R_1 \times R_n : x_0 = 0\}$. Let further

$$Af(x) = \frac{1}{2}\sum_{i=0}^{n} a_{ij}(x)\frac{\partial^2 f}{\partial x_i \partial x_j} + \sum_{i=1}^{n} b_i(x)\frac{\partial f_i}{\partial x_i} - \frac{\gamma}{x_0}\cdot\frac{\partial f}{\partial x_0} \qquad (x \in \Omega) \tag{2}$$

$$Lf(x) = \frac{\partial f(x)}{\partial x_0} + \sum_{i=1}^{n} \beta_i(x)\frac{\partial f(x)}{\partial x_i} = \frac{\partial f}{\partial n} + \langle\beta(x),\nabla f\rangle \qquad (x \in \partial\Omega) \tag{3}$$

Here $a_{ij}(x)$ are bounded, Lipschitz continuous functions on R^+_{n+1}, the matrix $(a_{ij}(x))$ is uniformly elliptic, i.e. there exists $0 < \lambda_0 < \lambda_1 < \infty$ such that $\sum \lambda_0 \xi_i \xi_j \le \sum a_{ij}\xi_i\xi_j \le \sum \lambda_1 \xi_i \xi_j$ for all $x \in R^+_{n+1}$ and $\xi \in R_{n+1}$. The functions $b_i(x)$ and $\beta_j(x)$ are assumed to be bounded, measurable on $R^+_{n+1} = \Omega$ and $R_n \triangleq \partial\Omega$ respectively, while γ is a constant with $0 < \gamma < 1$. (The signes $\langle\,,\rangle$ of scalar product and ∇ the gradient refer always to operations with vectors of n-components, i.e. to $\xi \in \partial\Omega$, we use also the notation $\alpha = \gamma + 1$.) In the case of $\beta \equiv 0$ the boundary condition $Lf = \partial f/\partial n = 0$ corresponds to the case of the normal reflection from the boundary. Hence $\beta(x)$ can be interpreted as the tangential component of a vector showing in the direction of the symmetry axis of a general (non-normal) reflexion at boundary point x. As it will be evidenced later $\beta(x)$ can also be called a tangential drift vector on the boundary.

We are looking for a Markov process X_t with transition semigroup $P_t f(x) = E_x f(X_t)$ such that

$$P_t f(x) - f(x) = \int_0^t P_s Af(x)\,ds \qquad (x \in \bar{\Omega}) \tag{4}$$

holds true for any $f \in C_0^\infty(\bar{\Omega})$ with $Lf(\xi) = 0$ on $\xi \in \partial\Omega$. (Here $C_0^\infty(\bar{\Omega})$ denotes the class of all C^∞ functions on Ω which tend to zero as $|x| \to \infty$ and whose derivatives can be continuously extended to $\bar{\Omega}$.) We call the process X_t (and also the semigroup P_t) to be generated by (A,L).

<u>Theorem 1:</u> There exists a Markov process on $\bar{\Omega}$ generated by (A,L) of type (2) - (3), and it is uniquely determined by the restriction of operators A and L to $C_0^\infty(\bar{\Omega})$.

Proof: a) We have to show that equations

$$(\lambda - A)u = f \quad ; \quad Lu = 0 \tag{5}$$

have unique solution in a sufficiently broad class $\mathcal{D}(A_L)$ for any $f \in C_0(\Omega)$ and $\lambda > 0$. (In fact $\mathcal{D}(A_L)$ has to be dense in $C_0(\bar{\Omega})$.) Denote R_λ the resolvent corresponding to the process in the interior Ω and killed at the first hitting of $\partial\Omega$. This minimal process is well defined by the system $(A,0)$ where 0 denotes the boundary operator $0u = 0$. R_λ is in fact the solution of the equations

$$(\lambda - A) R_\lambda f(x) = f(x) \quad \text{if} \quad x \in \Omega \; ; \quad R_\lambda f(x) = 0 \quad \text{if} \quad x \in \partial\Omega \qquad (6)$$
$$(f \in C_0(\bar{\Omega})) \; .$$

We denote by H_λ the λ-harmonic operator corresponding to the minimal process, i.e. $H_\lambda f$ is the solution of the equations

$$(\lambda - A) H_\lambda \varphi(x) = 0 \quad \text{if} \quad x \in \Omega \; ; \quad H_\lambda \varphi(x) = \varphi(x) \quad \text{if} \quad x \in \partial\Omega \qquad (7)$$
$$(\varphi \in C_0(\partial\Omega)) \; .$$

Suppose that LH_λ maps $C_0^\infty(\partial\Omega)$ into $C_0(\partial\Omega)$ and that it is a (pre-) generator of a non-conservative Markov semigroup. Then its closoure is invertible on $C_0(\partial\Omega)$ and a natural candidate for the solution $u = G_\lambda f$ of (5) is

$$G_\lambda f = R_\lambda f - H_\lambda (LH_\lambda)^{-1} LR_\lambda f \quad . \qquad (8)$$

In the rest of the proof we are going to show, that G_λ of the form (8) satisfies all conditions of the Hille-Yosida-Feller theorem and hence (A,L) generates a Markov semigroup.

b) First we turn to the analysis of LH_λ. For simplicity we suppose that (a_{ij}) is the unit matrix I and $b \equiv 0$. In this case a straightforward computation shows [1] that for any $\varphi \in C_0^2(\partial\Omega)$

$$LH_\lambda \varphi(x) = \int \left(\varphi(x+y) - \varphi(x) - \frac{\langle y, \nabla\varphi(x)\rangle}{1+|y|^2} \right) \frac{dy}{|y|^{n+\alpha}} - $$
$$- \lambda^{\alpha/2} \Gamma^{-1}(\alpha/2) + \langle \beta(x), \nabla\varphi(x)\rangle \quad . \qquad (9)$$

For general $a_{ij}(x)$ and $b_i(x)$ the only necessary change would be the occurance of an absolute continuous measure $h(x,y)dy$ with bounded, strictly positive density $0 < \delta \le h(x,y) \le M$ instead of dy in the integral term. This can be seen by majorating and minorating the positiv definite matrix (a_{ij}) by $\lambda_1 I$ and $\lambda_0 I$, where λ_1, λ_0 are the maximal and minimal eigenvalues of (a_{ij}). From the smoothness of $a(x)$, $b(x)$ follows that $h(x,y)$ is a continuous function of variable x.

Observe that the Fourier multiplier corresponding to the first (integral) term (denoted by B_0) of the right-hand side of (9) is $FB = -|\xi|^{\alpha}$. Introducing the Bessel potential spaces

$$H_p^{\alpha}(\partial\Omega) = \left\{ f \in L_p(\partial\Omega) : \|f\|_{\alpha,p} = \|F^{-1}(1+|\xi|^2)^{\alpha/2} Ff\|_{L_p} < \infty \right\} ,$$

we see that the operator $B_\lambda = B_0 - \lambda^{\alpha/2}\Gamma^{-1}(\alpha/2)$ maps $C_0^{\infty} \subset H_p^{\alpha}$ continuously into $H_p^0 = L_p$, and so it can be extended to all H_p^{α} . (Here and everywhere in the paper we use the old symbol for the extended operator. Once an operator have been shown to have an extension, its symbol refers always to the extended operator.) Moreover B_λ is boundedly invertible, and B_λ^{-1} as a mapping of L_p into L_p suffices

$$\|B_\lambda^{-1} f\|_{0,p} \le \lambda^{-\alpha/2}\,\Gamma(\alpha/2)\,\|f\|_{0,p} \quad . \tag{10}$$

By $FB^{\beta} = F\langle\beta(x),\nabla\rangle = i\langle\beta(x),\xi\rangle$ we see that B^{β} maps H_p^1 boundedly into $H_p^0 = L_p$, i.e. $\|B^{\beta} f\|_{0,p} \le K_1\|f\|_{1,p}$ for all $f \in H_p^1(\partial\Omega)$. As $\alpha > 1$ we have the interpolation inequality

$$\|u\|_{1,p} \le K_2\left(\varepsilon\|u\|_{\alpha,p} + \varepsilon^{-1/(\alpha-1)}\|u\|_{0,p}\right)$$

for all $u \in H_p^{\alpha}$ and for any $\varepsilon > 0$. Coupling the last two inequalities we obtain

$$\|B^{\beta} f\|_{0,p} \le K\left(\varepsilon\|f\|_{\alpha,p} + \varepsilon^{-1/(\alpha-1)}\|f\|_{0,p}\right) \tag{11}$$

for any $\varepsilon > 0$, $0 < p < \infty$ and $f \in H_p^{\alpha}(\partial\Omega)$. Here K depends only on p and $\sup|b_i(x)|$. Inequalities (10) and (11) give the estimate

$$\|B^{\beta} B_\lambda^{-1} f\|_{0,p} = \|\langle\beta,\nabla B_\lambda^{-1} f\rangle\|_{0,p} \le K\left(\varepsilon\|B^{-1} f\|_{\alpha,p} + \varepsilon^{-1/(\alpha-1)}\|B_\lambda^{-1} f\|_{0,p}\right)$$

$$\le K\left(2\varepsilon + \lambda^{-\alpha/2}\,\Gamma(\alpha/2)\,\varepsilon^{-1/(\alpha-1)}\right)\|f\|_{0,p} \quad .$$

From here we see that first choosing ε small and then choosing λ sufficiently large, say $\lambda > \lambda_\delta$ we can make $\|B^{\beta}B_\lambda^{-1}\|_{0,p} < \delta$ for arbitrary small $\delta > 0$.

Now we are able to construct $(LH_\lambda)^{-1} = (B_\lambda + B^{\beta})^{-1}$ by

$$(B_\lambda + B^{\beta})^{-1} = B_\lambda^{-1}(I + B^{\beta}B_\lambda^{-1}) = B_\lambda^{-1}\sum_{k=0}^{\infty}(B^{\beta} B_\lambda^{-1})^k \quad . \tag{12}$$

As $\|B^{\beta}B_\lambda^{-1}\|_p < \delta < 1$ the Neumann series on the right-hand side of (12) converge on all L_p . In other words $(LH_\lambda)^{-1}$ makes sense, it maps $L_p(\partial\Omega)$ into itself and its norm is

$$\|(LH_\lambda)^{-1}\|_p \le c\,\lambda^{-\alpha/2} \quad \text{with} \quad c = \Gamma(\alpha/2)/(1-\delta) \quad \text{if} \quad \lambda > \lambda_\delta \quad .$$

Moreover $(LH_\lambda)^{-1}$ can be extended to all values $\lambda > 0$ by the resolvent equation

$$(LH_\lambda)^{-1} - (LH_\mu)^{-1} + (\lambda^{\alpha/2} - \mu^{\alpha/2})\,\Gamma^{-1}(\alpha/2)\,(LH_\lambda)^{-1}\,(LH_\mu)^{-1} = 0 \quad .$$

Using the representation $(LH_\lambda)^{-1} = B_\lambda^{-1}(I + B^\beta B_\lambda^{-1})$ and the fact that B_λ is one to one between L_p and H_p^α we see that $(LH_\lambda)^{-1}$ maps $L_p(\partial\Omega)$ boundedly into $H_p^\alpha(\partial\Omega)$. As H_p^α is dense in L_p and LH_λ is a closed invertible operator from H_p^α onto L_p, by the Hille-Yosida theorem LH_λ generates an equi-continuous semigroup $S_t^{(\lambda)}$ on $L_p(\partial\Omega)$. Let us observe that $LH_\lambda | C_0^2(\partial\Omega)$ is a core of the generator $LH_\lambda | H_p^\alpha(\partial\Omega)$ and hence $S_t^{(\lambda)}$ is uniquely determined by $LH_\lambda | C_0^2(\partial\Omega)$.

Let us show that $S_t^{(\lambda)}$ is generated by a (non-conservative) Markov transition function, i.e. that

$$S_t^{(\lambda)} f(x) = \int P^{(\lambda)}(x,t,dy)\, f(y) \quad . \tag{13}$$

If $p > n/(\alpha - 1)$ then by Sobolev's imbedding theorem all functions $f \in H_p^\alpha$ are continuously differentiable. B_λ is a generator of a (sub-) Markov semigroup, hence it suffices the maximum principle. As $\nabla f(x) = 0$ at an interior maximum x for any $f \in H_p^\alpha \subset C^1$, we see that $B = B_\lambda + B^\beta$ inherites the maximum property. Hence by the Riesz representation theorem there exists a (sub-) Markov transition function such that (13) holds for any $f \in C_K(\partial\Omega)$.

Remark 1: We have implicitely proved, that $S_t^{(\lambda)}$ is a Feller-Dynkin semigroup, i.e. it leaves $C_0(\partial\Omega)$ invariant. In fact choose $p > n/\alpha$, then B_λ^{-1} maps $C_K \subset L_p$ into $H_p^\alpha \subset C_0$. As $\|(LH_\lambda)^{-1}\|_\infty \le K$ it follows that $(LH_\lambda)^{-1}$ maps C_0 into C_0, and so does $S_t^{(\lambda)}$ as well.

Remark 2: The domain of the L_p-generator of $S_t^{(\lambda)}$ is exactly $H_p^\alpha(\partial\Omega)$ independently of the drift term $\beta(x)$. This statement is true for all $\lambda \ge 0$ as they differ only in a term of multiplication with a constant, but it is in general not true for the C_0-infinitesimal generators.

c) To finish the proof of the theorem let us denote by $\mathcal{D}_L$ the class of all functions f from the domain of the closoure (in sup-norm) of the operator $A | C_0^\infty(\Omega)$, which can be represented in the form

$$f = \sum_{i=1}^{N} \left(R_{\lambda_i} u_i + H_{\lambda_i} v_i \right)$$

with positive reals λ_i, μ_i and $u_i \in C_0(\bar{\Omega})$, $v_i \in \mathcal{D}(LH_\lambda)$. By paragraph (b) Theorem 5.2. of [3] is applicable, which gives the following result: Denote $\mathcal{D}(A_L) = \{u \in \mathcal{D}_L : Lu = 0\}$, then the restriction of $A|\mathcal{D}_L$ to $\mathcal{D}(A_L)$ is infinitesimal generator of a $C_0(\bar{\Omega})$ semigroup P_t. In fact

$$G_\lambda u = R_\lambda u - H_\lambda (LH_\lambda)^{-1} LR_\lambda u \qquad (14)$$

maps $C_0(\bar{\Omega})$ into $\mathcal{D}(A_L)$ and $A_L = \lambda I - G_\lambda^{-1}$ possesses the maximum property. Moreover P_t is uniquely determined by the minimal process and the boundary semigroup $S_t^{(\lambda)}$ whose uniqueness has already been shown.

Q.e.d.

3. THE OPTIMAL CONTROL PROBLEM

Let u be a measurable mapping from $\partial\Omega$ into a compact set $D \subset R^d$ we call u an admissible control strategy, and we call a Markov process X_t^u controlled according u, if X_t^u is the unique process determined by a generator (A, L^u) of type (2) - (3) where

$$Lf(x) = \frac{\partial f}{\partial n} + \langle \beta(x, u(x)), \nabla f(x) \rangle \qquad (x \in \partial\Omega) . \qquad (15)$$

Here b is a bounded measurable mapping from $R^n \times R^d$ into R^n. In other words a control strategy u determines at every point of the boundary the angle of refexion of the process. As we can observe the process only on the boundary the control $u(x)$ too may depend only on the boundary values of the process.

Assume that the process gets killed when it attains the smooth target set $\partial\Omega_0 \subset \partial\Omega$, and that at the termination a final cost $p(x)$ arises $(\xi \in \partial\Omega_0)$. The evolution of the process until its death gives rise to a state dependent differential cost $q(x_t)$. The functions p and q are Lipschitz continuous and bounded. We call a strategy u^* optimal if it minimizes the expected cost of the form:

$$\phi^u(x) = E_x^u \left\{ p(x_\tau)\, e^{-\lambda\tau} + \int_0^\tau e^{-\lambda t} q(x_t)\, dt \right\} \qquad (16)$$

for any boundary point $x \in \partial\Omega$ as initial state. (Here $\lambda \geq 0$ is a fixed discount rate and τ denotes the first hitting time of the target set $\partial\Omega_0$.) In other words u^* is optimal iff

$$\phi^{u^*}(x) = \min_u \phi^u(x) \qquad \text{for all} \quad x \in \partial\Omega .$$

Now we are able to state a necessary and sufficient optimality condition in the form of a Hamilton - Jacobi - Bellman equation:

<u>Theorem 2:</u> A strategy u^* is optimal if and only if there exists a function $\psi \in H_p^2(\Omega) \cap H_p^\alpha(\partial\Omega)$ such that it is the solution of

$$A\psi(x) - \lambda\psi(x) + q(x) = 0 \qquad \text{if} \qquad x \in \Omega , \qquad (17)$$

$$L^{u^*}\psi(x) = \inf_{d \in D} L^d \psi(x) = 0 \qquad \text{if} \qquad x \in \partial\Omega \setminus \partial\Omega_0 = \partial\Omega_1 \ , \tag{18}$$

$$\psi(x) = p(x) \qquad \text{if} \qquad x \in \partial\Omega_0 \ . \tag{19}$$

Moreover if $p > n/(\alpha - 1)$ then ψ is the unique solution of (17), (18), (19), and $\psi(x) = \phi^{u^*}(x)$ for all $x \in \bar{\Omega}$.

Proof: Observe that by (17) $\psi(x) = R_\lambda q + H_\lambda[\psi]_{\partial\Omega}$. As $[\psi]_{\partial\Omega} = [H_\lambda \psi]_{\partial\Omega}$, we can write $L^u \psi$ in the form

$$L^u \psi = (L^u H_\lambda)\psi + L^u R_\lambda q \quad .$$

In the second paragraph we have shown that $L = L^u H_\lambda$ is an infinitesimal generator of a Markov process. Denoting $f^u(x) = (L^u R_\lambda q)(x)$ $(x \in \partial\Omega)$ we find that (17) - (19) can be rewritten in the form

$$L^{u^*}\psi(\xi) + f^{u^*}(\xi) = \inf_{d \in D}\left[L^d \psi(\xi) + f^d(\xi)\right] = 0 \qquad \text{if} \qquad \xi \in \partial\Omega_1 \ , \tag{20}$$

$$\psi(\xi) = p(\xi) \qquad \text{if} \qquad \xi \in \partial\Omega_0 \ . \tag{21}$$

But this is the Hamilton - Jacobi - Bellman equation of a general Markovian control problem treated in [4]. By Remark 2 the condition on the coincidence of the domains of the generators is satisfied, and the main theorem of [4] becomes applicable, and this gives the statement of our Theorem 2.

Remark 3: The same method would give an analogous result for the more general cost functional

$$\phi^u(x) = E_x^u \left\{ p(x_\tau) + \int_0^\tau q(x_t)\,dt + \int_0^{\Lambda_\tau} Q(x_t, u(x_t))\,d\Lambda_t \right\}$$

where Λ_t is the local time at the boundary and Q is a bounded function $\partial\Omega \times R^d \to R^1$. The only necessary change is the inclusion of an additional term $Q(x,u(x))$ to the left - hand side of (18) and (20).

Remark 4: If $q \equiv 0$ then our control problem can be directly reduced to a control problem on the boundary. In this case we would get a HJB - equation of the form (20) - (21), as a special case of Pragarauskas' more general results [2] or those of the author [5]. But one has no idea how should one effectively solve the nonlinear singular integro - differential equation (20) - (21). The solution of the elliptic boundary value problem (17) with reflecting and Dirichlet boundary data (18) - (19) is quite standard.

Remark 5: The inclusion of the drift term becoming infinit near the boundary is necessited by the technique of the proof of Theorem 1. For the same problem, but with bounded drift and diffusion coefficients one could prove Theorem 2 under the additional hypothesis that the set

$$I(x) = \left\{ y \in R^n : y = \beta(x,d),\ d \in D \right\}$$

is strictly convex. But the proof of the substitute of Theorem 1 would be entirely different.

Acknowledgement: The results of this paper were obtained while the author was staying with Université de Paris - Sud (Orsay). The author is very indepted to Prof. J.M. Bismut for his permanent interest, support and the numeruous motivating discussions on the subject.

REFERENCES

[1] Molchanov, S.A., and Ostrovskii, E.: Symmetric stable process as traces of degenerate diffusion process. Theory Probab. Applications 14 (1969) pp. 128 - 131.

[2] Pragarauskas, H.: On the optimal control of discontinuous random processes (Russian). In: "Trudi Skoli - seminara po teorii Sluchaynih Processov", Vilnius 1975.

[3] Sato, K., and Ueno, T.: Multi - dimensional diffusion and the Markov process on the boundary. J. Math. Kyoto Univ. 4 (1965) pp. 529 - 605.

[4] Vermes, D.: A necessary and sufficient condition of optimality for Markovian control problems. Acta Sci. Math. (Szeged) 34 (1973) pp. 401 - 413.

[5] Vermes, D.: On the semigroup theory of stochastic control. In: "Stochastic Differential Systems, Vilnius 1978", Lecture Notes in Control and Information Sciences, Vol. 25, pp. 91 - 102, Springer Verlag.

[6] Vermes, D.: Extremality properties of the optimal strategy in Markovian control problems. In: "Analysis and Optimisation of Stochastic Systems", Oxford,1978, pp. 35 - 47. Academic Press 1980.

Domokos VERMES
Bolyai Institute
University of Szeged
6720 Szeged
Aradi vèrtanuk tere 1
HUNGARY

Lecture Notes in Control and Information Sciences

Edited by A. V. Balakrishnan and M. Thoma

Vol. 22: Optimization Techniques
Proceedings of the 9th IFIP Conference on Optimization Techniques,
Warsaw, September 4–8, 1979
Part 1
Edited by K. Iracki, K. Malanowski, S. Walukiewicz
XVI, 569 pages. 1980

Vol. 23: Optimization Techniques
Proceedings of the 9th IFIP Conference on Optimization Techniques,
Warsaw, September 4-8, 1979
Part 2
Edited by K. Iracki, K. Malanowski, S. Walukiewicz
XV, 621 pages. 1980

Vol. 24: Methods and Applications in Adaptive Control
Proceedings of an International Symposium
Bochum, 1980
Edited by H. Unbehauen
VI, 309 pages. 1980

Vol. 25: Stochastic Differential Systems – Filtering and Control
Proceedings of the IFIP-WG7/1 Working Conference
Vilnius, Lithuania, USSR, Aug. 28 – Sept. 2, 1978
Edited by B. Grigelionis
X, 362 pages. 1980

Vol. 26: D. L. Iglehart, G. S. Shedler
Regenerative Simulation of Response Times in Networks of Queues
XII, 204 pages. 1980

Vol. 27: D. H. Jacobson, D. H. Martin, M. Pachter, T. Geveci
Extensions of Linear-Quadratic Control Theory
XI, 288 pages. 1980

Vol. 28: Analysis and Optimization of Systems
Proceedings of the Fourth International Conference on Analysis and Optimization of Systems
Versailles, December 16–19, 1980
Edited by A. Bensoussan and J. L. Lions
XIV, 999 pages. 1980

Vol. 29: M. Vidyasagar,
Input-Output Analysis of Large-Scale Interconnected Systems –
Decomposition, Well-Posedness and Stability
VI, 221 pages. 1981

Vol. 30: Optimization and Optimal Control
Proceedings of a Conference Held at Oberwolfach, March 16–22, 1980
Edited by A. Auslender, W. Oettli, and J. Stoer
VIII, 254 pages. 1981

Vol. 31: Berc Rustem
Projection Methods in Constrained Optimisation and Applications to Optimal Policy Decisions
XV, 315 pages. 1981

Vol. 32: Tsuyoshi Matsuo,
Realization Theory of Continuous-Time Dynamical Systems
VI, 329 pages, 1981

Vol. 33: Peter Dransfield
Hydraulic Control Systems –
Design and Analysis of Their Dynamics
VII, 227 pages, 1981

Vol. 34: H.W. Knobloch
Higher Order Necessary Conditions in Optimal Control Theory
V, 173 pages, 1981

Vol. 35: Global Modelling
Proceedings of the IFIP-WG 7/1 Working Conference Dubrovnik, Yugoslavia,
Sept. 1–5, 1980
Edited by S. Krčevinac
VIII, 232 pages, 1981

Vol. 36: Stochastic Differential Systems
Proceedings of the 3rd IFIP-WG 7/1 Working Conference
Visegrád, Hungary, Sept. 15–20, 1980
Edited by M. Arató, D. Vermes, A.V. Balakrishnan
VI, 238 pages, 1981

Vol. 37: Rüdiger Schmidt
Advances in Nonlinear Parameter Optimization
VI, 159 pages, 1982

Vol. 38: System Modeling and Optimization
Proceedings of the 10th IFIP Conference
New York City, USA, Aug. 31 – Sept. 4, 1981
Edited by R.F. Drenick and F. Kozin
XI, 894 pages. 1982

Vol. 39: Feedback Control of Linear and Nonlinear Systems
Proceedings of the Joint Workshop on Feedback and Synthesis of Linear and Nonlinear Systems
Bielefeld/Rom
XIII, 284 pages. 1982

Vol. 40: Y.S. Hung, A.G.J. MacFarlane
Multivariable Feedback:
A Quasi-Classical Approach
X, 182 pages. 1982

Vol. 41: M. Gössel
Nonlinear Time-Discrete Systems –
A General Approach by Nonlinear Superposition
VIII, 112 pages. 1982

Vol. 42: Advances in Filtering and Optimal Stochastic Control
Proceedings of the IFIP-WG 7/1 Working Conference
Cocoyoc, Mexico, February 1–6, 1982
VIII, 391 pages. 1982